
Function Spaces

Proceedings of the Second International Conference,

Poznań 1989, August 28–September 2

Edited by Julian Musielak, Henryk Hudzik, Ryszard Urbański

B. G. Teubner Verlagsgesellschaft
Stuttgart · Leipzig 1991

This volume contains papers from selected topics in the theory of function spaces. Part I contains papers on Orlicz spaces with special emphasis on geometric properties. Part II is devoted to Banach algebras, operators, vector-valued functions and convex analysis. In Part III there are presented results on approximation and integral theory. Finally, Part IV contains papers on interpolation spaces and spaces of differentiable functions on manifolds.

Dieser Band enthält Vorträge über ausgewählte Gebiete der Theorie der Funktionenräume. In Teil I findet der Leser Arbeiten über Orliczräume mit spezieller Beachtung geometrischer Eigenschaften. Teil II enthält Resultate über Banachalgebren, Operatoren, Vektorfunktionen und konvexe Analysis. In Teil III findet man Arbeiten aus der Approximationstheorie und Integraltheorie. Teil IV enthält Resultate über Interpolationsräume und über Räume differenzierbarer Funktionen auf Mannigfaltigkeiten.

Ce volume contient un choix des articles relatifs â la thêorie des espaces fonctionnels. Dans la partie I on y trouve les traveaux sur les espaces de Orlicz, en particulier sur les propriêtes géométriques. La partie II contient les traveaux sur Banach algêbras, opêrateurs, fonctions vectorielles et analyse convexe. Dans la partie III les auteurs presentent les resultats en la thêorie de l'approximation et la thêorie de l'integral. La partie IV contient les traveaux sur les espaces d'interpolation et sur les espaces de fonctions differentiables sur les variêtes.

В сборнике помещено работы по теории функциональных пространств. Часть I содержит работы по пространствам Орлича в особенности по их геометрическим своиствам. Часть II посвящена Банаховым алгебрам, оператором, векторным функциям и выпуклой анализе. В Части III помещены работы по аппроксимации и интегральной теории. Часть IV содержит работы по интерполации и пространствах дифференцирыемых функции на многообразиях.

PREFACE

Modern development of functional analysis and its applications has proved the importance of investigations in the field of the theory of function spaces. This concerns both the general theory and the consideration of particular function spaces and sequence spaces. The Second International Conference "Function Spaces" held in Poznań, August 28th – September 2nd, was a contribution in this direction. There were held lectures both on plenary sessions and in two sections. These Proceedings contain selected papers devoted to modern directions in the theory of function spaces and their applications.

Orlicz spaces and their generalizations, which form a special family of function spaces, are considered in Part I. There are investigated topologic and geometric properties of these spaces as imbeddings, roughness, stability of unit ball etc. There is given also a probabilistic interpretation.

Part II is devoted to various problems concerning convex analysis, spaces of vector-valued functions, operators and multi-valued functions with applications. There are considered also ideal spaces, rearrangement invariant spaces and quasidifferentiable functions.

Function spaces, from the point of view of special inequalities and the theory of approximation, are dealt with in Part III, together with some problems of Henstock and Denjoy-Perron integrals.

Part IV contains investigations in the theory of interpolation spaces and related problems. The last paper presents a theory of anisotropic Sobolev spaces on Riemannian symmetric manifolds with special emphasis on local smoothness.

The credit of organization of the Conference goes mainly to Doc. Dr. Magdalena Jaroszewska, Dr. Mirosława Mikosz, Dr. Marian Nowak, Dr. Ryszard Płuciennik and Dr. Marek Wisła. The final text was prepared by Dr. Marek Wisła.

Thanks are due to the staff of the Publishing House TEUBNER, Leipzig, for the publication of these Proceedings.

The Editors

CONTENTS

Part III. Approximation and Integration Theory

Part IV. Interpolation Spaces and Spaces of Infinitely Differentiable Functions

NEAR CONVEXITY IN BANACH SPACES

JÓZEF BANAŚ

Rzeszów, Poland

ABSTRACT. Some basic facts concerning the concept of near convexity in Banach spaces are presented. In particular, with help of this concept we extend the Day's classification of some product spaces.

1. Introduction. In the last years a lot of concepts of classical geometry of Banach spaces such as strict convexity, uniform convexity, smoothness and uniform smoothness have been generalized in the direction involving compactness. R.Huff seems to be the first who initiated this direction. Namely, in 1980 he introduced the concept of a nearly uniformly convex Banach space [11]. Independently this definition was also given by Goebel and Sękowski [10]. Subsequently the concept of near smoothness in Banach spaces has been formulated in [20,16,4].

It turned out that the mentioned concepts are useful tools in the geometric theory of Banach spaces [4,10,11,13,17,18] and find also applications in other branches of mathematics, especially in fixed point theory and optimal control theory (cf. [6,10,15,16,19] for example).

The aim of this paper is to present some basic facts concerning the concept of near convexity of Banach spaces. This concept will be used in extending of the well known Day's classification of the so-called product spaces.

AMS Subject Classification: 46B20

2. Near convexity.

Let E be an infinite dimensional Banach space with the norm $\|\cdot\|$ and the zero element θ and let E^* denote the dual of E.

Throughout this paper we shall use the following notation: $K(x,r)$ and $S(x,r)$ will denote the closed ball and the sphere centered at x and of radius r. We write $B = K(\theta,1)$, $S = S(\theta,1)$. The symbol S^* stands for the unit sphere in E^*. The closed and convex hull of a set X will be denoted by Conv X.

For a given nonempty and bounded subset X of E let $\mu(X)$ and $\alpha(X)$ denote the Hausdorff and the Kuratowski measure of noncompactness of a set X, respectively (cf. [2]).

The concept of a nearly uniformly convex (NUC) Banach space was introduced by Huff [11] in the following way:

> A space E is called *NUC* if for any $\varepsilon>0$ there exists $\delta>0$ such that whenever a closed and convex subset X of the unit ball has $\mathrm{dist}(\theta,X) \ge 1-\delta$ then $\mu(X) \le \varepsilon$.

In connection with the above concept we can define the function $\beta_E:[0,1] \longrightarrow [0,1]$,

$$\beta_E(\varepsilon) = \sup \{\mu(X)\colon X \subset B,\ X = \mathrm{Conv}\ X,\ \mathrm{dist}(\theta,X) \ge 1-\varepsilon\}.$$

This function will be called *the modulus of near convexity of E* [5]. It is an easy matter to check that β_E is nondecreasing on [0,1], so there exists the limit $\beta_0(E) = \lim_{\varepsilon\to 0} \beta_E(\varepsilon)$. Obviously E is NUC if and only if $\beta_0(E) = 0$. A space E for which $\beta_E(0) = 0$ will be called *nearly strictly convex (NSC)*.

As was mentioned in Introduction, the concept of near uniform convexity has been defined independently be Goebel and Sękowski in [11]. Namely, they defined the function

$$\tilde{\Delta}_E(\varepsilon) = \inf \{1-\mathrm{dist}(\theta,X)\colon X \subset B,\ X = \mathrm{Conv}\ X,\ \alpha(X) \ge\varepsilon\}$$

for $\varepsilon \in [0,2]$. This function is said to be *the modulus of noncompact convexity of E*. If we denote by $\tilde{\varepsilon}_1(E) = \sup \{\varepsilon\ge 0\colon \tilde{\Delta}_E(\varepsilon) = 0\}$ then it is easily seen that E is NUC iff $\tilde{\varepsilon}_1(E) = 0$.

In what follows we shall also use the function Δ_E defined in the same way as $\tilde{\Delta}_E$ but with help of the measure μ. Similarly, $\varepsilon_1(E) = \sup \{\varepsilon\ge 0\colon \Delta_E(\varepsilon) = 0\}$ (cf. [3]).

Now, let us review some basic results obtained up to now and concerning the above described concept.

Theorem 1. *([5]) $\beta_0(E) = \varepsilon_1(E)$.*

Moreover, it may be shown that the functions Δ_E and β_E are, in a certain sense, inverse of one another [5].

Theorem 2. *([10]) If $\varepsilon_1(E) < 1/2$ (or $\tilde{\varepsilon}_1(E) < 1$) then E is reflexive and has normal structure.*

It is worthwhile to mention that the function $\tilde{\Delta}_E$ is a generalization of the classical modulus of convexity δ_E of a space E [9]. Moreover, it is easy to see that $\tilde{\Delta}_E(\varepsilon) \geq \delta_E(\varepsilon)$ and for some Banach spaces this inequality is sharp [10].

Now we illustrate our considerations by a few examples.

Example 1. Consider the space $C = C[a,b]$ with the standard maximum norm. Then we can easily calculate that $\Delta_C(\varepsilon) = 0$ for $\varepsilon \in [0,1]$ and $\tilde{\Delta}_C(\varepsilon) = 0$ for $\varepsilon \in [0,2]$. Apart from this $\beta_C(\varepsilon) = 1$ on $[0,1]$.

Example 2. Let H be a Hilbert space with the norm induced by scalar product in H. Then we have the following formulas [5]:

$$\Delta_H(\varepsilon) = 1-(1-\varepsilon^2)^{1/2},$$

$$\tilde{\Delta}_H(\varepsilon) = 1-(1-(\varepsilon/2)^2)^{1/2},$$

$$\beta_H(\varepsilon) = (1-(1-\varepsilon)^2)^{1/2}.$$

Example 3. Take the space ℓ^p $(1<p<\infty)$ with the classical norm. It was shown in [10] (cf. also [5]) that

$$\Delta_{l^p}(\varepsilon) = 1-(1-\varepsilon^p)^{1/p},$$

$$\tilde{\Delta}_{l^p}(\varepsilon) = 1-(1-(\varepsilon/2)^p)^{1/p},$$

$$\beta_{l^p}(\varepsilon) = (1-(1-\varepsilon)^p)^{1/p}.$$

Let us pay attention to the fact that $\tilde{\Delta}_{l^p} = \delta_{l^p}$ provided $2 \leq p < \infty$ but $\tilde{\Delta}_{l^p}(\varepsilon) > \delta_{l^p}(\varepsilon)$ if p belongs to the interval $(1,2)$ (cf. [10]).

It what follows we will need some characterization of the notion of near uniform convexity given by Huff [11].

First we recall the following definition.

Definition 1. A Banach space E is called *uniformly Kadec-Klee* (UKK in short) if for every $\varepsilon>0$ there is $\delta>0$ such that whenever x is in B and (x_n) is a sequence in B such that $x_n \to x$ weakly and $\|x_n-x_m\| \geq \varepsilon$ for $n \neq m$ then $\|x\| \leq 1-\delta$.

We have the following theorem.

Theorem 3. *([11]) A Banach space E is NUC iff it is UKK and reflexive.*

Based on Definition 1 Partington [14] defined the modulus of UKK-ness as the function $P_E:[0,1] \longrightarrow [0,1]$ given by

$$P_E(\varepsilon) = \inf \{1-\|x\|: x \in B \text{ and there exists } (x_n) \subset B \text{ such that } x_n \rightarrow x \text{ weakly and } \inf\{\|x_n-x_m\|: n\neq m\} \geq \varepsilon\}.$$

Observe that E is UKK iff $P_E(\varepsilon) > 0$ whenever $\varepsilon>0$. The below given theorem provide the quantitative translation of the results from Theorem 3.

Theorem 4. *([5]) If E is an arbitrary Banach space then $\Delta_E(\varepsilon) = P_E(\varepsilon)$ for each $\varepsilon \in [0,1]$.*

Proof. Fix $\varepsilon>0$ and take an arbitrary sequence $(x_n) \subset B$ such that $\|x_n-x_m\| \geq \varepsilon$ for $n,m = 1,2,\ldots$, $n \neq m$. Assume that (x_n) converges weakly to x. In view of weak lower semicontinuity of the norm [9] we have that $x \in B$. Next, let us take $\eta>0$ small enough. Choose a functional $f \in S^*$ such that $f(x) = \|x\|$. Consider the set $D(f,\eta) = \{y\in B: f(y) \geq \|x\|-\eta\}$. Obviously the set $D(f,\eta)$ is convex and $\text{dist}(\theta,D(f,\eta)) \geq \|x\|-\eta$. In virtue of the fact that $f(x_n) \rightarrow f(x) = \|x\|$ we can find n_o with the property that $f(x_n) \geq \|x\|-\eta$ for $n\geq n_o$. This implies that the set $(x_n)_{n\geq n_o}$ is contained in $D(f,\eta)$ what allows us to deduce that $\text{Conv}(x_n)_{n\geq n_o} \subset D(f,\eta)$. Hence

$$\text{dist}(\theta,\text{Conv}(x_n)_{n\geq n_o}) \geq \|x\|-\eta. \tag{1}$$

Since $\alpha((x_n)_{n\geq n_o}) \geq \varepsilon$ then

$$\Delta_E(\varepsilon) \leq 1-\text{dist}(\theta,\text{Conv}(x_n)_{n\geq n_o}).$$

Hence, in view of (1) we get $\|x\|-\eta \leq 1-\Delta_E(\varepsilon)$. Consequently

$$1-\|x\| \geq \Delta_E(\varepsilon)-\eta.$$

Thus $P_E(\varepsilon) \geq \Delta_E(\varepsilon)-\eta$. Taking into account the arbitrariness of η we finish the proof.

Corollary 1. *If E is NUC then E is UKK.*

Theorem 5. *([5]) If E is a reflexive Banach space then $\Delta_E(\varepsilon) \geq P_E(\varepsilon/2)$ for any $\varepsilon \in [0,2]$.*

Proof. Assume that $\varepsilon>0$ is a fixed number from [0,2]. Take a convex and closed subset X of B such that $\alpha(X) \geq \varepsilon$. Hence, there exists a sequence $(x_n) \subset X$ such that $\|x_n - x_m\| \geq \varepsilon/2$ for $n \neq m$ [1]. By reflexivity of E there is a subsequence (x_{k_n}) of (x_n) which converges weakly to some $x \in B$. Obviously $\|x_{k_n} - x_{k_m}\| \geq \varepsilon/2$ for $n \neq m$ what gives $\|x\| \leq 1 - P_E(\varepsilon/2)$. On the other hand $x \in \mathrm{Conv}(x_{k_n}) \subset \mathrm{Conv}(x_n)$ what implies

$$(2) \qquad \mathrm{dist}(\theta, \mathrm{Conv}(x_n)) \leq \mathrm{dist}(\theta, \mathrm{Conv}(x_{k_n})) \leq \|x\| \leq 1 - P_E(\varepsilon/2).$$

Moreover, $\mathrm{dist}(\theta, X) \leq \mathrm{dist}(\theta, \mathrm{Conv}(x_n))$. Hence, in view of (2) we get

$$\mathrm{dist}(\theta, X) \leq 1 - P_E(\varepsilon/2)$$

or equivalently

$$1 - \mathrm{dist}(\theta, X) \geq P_E(\varepsilon/2).$$

But this gives $\Delta_E(\varepsilon) \geq P_E(\varepsilon/2)$ and the proof is complete.

Corollary 2. *If E is UKK and reflexive then E is NUC.*

3. Classification of product spaces.

Let $(E_i, \|\cdot\|_i)$, $i = 1,2,\ldots$, be a sequence of infinite dimensional Banach spaces. Consider the so-called product space $\ell^p(E_i) = \ell^p(E_1, E_2, \ldots)$, ($p \in (1,\infty)$ is fixed) which consists of all sequences $x = (x_i)$, $x_i \in E_i$ for $i = 1,2,\ldots$ and $\sum_{i=1}^{\infty} \|x_i\|_i^p < \infty$. It is well known that $\ell^p(E_i)$ forms a Banach space under the norm

$$\|x\| = \|(x_i)\| = \left(\sum_{i=1}^{\infty} \|x_i\|_i^p\right)^{1/p}$$

(cf. [12]).

Now, let α be the Kuratowski measure of noncompactness in $\ell^p(E_i)$ and α_i be such a measure in E_i, $i = 1,2,\ldots$. Taking a nonempty and bounded subset X of E_i we have $\alpha(X) = \alpha_i(X)$. Consequently

$$\Delta_{l^p(E_i)}(\varepsilon) \leq \Delta_{E_i}(\varepsilon)$$

for any $\varepsilon \in [0,2]$, $i = 1,2,\ldots$.

Further, let us denote by $m(\varepsilon)$ the function $m:[0,2] \longrightarrow [0,1]$ defined by

$$(3) \qquad m(\varepsilon) = \inf \{\Delta_{E_i}(\varepsilon) : i = 1,2,\ldots\}.$$

Then

$$\Delta_{l^p(E_i)}(\varepsilon) \leq m(\varepsilon), \qquad \varepsilon \in [0,2].$$

Hence we obtain

Lemma 1. *If there exists a number* $a \in [0,2]$ *such that* $\liminf_{i\to\infty} \tilde{\Delta}_{E_i}(\varepsilon) = 0$ *for* $\varepsilon \in [0,a)$ *then* $\tilde{\Delta}_{l^p(E_i)}(\varepsilon) = 0$ *on the interval* $[0,a)$ *and the space* $\ell^p(E_i)$ *is not NUC.*

In the sequel we shall need the following theorem.

Theorem 6. *If there exists a function* $r:(0,1] \longrightarrow (0,1]$ *such that* $P_{E_i}(\varepsilon) \geq r(\varepsilon)$ *for* $\varepsilon \in (0,1]$ *and for* $i = 1,2,\ldots$ *then the product space* $\ell^p(E_i)$ *is UKK.*

The proof of this theorem for the case $E_i = E$ ($i = 1,2,\ldots$) was given by Partington [14]. Smith observed in [21] that this proof is also valid in the situation considered in the above theorem.

Based on the above theorem we are able to prove

Theorem 7. *Let* $(E_i,\|\cdot\|_i)$ *be a sequence of NUC Banach spaces such that the function* $m(\varepsilon)$ *(defined by (3)) is positive on the interval* $(0,2]$*. Then the product space* $\ell^p(E_i)$ *is also NUC.*

Proof. In view of Theorem 4 we have

$$P_{E_i}(\varepsilon) \geq m(\varepsilon)$$

for $\varepsilon \in [0,1]$ and for $i = 1,2,\ldots$. Hence, by Theorem 6 we obtain that $\ell^p(E_i)$ is UKK. On the other hand, taking into account that all E_i are reflexive (Theorem 2) we deduce [12] that $\ell^p(E_i)$ is reflexive. Now applying Theorem 3 we finish the proof.

Combining Lemma 1 and Theorem 7 we get

Corollary 3. *The product space* $\ell^p(E_i)$ *is NUC if and only if the function* $m(\varepsilon)$ *defined by (3) is positive on the interval* $(0,2]$*.*

This result extends the classification due to Day [7,8] and concerning the concept of uniform convexity in Clarkson sense.

Now we pay our attention to the particular case of a product space, namely to the product space $\ell^p(\ell^{p_i})$, where ℓ^{p_i} ($i = 1,2,\ldots$) is furnished with the classical norm. We shall always assume that (p_i) is a sequence of real numbers greater than 1.

First we recall that in view of results of Day [7,8] the space $\ell^p(\ell^{p_i})$ is uniformly convex if and only if the sequence (p_i) is bounded and bounded away from 1. We extend this result using the concept of near uniform convexity.

Lemma 2. *If* (p_i) *is a sequence such that* $\limsup_{i\to\infty} p_i = \infty$ *then* $\chi_{\ell^p(\ell^{p_i})}(\varepsilon) = 0$ *on the interval* $[0,2)$ *what means that in such case the space* $\ell^p(\ell^{p_i})$ *is not NUC.*

Indeed, according to the formula from Example 3 we have

$$\chi_{\ell^{p_i}}(\varepsilon) = 1-(1-(\varepsilon/2)^{p_i})^{1/p_i}, \qquad \varepsilon \in [0,2].$$

This easily implies that $\chi_{\ell^p(\ell^{p_i})}(\varepsilon) = 0$ for all $\varepsilon \in [0,2]$.

Theorem 8. *If a sequence* (p_i) *is bounded then* $\ell^p(\ell^{p_i})$ *is NUC.*

Proof. Denote $M = \sup\{p_i: i = 1,2,\dots\} < \infty$. Then we have two possibilities. First, suppose that $M \le 2$. Then it is easily seen that

$$(4) \qquad \chi_{\ell^{p_i}}(\varepsilon) \ge \chi_{\ell^2}(\varepsilon) = 1-(1-\varepsilon^2/4)^{1/2}, \qquad \varepsilon \in [0,2].$$

Second, assume that $M > 2$. In such case we can show that the following inequality holds

$$(5) \qquad \chi_{\ell^{p_i}}(\varepsilon) \ge \chi_{\ell^M}(\varepsilon), \qquad \varepsilon \in [0,2].$$

Thus, combining (4), (5) and Theorem 7 we complete the proof.

Summing up the results from Lemma 2 and Theorem 8 we infer the following characterization of the spaces $\ell^p(\ell^{p_i})$.

Theorem 9. *The space* $\ell^p(\ell^{p_i})$ *is NUC if and only if the sequence* (p_i) *is bounded.*

REFERENCES

[1] R.R.Akmerov, M. I. Kamenskii, A. S. Potapov, A. E. Rodkina, B. N. Sadovskii, *Measures of noncompactness and condensing operators*, Nauka, Novosibirsk 1986 (in Russian).

[2] J.Banaś, K. Goebel, *Measures of noncompactness in Banach spaces*, Lect. Notes in Pure and Applied Math., Vol. 60, M. Dekker 1980.

[3] J.Banaś, *On modulus of noncompact convexity and its properties*, Canad. Math. Bull. 30 (1987), 186-192.

[4] J.Banaś, *On drop property and nearly uniformly smooth Banach spaces*, Nonlin. Analysis, Theory, Methods, Appl. (to appear).

[5] J.Banaś, *On nearly uniformly convex and nearly uniformly smooth Banach spaces*, Trans. Amer. Math. Soc. (preprint).

[6] J.Daneš. *A geometric theorem useful in nonlinear functional analysis*, Boll. Un. Mat. Ital. 6 (1972), 369–372.

[7] M.M.Day. *Some more uniformly convex spaces*, Bull. Amer. Math. Soc. 47 (1941), 504–507.

[8] M.M.Day. *Uniform convexity III*. Bull. Amer. Math. Soc. 49 (1943), 745–750.

[9] M.M.Day. *Normed Linear Spaces*. Springer Verlag 1973.

[10] K.Goebel, T. Sękowski. *The modulus of noncompact convexity*, Ann. Univ. Mariae Curie-Skłodowska, Sect. A, 38 (1984), 41–48.

[11] R.Huff. *Banach spaces which are nearly uniformly convex*, Rocky Mountain J. Math. 10 (1980), 743–749.

[12] G.Köthe. *Topological Vector Spaces I*, Springer Verlag 1969.

[13] V.Montesinos. *Drop property equals reflexivity*, Studia Math. 87 (1987), 93–100.

[14] J.R.Partington. *On nearly uniformly convex Banach spaces*, Math. Proc. Camb. Phil. Soc. 93 (1983), 127–129.

[15] J.P.Penot. *The drop theorem, the petal theorem and Ekeland's variational principle*, (preprint).

[16] S.Prus. *Nearly uniformly smooth Banach spaces*, (preprint).

[17] S.Rolewicz. *On drop property*, Studia Math. 85 (1987), 27–35.

[18] S.Rolewicz. *On Δ-uniform convexity and drop property*, Studia Math. 87 (1987), 181–191.

[19] T.Sekowski. *On normal structure, stability of fixed point property and the modulus of noncompact convexity*, Rend. Sem. Mat. Fis. di Milano, 56 (1986), 147–153.

[20] T.Sekowski, A.Stachura. *Noncompact smoothness and noncompact convexity*, Atti Sem. Mat. Fis. Univ. Modena, 36 (1988), 329–338.

[21] M.A.Smith. *Rotundity and extremity in* $\ell^p(X_i)$ *and* $L^p(\mu,X)$, Contemp. Math. 52 (1986), 143–162.

ORLICZ SPACES CONTAINING SINGULAR ℓ^p-COMPLEMENTED COPIES

FRANCISCO L. HERNÁNDEZ, B. RODRIGUEZ-SALINAS

Madrid, Spain

ABSTRACT. The main purpose of this paper is to show the existence of Orlicz function spaces $L^F[0,1]$ containing singular ℓ^p-complemented copies for p>1. We give also several properties of the disjointly singular operators between Orlicz spaces.

Given $p\geq 1$, we say that a Banach function space $X([0,1])$ contains a *singular ℓ^p-complemented copy* if there is a complemented subspace in $X([0,1])$ isomorphic to ℓ^p and it does not exist any sequence of mutually disjoint characteristic functions (χ_{A_n}) which spans a subspace isomorphic to ℓ^p.

For the case p=2 there are easy examples of function spaces containing singular ℓ^2-complemented copies: the $L^q[0,1]$ spaces for $1 < q \neq 2 < \infty$. Indeed, the span of Rademacher functions in $L^q[0,1]$ is a complemented subspace isomorphic to ℓ^2 and every pairwise disjoint characteristic function sequence (χ_{A_n}) spans a subspace isomorphic to ℓ^q (cf.[7]).

The existence of Orlicz sequence spaces ℓ^F containing singular ℓ^p-complemented copies was proved by N.J.Kalton ([8]) (answering a question in [10]):Given $p > 3/2$ there exist an Orlicz sequence space ℓ^F with indices $\alpha_F = p - \frac{1}{2}$, $\beta_F = p + \frac{1}{2}$, which contains a complemented subspace isomorphic to ℓ^p and the span of any block basis with constant coefficients of the unit vector basis of ℓ^F is not isomorphic to ℓ^p.

Given an Orlicz function F satisfying the Δ_2-condition at ∞, we will consider the compact sets E_F^∞ and C_F^∞ of functions in $C(0,1)$

Supported in part by CAICYT grant 0338-84

defined by

$$E_F^\infty = \left\{G:\ \exists\ (s_n)\longrightarrow\infty,\ \ G(x) = \lim_{n\to\infty} \frac{F(s_n x)}{F(s_n)}\ \text{ uniformly on } (0,1)\right\}$$

and $C_F^\infty = \overline{\text{conv}}\ E_F^\infty$ (for other definitions we remit to [11],[12] and [13]).

In general the problem of characterizing when an ℓ^p-space is isomorphic to a complemented subspace of $L^F[0,1]$ does not have still solution (cf. [11],[12],[6]). A criterium for that is that the function x^p is, up to equivalence at 0, in the set E_F^∞ ([10]):

Proposition 1. *If $x^p \underset{\sim}{\in} E_F^\infty$ then there exists a complemented subspace in $L^F[0,1]$ isomorphic to ℓ^p.*

Under the above condition we can take in a very natural way a complemented subspace H in $L^F[0,1]$ isomorphic to ℓ^p. Indeed, H is the subspace generated by certain sequence of mutually disjoint characteristic functions (χ_{A_n}) in $L^F[0,1]$, being the projection from $L^F[0,1]$ onto $[\chi_{A_n}] = H \approx \ell^p$ the contractive averaging projection

$$P(f) = \sum_{n=1}^{\infty} \frac{\left(\int f\chi_{A_n}\right)}{\mu(A_n)}\cdot\chi_{A_n}.$$

It is known ([3], Corollary 3) that the condition $x^p \underset{\sim}{\in} E_F^\infty$ is equivalent to the existence of a sequence of disjoint characteristic functions (χ_{A_n}) verifying that the span $[\chi_{A_n}]$ in $L^F[0,1]$ is isomorphic to ℓ^p. Using this, we get for Orlicz spaces an equivalent definition of singular ℓ^p-complemented copies: $L^F[0,1]$ contains *a singular ℓ^p-complemented copy* if and only if there exists a complemented subspace isomorphic to ℓ^p and $x^p \underset{\sim}{\notin} E_F^\infty$.

By proving the existence of Orlicz spaces $L^F[0,1]$ containing singular ℓ^p-complemented copies (Theorem 7) we show that the converse of Proposition 1 does not hold for every p>1.

Let us also mention that the study of Orlicz spaces containing singular ℓ^p-complemented copies is related to the open question of whether there exist non-trivial minimal Orlicz spaces having ℓ^p-complemented subspaces, since it follows from the definition of minimality (see [10],[11],[2]) that minimal Orlicz spaces could only contain singular ℓ^p-complemented copies.

The following notion will be useful in order to find projections:

Definition. An operator $T:L^F[0,1] \longrightarrow L^G[0,1]$ is said to be *disjointly singular* if there is no a disjoint sequence of non-null functions (f_n) in $L^F[0,1]$ such that the restriction of T to the subspace $[f_n]$ is an isomorphism.

Clearly every strictly singular operator is disjointly singular. However the converse is not true as the following example shows:

The inclusion map J: $L^q[0,1] \hookrightarrow L^p[0,1]$ for $1 \leq p < q$ is not strictly singular. Indeed, its restriction to the subspace $[r_n]_q$ generated by the Rademacher functions is an isomorphism since by Khincthine inequality there exist positive constants K_1 and K_2 such that

$$K_1 \|\sum a_n r_n\|_p \leq \|\sum a_n r_n\|_q \leq K_2 \|\sum a_n r_n\|_p.$$

However, the operator J is disjointly singular because for any sequence of mutually disjoint support functions (f_n) in $L^q[0,1]$ we have $[f_n]_p \cong \ell^p$ and $[f_n]_q \approx \ell^q$.

Our interest here in this class of operators is due to the following fact:

Proposition 2. *If T is a positive operator $T: L^F[0,1] \to L^p[0,1]$ which is not disjointly singular then there exists a complemented subspace in $L^F[0,1]$ isomorphic to ℓ^p.*

Proof. Let us suppose that there exists a sequence of normalized functions (f_n) in $L^F[0,1]$ with pairwise disjoint support, such that the restriction of T on $[f_n] = X$ is an isomorphism. Then the sequence of disjoint functions (g_n), for $g_n = \dfrac{Tf_n}{\|Tf_n\|_p}$, spans a complemented subspace $T(X) = [g_n]$ in L^p isomorphic to ℓ^p (see [7], Theorem 5]. The projection P from $L^p[0,1]$ on $T(X)$ is defined by

$$P(h) = \sum_{n=1}^{\infty} \left(\int_{A_n} h\, g_n' \right) g_n,$$

where $A_n = \operatorname{supp} g_n$ for every $n \in \mathbb{N}$ and (g_n') is an orthonormal sequence to (g_n) in $L^q[0,1]$ ($\frac{1}{p} + \frac{1}{q} = 1$). Then, considering the composition operator $T^{-1}_{|T(X)} \circ P \circ T = Q$ we get a projection Q of $L^F[0,1]$ onto $X \approx \ell^p$.

A necessary condition for a positive operator between Orlicz spaces to be not disjointly singular is that the associated intervals defined by the indices are not disjoint:

Proposition 3. *If T is a positive operator $T: L^F[0,1] \to L^G[0,1]$ and $[\alpha_F^\infty, \beta_F^\infty] \cap [\alpha_G^\infty, \beta_G^\infty] = \emptyset$, then T is a disjointly singular operator.*

Proof. Let us assume that there exists a normalized sequence (f_n) of pairwise disjoint support in $L^F[0,1]$ such that $T_{|[f_n]_F}$ is an isomorphism. It is clear that the sequence (f_n) is not contained in the set

$$A_\varepsilon^F = \left\{ f \in L^F[0,1]: \mu(|f| > \varepsilon \cdot \|f\|_F) > \varepsilon \right\}$$

for every $\varepsilon>0$. Then, by ([10] Proposition 4.3) there exists a subsequence (f_{n_k}) of (f_n) which is equivalent to the unit vector basis in ℓ^H for some function $H \in C_F^\infty$. But this implies that $[\alpha_H,\beta_H] \subset [\alpha_F^\infty,\beta_F^\infty]$. Now, doing the same with the sequence $\left(\frac{Tf_{n_k}}{\|Tf_{n_k}\|_G} \right) = (g_k)$ in $L^G[0,1]$ we find a subsequence of (g_k) which is equivalent to the unit vector basis in ℓ^H for some function $M \in C_G^\infty$. Hence $[\alpha_H,\beta_H] \subset [\alpha_G^\infty,\beta_G^\infty]$. Then, the space $\ell^H \approx [f_{n_k}] \approx [Tf_{n_k}]$ contains a subspace isomorphic to ℓ^M, which implies that $\ell^p \underset{\sim}{\subset} \ell^H$ for some $p \in [\alpha_G^\infty,\beta_G^\infty]$, which is a contradiction.

Corollary 4. *If $L^F[0,1] \subset L^p[0,1]$ (resp. $L^p[0,1] \subset L^F[0,1]$) and the function F is supermultiplicative (resp. submultiplicative) then the inclusion operator is disjointly singular.*

Proof. Let as assume that $L^F[0,1] \subset L^p[0,1]$. So $p \le \beta_F^\infty$. From above Proposition we need only to show that $p < \alpha_F^\infty$. Indeed, since $t^p \le F(t)$ at ∞ we have $F_{(-1)}(t) = 1/F(1/t) \le t^p$ at 0 and $F_{(-1)}$ is submultiplicative at 0. Now, by ([14] pp. 26), there exists $\varepsilon>0$ and $M>0$ such that $F_{(-1)}(t) \le Mt^{p+\varepsilon}$ at 0. Hence $F(t) \ge \frac{1}{M}\cdot t^{p+\varepsilon}$ at ∞ and F is $(p+\varepsilon')$-convex at ∞ for $0 < \varepsilon' < \varepsilon$. Thus $p < \alpha_F^\infty$.

The proof in the remaining case of F submultiplicative is similar.

We give a characterization of when the inclusion map between Orlicz function spaces is not a disjointly singular operator.

Let $L^F[0,1] \subset L^G[0,1]$, so the function $W(t) = \frac{G(t)}{F(t)}$, for $t\ge1$ is bounded on $I_\infty = [1,\infty)$. Let us denote by $W(t)$ its unique extension to the Stone-Čech compactification βI_∞ of I_∞. Similarly we denote by $F_\tau(x)$ the extension to βI_∞ of the function F_τ defined by $F_\tau(x) = \frac{F(\tau x)}{F(\tau)}$, for $\tau \in I_\infty$. Using the method in [8] for Orlicz sequence spaces we can prove the following characterization ([5]):

Proposition 5. *The inclusion map $L^F[0,1] \subset L^G[0,1]$ is not a disjointly singular operator if and only if there exists a constant $K>0$ and a propability measure μ on βI_∞ such that $\mu(I_\infty)=0$ and*

$$\int_{\beta I_\infty} F_\tau(x)d\mu(\tau) \le K \cdot \int_{\beta I_\infty} W(\tau)G_\tau(x)d\mu(\tau)$$

for $0 \le x \le 1$.

In the special cases of considering L^p-spaces in one side of the inclusions we can get suitable analytic criteria:

Proposition 6. *(a) The inclusion $L^p[0,1] \subset L^F[0,1]$ is not a disjointly singular operator if and only if*

(+) $$\lim_{a\to\infty} \sup_{s\ge 1} \frac{1}{\log a} \int_1^a \frac{F(su)}{s^p u^{p+1}}\, du > 0.$$

(b) The inclusion $L^F[0,1] \subset L^p[0,1]$ is not a disjointly singular operator if and only if

(++) $$\lim_{a\to\infty} \inf_{s\ge 1} \frac{1}{\log a} \int_1^a \frac{F(su)}{s^p u^{p+1}}\, du < \infty.$$

Examples. The inclusions $L^{x^p \log(x)} \subset L^p$ and $L^p \subset L^{x^p/\log(1+x)}$ are disjointly singular operators. This follows easily from the above criteria. Let us note that however both spaces $L^{x^p \log(x)}$ and $L^{x^p/\log(1+x)}$ have ℓ^p-complemented subspaces in virtue of Proposition 1 for $p>1$.

Notice also that these examples show that the converse of Proposition 3 does not hold.

We can now present the main result:

Theorem 7. *For any $p>1$ there exists an Orlicz function space $L^F[0,1]$ with indices $\alpha_F^\infty = \beta_F^\infty = p$, containing a singular ℓ^p-complemented copy.*

Sketch of the proof: Let us consider the sequence of positive periodic functions (f_n) on $[0,\infty)$ defined in the following way: for each $n\in\mathbb{N}$, f_n is the function of period $P_n = 2^{2^{2n}}$ defined by

$$f_n(t) = \begin{cases} 0 & \text{if } 0 \le t \le P_n - 4\cdot 2^n \\ \sum_{k=1}^{n} (1-\cos \frac{\pi t}{2^k}) & \text{if } P_n - 4\cdot 2^n \le t \le P_n. \end{cases}$$

Now, if $f(t) = \sup_n f_n(t)$ we define the Orlicz function

$$F(t) = t^p \cdot \exp\,(qf(\log t)),$$

for $t>1$ and $0 < q < \frac{(p-1)}{3\pi}$.

We have the inclusion $L^F[0,1] \subset L^p[0,1]$ since $\overline{\lim\limits_{t\to\infty}}\, t^p/F(t) < \infty$. Now, it can be checked that the function F satisfies the condition (++) given in Proposition 6(b). Hence the inclusion operator $J{:}L^F[0,1] \subset L^p[0,1]$ is not a disjointly singular operator. So, using Proposition 2, we deduce that the space $L^F[0,1]$ contains a complemented subspace generated by simple functions isomorphic to ℓ^p.

It happens that the function x^p is not equivalent at 0 to any function in E_F^∞. Indeed, if $\bar{F} \in E_F^\infty$ there exists a sequence $(s_k) \to \infty$ such that

$$\bar{F}(t) = \lim_{k\to\infty} \frac{F(e^{s_k}\cdot t)}{F(e^{s_k})} = t^p \cdot \lim_{k\to\infty} e^{q(f(s_k+\log t)-f(s_k))}$$
$$= t^p e^{q\bar{f}(\log t)}$$

for $0 < t < 1$, and where the function $\bar{f}$ is defined by

$$\bar{f}(x) = \lim_{k\to\infty} [f(x+s_k)-f(s_k)].$$

Now, it can be shown that the function $\bar{f}$ is unbounded (see [5]) so we get that $\bar{F}$ is not equivalent to x^p at 0.

Finally, proving that for every $\varepsilon>0$ there is a constant $K_\varepsilon>0$ such that

$$|f(x+t)-f(x)| < \varepsilon t + K_\varepsilon \qquad \text{for } t\geq 0,$$

we can deduce easily that $\alpha_F^\infty = \beta_F^\infty = p$.

Notice that the Young conjugate function $\hat{F}$ of the above function verifies also that the Orlicz space $L^{\hat{F}}[0,1]$ contains a singular ℓ^r-complemented copy (for $\frac{1}{r} + \frac{1}{p} = 1$, $p>1$). This follows from the fact that a function G is equivalent at 0 to a function of E_F^∞ if and only if $\hat{G}$ is equivalent to a function in $E_{\hat{F}}^\infty$.

In the case $p=1$ it is not possible to find Orlicz spaces $L^F[0,1]$ containing singular ℓ^1-complemented copies since every Orlicz space $L^F[0,1]$ with $\alpha_F^\infty = 1$ has a complemented subspace isomorphic to ℓ^1, which is generated by a sequence of pairwise disjoint characteristic functions (cf. [1] Theorem 2.2).

Remark. We do not know whether there exists a separable Banach function space $X([0,1])$ containing a singular ℓ^1-complemented copy.

Using the above result we can find Orlicz function spaces $L^F[0,1]$ with different indices $\alpha_F^\infty \neq \beta_F^\infty$ containing singular ℓ^p-complemented copies for several values of p. More generally, we can consider the following problem:

Given an arbitrary set H of real numbers p>1, find an Orlicz function space $L^F[0,1]$ containing only singular ℓ^p-complemented copies and such that the set of values p>1 for which ℓ^p is complementably embedded into $L^F[0,1]$ is exactly the set $H \cup \{2\}$.

This problem in the case of considering "natural" ℓ^p-complemented copies (i.e. when the functions $x^p \underset{\sim}{\in} E_F^\infty$) has been solved for closed sets H in ([4] Theorem 7). Following the method used in the proof of this Theorem and taking now the functions $F_p(t) = t^p \cdot \exp(qf(\log t))$ defined in Theorem 7 we can obtain the following

Theorem 8. *Let $1 < \alpha \leq \beta < \infty$ and let H be an arbitrary closed subset of the interval $[\alpha,\beta]$. Then there exists an Orlicz function space $L^F[0,1]$ with indices $\alpha_F^\infty = \alpha$ and $\beta_F^\infty = \beta$, which contains an ℓ^p-complemented copy if and only if $p \in H \cup \{2\}$. Furthermore for each $p \in H$, $L^F[0,1]$ has a singular ℓ^p-complemented copy.*

It would be interesting to know whether or not the above result is the best possible. In other words, is the set of values p>1 for which an Orlicz space $L^F[0,1]$ contains a singular ℓ^p-complemented copy a closed set?

REFERENCES

[1] F.L.Hernández, *On the galb of weighted Orlicz sequence spaces II*, Arch. Math. 45 (1985), 158–168.

[2] F.L.Hernández, V.Peirats, *Orlicz function spaces without complemented copies of ℓ^p*, Israel J. Math. 56 (1986), 355–360.

[3] F.L.Hernández, V.Peirats, *Weighted sequence subspaces of Orlicz function spaces isomorphic to ℓ^p*, Arch. Math. 50 (1988), 270–280.

[4] F.L.Hernández, B.Rodriguez-Salinas, *On ℓ^p-complemented copies in Orlicz spaces*, Israel J. Math. 62 (1988), 37–55.

[5] F.L.Hernández, B.Rodriguez-Salinas, *On ℓ^p-complemented copies in Orlicz spaces II*, Israel J. Math. (to appear).

[6] W.B.Johnson, B.Maurey, G.Schechtmann, L.Tzafriri, *Symmetric structures in Banach spaces*, Memoirs Am. Math. Soc. N. 217 (1979).

[7] M.Kadec, A.Pełczyński, *Bases, lacunary sequences and complemented subspaces in the space* L^p, Studia Math 21 (1962), 161–176.

[8] N.J.Kalton, *Orlicz sequence spaces without local convexity*, Math. Proc. Cambridge Philos. Soc. 81 (1977), 253–277.

[9] J.Lindenstrauss, L. Tzafriri, *On Orlicz sequence spaces II*, Israel J. Math. 11 (1972), 355–379.

[10] J.Lindenstrauss, L. Tzafriri, *On Orlicz sequence spaces III*, Israel J. Math. 14 (1973), 368–389.

[11] J.Lindenstrauss, L. Tzafriri, *Classical Banach spaces I. Sequence spaces*, Springer-Verlag 1977.

[12] J.Lindenstrauss, L. Tzafriri, *Classical Banach spaces II. Function spaces*, Springer-Verlag 1979.

[13] J.Musielak, *Orlicz spaces and modular spaces*, Lect. Notes in Math. 1034, Springer-Verlag 1983.

[14] Ph.Turpin, *Convexités dans les espaces vectoriels topologiques generaux*, Dissertationes Math. 131 (1976).

NOTES ON ORLICZ SPACES

HENRYK HUDZIK

Poznań, Poland

ABSTRACT. For any measure μ and any Orlicz function φ, the Orlicz space $L^{\Psi_\varphi}(\mu)$ is characterized in terms of the function φ, where $\Psi_\varphi(u) = \varphi(u)/(1+\varphi(u))$. The convergence with respect to the F-norm $\|\cdot\|_{\Psi_\varphi}$ is also described. There are also given some sufficient conditions for the equalities $L^{\Phi}(\mu) = L^{\varphi}(\mu) \cap L^{\psi}(\mu)$ and $L^{\Phi}(\mu) = L^{\varphi}(\mu) + L^{\psi}(\mu)$ to hold for a triple of Orlicz functions Φ, φ and ψ. Finally, the modular convergence is discussed.

In the sequel, (T,Σ,μ) denotes a measure space, $\mathbb{R} = (-\infty,+\infty)$, $\mathbb{R}_+ = [0,+\infty)$, $\mathbb{R}_+^e = [0,+\infty]$, and $\mathbb{N}$ denotes the set of all natural numbers. By $L^0(\mu)$ we shall denote the space of all Σ-measurable functions $x:T \to \mathbb{R}$ and $\tilde{L}(\mu)$ will stand for the space of all $x \in L^0(\mu)$ such that $\mu(\text{supp } x) < +\infty$, where $\text{supp } x = \{t\in T: x(t)\neq 0\}$. For any sequence (x_n) in $L^0(\mu)$, the notation $x_n \to 0(\mu)$ as $n \to +\infty$ means that x_n tends to 0 with respect to the measure μ as $n \to +\infty$.

A mapping $\Phi:\mathbb{R} \to \mathbb{R}_+^e$ is said to be *an Orlicz function* if $\Phi(0) = 0$, Φ is nondecreasing in $\mathbb{R}_+$, even, continuous at 0 and not identically equal to 0 (cf. [3],[4],[5] and [6]).

Given an Orlicz function φ define a new Orlicz function Ψ_φ by the formula

$$\Psi_\varphi(u) = \varphi(u)/(1+\varphi(u))$$

for any $u\in\mathbb{R}$. Every Orlicz function φ and every measure space (T,Σ,μ) generate *an Orlicz space* $L^\varphi(\mu)$ as follows

$$L^\varphi(\mu) = \left\{x\in L^0(\mu)\colon I_\varphi(\lambda x) = \int_T \varphi(\lambda x(t))d\mu < +\infty \text{ for some } \lambda>0 \right\}.$$

This space can be endowed with the F-norm $\|\cdot\|_\varphi$ defined by

$$\|x\|_\varphi = \inf \{\lambda>0\colon I_\varphi(x/\lambda) \le \lambda \}$$

(cf. [4]). It is well known that for any sequence (x_n) in $L^\varphi(\mu)$ we have $\|x_n\|_\varphi \to 0$ as $n \to +\infty$ if and only if $I_\varphi(\lambda x_n) \to 0$ as $n \to +\infty$ for any $\lambda > 0$ (cf. [4],[5] and [6]).

Let φ and ψ be Orlicz functions. φ is said to be *weaker than* ψ *at* $+\infty$, briefly $\varphi \overset{l}{\prec} \psi$, iff there exist positive numbers a,b and c such that $\varphi(c) > 0$ and

$$\text{(1)} \qquad \varphi(u) \le a\psi(bu) \qquad (\forall\ u \ge c).$$

Two Orlicz functions φ and ψ are called to be *equivalent at* $+\infty$, briefly $\varphi \overset{l}{\sim} \psi$, iff $\varphi \overset{l}{\prec} \psi$ and $\psi \overset{l}{\prec} \varphi$ (see [3]). Replacing "$u \ge c$" by "$0 \le u \le c$" in the inequality (1), we define the relation $\varphi \overset{s}{\prec} \psi$ which means that φ *is weaker than* ψ *at* 0. Analogously, $\varphi \overset{s}{\sim} \psi$ iff $\varphi \overset{s}{\prec} \psi$ and $\psi \overset{s}{\prec} \varphi$.

Note. If μ is finite then $L^{\Psi_\varphi}(\mu) = L^\circ(\mu)$ for any Orlicz function φ. Therefore, for any two Orlicz functions φ and ψ, the Orlicz functions Ψ_φ and Ψ_ψ are equivalent at $+\infty$ (see [3]). This follows also immediately by the inequalities

$$\frac{\lambda}{1+\lambda}\,\frac{|u|}{1+|u|} \le \frac{\lambda}{1+\lambda} \le \frac{\varphi(u)}{1+\varphi(u)} = \Psi_\varphi(u) \le 1 = \frac{1+\lambda}{\lambda}\,\frac{\lambda}{1+\lambda} \le \frac{1+\lambda}{\lambda}\,\frac{|u|}{1+|u|}$$

for any Orlicz function φ whenever $|u| \ge \max(a,\lambda)$, where a and λ are positive numbers such that $\varphi(a) = \lambda$.

For every purely atomic measure such that the sequence of measures of all atoms is bounded from below by a positive number, we have $L^{\Psi_\varphi}(\mu) = L^\varphi(\mu)$. This follows by the relation $\Psi_\varphi \overset{s}{\sim} \varphi$. To prove this relation, choose $a>0$ in such a manner that $0 < \lambda = \varphi(a) < +\infty$. Then, for every $u \in \mathbb{R}$ with $|u| \le a$, we have

$$\frac{\varphi(u)}{1+\lambda} \le \frac{\varphi(u)}{1+\varphi(u)} \le \varphi(u),$$

which proves the desired relation.

It is well known that, in the case of a finite measure μ, $L^\circ(\mu)$ is a Frechet space endowed with the F-norm $\|\cdot\|$ defined by

$$\|x\| = \int_T \frac{|x(t)|}{1+|x(t)|}\,d\mu,$$

and, for any sequence (x_n) in $L^\circ(\mu)$ we have $\|x_n\| \to 0$ as $n \to +\infty$ iff $x_n \to 0(\mu)$ as $n \to +\infty$ (cf. [1] and [7]). Therefore, $L^\circ(\mu)$ is an Orlicz space generated by the Orlicz function $\Psi_o(u) = |u|/(1+|u|)$ for any $u \in \mathbb{R}$, provided μ is finite. It seems to be interesting to describe the space $L^{\Psi_o}(\mu)$ for every measure μ. That description can be deduced from the following theorem.

Theorem 1. *For every measure μ and every Orlicz function φ we have*

$$L^{\Psi_\varphi}(\mu) = L^{\varphi}(\mu) + \tilde{L}(\mu).$$

Proof. Let $x \in L^{\Psi_\varphi}(\mu)$. There exists $\alpha>0$ such that $\int_T \Psi_\varphi(\alpha x(t))d\mu < +\infty$. Let $a>0$ be such that $0 < \lambda = \varphi(a) < +\infty$. Define

$$A = \{t \in T: \alpha|x(t)| > a\}.$$

Obviously, $\mu(A) < +\infty$, so $x\chi_A \in \tilde{L}(\mu)$. We have

$$\frac{\varphi(u)}{1+\lambda} \le \frac{\varphi(u)}{1+\varphi(u)}$$

whenever $|u| \le a$. Therefore,

$$\frac{\lambda}{1+\lambda}\int_{T\setminus A} \varphi(\alpha x(t))d\mu \le \int_{T\setminus A} \frac{\varphi(\alpha x(t))}{1+\varphi(\alpha x(t))}\, d\mu < +\infty,$$

whence

$$\int_{T\setminus A} \varphi(\alpha x(t))\, d\mu < +\infty, \quad \text{i.e.,} \quad x\chi_{T\setminus A} \in L^{\varphi}(\mu).$$

Hence, $x = x\chi_A + x\chi_{T\setminus A} \in \tilde{L}(\mu) + L^{\varphi}(\mu)$.

Assume now that $x = y+z$, where $y \in \tilde{L}(\mu)$ and $z \in L^{\varphi}(\mu)$. Since $\Psi_\varphi(u) \le \varphi(u)$ for every $u \in \mathbb{R}$, we infer that $z \in L^{\Psi_\varphi}(\mu)$. Further, it is obvious that $y \in L^{\Psi_\varphi}(\mu)$. By the linearity of $L^{\Psi_\varphi}(\mu)$, $x \in L^{\Psi_\varphi}(\mu)$ and the proof is finished.

Remark. $\tilde{L}(\mu) + L^{\varphi}(\mu)$ is the space of all functions from $L^{0}(\mu)$ which belong to $L^{\varphi}(\mu)$ outside a set of finite measure.

Now, we shall give a characterization of the convergence in $L^{\Psi_\varphi}(\mu)$.

Theorem 2. *Let (x_n) be a sequence in $L^{\Psi_\varphi}(\mu)$. Then $\|x_n\|_{\Psi_\varphi} \to 0$ as $n \to +\infty$ iff $x_n \to 0(\mu)$ as $n \to +\infty$ and for every subsequence (x_{n_k}) of (x_n) there exists a subsequence $(x_{n_{k_l}})$ of (x_{n_k}) and a set $A \in \Sigma$ of finite measure such that $\|x_{n_{k_l}}\chi_{T\setminus A}\|_\varphi \to 0$ as $l \to +\infty$.*

Proof. Sufficiency. Assume that (x_{n_k}) is an arbitrary subsequence of (x_n), let $(x_{n_{k_l}})$ and $A \in \Sigma$ with $\mu(A) < +\infty$ be such that $\|x_{n_{k_l}}\chi_{T\setminus A}\|_\varphi \to 0$ as $l \to +\infty$. In virtue of the double extract of subsequences theorem it suffices to prove that

$$\int_T \Psi_\varphi(\alpha x_{n_{k_l}}(t))\, d\mu \to 0 \quad \text{as} \quad l \to +\infty, \tag{2}$$

for every $\alpha>0$. Take an arbitrary $\varepsilon>0$ and choose $\delta>0$ in such a manner that $\varphi(\alpha\delta)\mu(A) < \varepsilon/3$. Let $l_0 \in \mathbb{N}$ be such that $I_\varphi(\alpha x_{n_{k_l}}\chi_{T\setminus A}) < \varepsilon/3$ for every $l \geq l_0$. Then we have, for any $l \geq l_0$,

$$(3) \qquad \int_T \Psi_\varphi(\alpha x_{n_{k_l}}(t))d\mu = \int_A \Psi_\varphi(\alpha x_{n_{k_l}}(t))d\mu + \int_{T\setminus A} \Psi_\varphi(\alpha x_{n_{k_l}}(t))d\mu$$

$$\leq \int_A \Psi_\varphi(\alpha x_{n_{k_l}}(t))d\mu + \int_{T\setminus A} \varphi(\alpha x_{n_{k_l}}(t))d\mu \leq \int_A \Psi_\varphi(\alpha x_{n_{k_l}}(t))d\mu + \varepsilon/3.$$

Define

$$(4) \qquad A_n^\delta = \{t\in T: |x_n(t)| > \delta\}, \qquad n=1,2,\ldots;\ \delta>0.$$

We have

$$(5) \quad \int_A \Psi_\varphi(\alpha x_{n_{k_l}}(t))d\mu = \int_{A\cap A^\delta_{n_{k_l}}} \Psi_\varphi(\alpha x_{n_{k_l}}(t))d\mu + \int_{A\setminus A^\delta_{n_{k_l}}} \Psi_\varphi(\alpha x_{n_{k_l}}(t))d\mu$$

$$\leq \mu(A^\delta_{n_{k_l}}) + \Psi_\varphi(\alpha\delta)\mu(A) \leq \varepsilon/3 + \varepsilon/3 = 2\varepsilon/3$$

for every $l\in\mathbb{N}$ large enough. Combining (3) and (5) we get $I_{\Psi_\varphi}(\alpha x_{n_{k_l}}) < \varepsilon$ for large $l\in\mathbb{N}$, i.e., condition (2) is proved.

Necessity. Without loss of generality, we can consider the sequence (x_n) instead of its subsequence (x_{n_k}). Assume that

$$(6) \qquad \|x_n\|_{\Psi_\varphi} \longrightarrow 0 \quad \text{as} \quad n \longrightarrow +\infty.$$

Define, for every $\varepsilon>0$ and $n\in\mathbb{N}$, the sets A_n^ε as in formula (4). We have

$$(7) \qquad \mu(A_n^\varepsilon)\Psi_\varphi(\alpha\delta) \leq \int_{A_n^\varepsilon} \Psi_\varphi(\alpha x_n(t))d\mu \longrightarrow 0 \quad \text{as} \quad n \longrightarrow +\infty,$$

for every $\alpha>0$. Now, choose $\alpha>0$ in such a manner that $\varphi(\alpha\varepsilon) > 0$. By (7), $\mu(A_n) \longrightarrow 0$ as $n \longrightarrow +\infty$ for every $\varepsilon>0$, i.e., $x_n \longrightarrow 0(\mu)$ as $n \longrightarrow +\infty$.

Let $a>0$ be such that $\lambda = \varphi(a) > 0$. We have $\mu(A_n^a) \longrightarrow 0$ as $n \longrightarrow +\infty$. Thus, there exists a sequence (n_k) of natural numbers such that $\mu(A^a_{n_k}) \leq 2^{-k}$ for $k=1,2,\ldots$. Define $A = \bigcup_{k=1} A^a_{n_k}$. We have $\mu(A) \leq \sum_{k=1} 2^{-k} = 1$, and $|x_{n_k}(t)| \leq a$ for every $k\in\mathbb{N}$ and $t \in T\setminus A$. Therefore, in virtue of assumption (6), we have

$$\frac{1}{1+\varphi(a)} \int_{T\setminus A} \varphi(\alpha x_{n_k}(t))d\mu \leq \int_{T\setminus A} \frac{\varphi(\alpha x_{n_k}(t))}{1+\varphi(\alpha x_{n_k}(t))}\, d\mu \longrightarrow 0$$

as $n \longrightarrow +\infty$ for every $\alpha>0$, i.e., $\|x_{n_k}\chi_{T\setminus A}\|_\varphi \longrightarrow 0$ as $k \longrightarrow +\infty$. The proof is finished.

THEOREM 3. *Let μ be an arbitrary measure and Φ, φ, ψ be a triple of Orlicz functions such that $\Phi \overset{s}{\sim} \varphi$, $\Phi \overset{l}{\sim} \psi$, $\varphi \overset{s}{\prec} \psi$ and $\psi \overset{l}{\prec} \varphi$. Then $L^\Phi(\mu) = L^\varphi(\mu) + L^\psi(\mu)$.*

Proof. Take $x \in L^{\Phi}(\mu)$ and define

$$A = \{t \in T: |x(t)| \le 1\}, \qquad B = T \setminus A.$$

We have $x\chi_A \in L^{\Phi}(\mu)$ and $x\chi_B \in L^{\Phi}(\mu)$. By the relations $\Phi \overset{s}{\sim} \varphi$, $\Phi \overset{l}{\sim} \psi$, we get $x\chi_A \in L^{\varphi}(\mu)$ and $x\chi_B \in L^{\psi}(\mu)$; so $x \in L^{\varphi}(\mu) + L^{\psi}(\mu)$. Therefore

$$L^{\Phi}(\mu) \subset L^{\varphi}(\mu) + L^{\psi}(\mu). \tag{8}$$

Now, assume that $x \in L^{\varphi}(\mu) + L^{\psi}(\mu)$, i.e., $x = x_1 + x_2$, where $x_1 \in L^{\varphi}(\mu)$, $x_2 \in L^{\psi}(\mu)$. Define

$$A_i = \{t \in T: |x_i(t)| \le 1\}; \quad i = 1,2.$$

We have $x_1\chi_{A_1}$, $x_2\chi_{A_2} \in L^{\Phi}(\mu)$. Since $x_1\chi_{T\setminus A_1} \in L^{\varphi}(\mu)$ and $\psi \overset{l}{\prec} \varphi$, we have $x_1\chi_{T\setminus A_1} \in L^{\psi}(\mu)$. Now, in virtue of the relation $\Phi \overset{l}{\sim} \psi$, we get $x_1\chi_{T\setminus A_1} \in L^{\Phi}(\mu)$. We have $x_2\chi_{A_2} \in L^{\psi}(\mu)$. Therefore, the relation $\varphi \overset{s}{\sim} \psi$ yields $x_2\chi_{A_2} \in L^{\varphi}(\mu)$. Now, in view of the relation $\varphi \overset{s}{\sim} \Phi$, we get $x_2\chi_{A_2} \in L^{\Phi}(\mu)$. Thus, $x_1, x_2 \in L^{\Phi}(\mu)$. By the linearity of $L^{\Phi}(\mu)$, $x \in L^{\Phi}(\mu)$. Therefore,

$$L^{\varphi}(\mu) + L^{\psi}(\mu) \subset L^{\Phi}(\mu). \tag{9}$$

Inclusions (8) and (9) yield the desired equality.

Theorem 4. *Let μ be an arbitrary measure and Φ, φ, ψ be a triple of Orlicz functions such that $\Phi \overset{s}{\sim} \varphi$, $\Phi \overset{l}{\sim} \psi$, $\psi \overset{s}{\prec} \varphi$ and $\varphi \overset{l}{\prec} \psi$. Then $L^{\Phi}(\mu) = L^{\varphi}(\mu) \cap L^{\psi}(\mu)$.*

Proof. Assume that $x \in L^{\Phi}(\mu)$ and define

$$A = \{t \in T: |x(t)| \le 1\}.$$

Then $x\chi_A \in L^{\Phi}(\mu)$ and, in virtue of $\Phi \overset{s}{\sim} \varphi$, we get $x\chi_A \in L^{\varphi}(\mu)$. Now, the relation $\psi \overset{s}{\prec} \varphi$ yields $x\chi_A \in L^{\psi}(\mu)$, i.e., $x\chi_A \in L^{\varphi}(\mu) \cap L^{\psi}(\mu)$.

We have $x\chi_{T\setminus A} \in L^{\Phi}(\mu)$ and, in virtue of $\Phi \overset{l}{\sim} \psi$, we have $x\chi_{T\setminus A} \in L^{\psi}(\mu)$. Next, in view of $\varphi \overset{l}{\prec} \psi$, we get $x\chi_{T\setminus A} \in L^{\varphi}(\mu)$, i.e., $x\chi_{T\setminus A} \in L^{\varphi}(\mu) \cap L^{\psi}(\mu)$.

The linearity of $L^{\varphi}(\mu) \cap L^{\psi}(\mu)$ yields now $x \in L^{\varphi}(\mu) \cap L^{\psi}(\mu)$, whence the inclusion $L^{\Phi}(\mu) \subset L^{\varphi}(\mu) \cap L^{\psi}(\mu)$ follows.

Assume now that $x \in L^{\varphi}(\mu) \cap L^{\psi}(\mu)$ and define the set A as above. Then $\Phi \overset{s}{\sim} \varphi$ yields $x\chi_A \in L^{\Phi}(\mu)$ and $\Phi \overset{l}{\sim} \psi$ yields $x\chi_{T\setminus A} \in L^{\Phi}(\mu)$. Therefore, by the linearity of $L^{\Phi}(\mu)$, $x \in L^{\Phi}(\mu)$. Hence the inclusion $L^{\varphi}(\mu) \cap L^{\psi}(\mu) \subset L^{\Phi}(\mu)$ follows and the proof is finished.

Example. Assume that $0 < p \le q \le +\infty$. Define $\varphi(u) = |u|^p$, $\psi(u) = |u|^q$, $\Phi(u) = \max(|u|^p, |u|^q)$, $\tilde{\Phi}(u) = \min(|u|^p, |u|^q)$. Then $\Phi \overset{s}{\sim} \varphi$, $\Phi \overset{l}{\sim} \psi$, $\psi \overset{s}{\prec} \varphi$, $\varphi \overset{l}{\prec} \psi$, $\tilde{\Phi} \overset{s}{\sim} \psi$, $\tilde{\Phi} \overset{l}{\sim} \varphi$. Then, in virtue of Theorems 3 and 4, we get

$$L^{\Phi}(\mu) = L^p(\mu) \cap L^q(\mu) \quad \text{and} \quad L^{\tilde{\Phi}}(\mu) = L^p(\mu) + L^q(\mu).$$

The last equalities follow also by the results of [2].

An Orlicz function φ is said *to satisfy the condition Δ_2 for all $u \in \mathbb{R}$ (at infinity) [at zero]* if there exist positive constants K and a such that $0 < \varphi(a) < +\infty$ and the inequality $\varphi(2u) \le K\varphi(u)$ holds for all $u \in \mathbb{R}$ (for $|u| \ge a$) [for $|u| \le a$].

We shall say that an Orlicz function φ satisfies the suitable condition Δ_2 if it satisfies the condition Δ_2 for all $u \in \mathbb{R}$ (at infinity) [at zero] when μ is a nonatomic infinite (a nonatomic finite) [the counting] measure.

For a given Orlicz function φ, let us define

$$E^{\varphi}(\mu) = \{x \in L^{0}(\mu) : I_{\varphi}(\lambda x) < +\infty \text{ for every } \lambda > 0\},$$

and endow this space with the F-norm $\|\cdot\|_{\varphi}$ induced from $L^{\varphi}(\mu)$.

Recall that a sequence (x_n) in $L^{\varphi}(\mu)$ is *modularly convergent to 0* if $I_{\varphi}(\lambda x_n) \to 0$ as $n \to +\infty$ for some $\lambda > 0$. It is well known (cf. [5]) that the modular convergence and the norm convergence coincide in $L^{\varphi}(\mu)$ if and only if φ satisfies the suitable condition Δ_2.

Now, we shall prove a little more.

Theorem 5. *Let φ be a finite Orlicz function not satisfying the suitable condition (Δ_2). Then, for every $\alpha > 0$ there is a sequence (x_n) in $E^{\varphi}(\mu)$ such that $I_{\varphi}(\alpha x_n) \to 0$ but $I_{\varphi}(\beta x_n) \not\to 0$ for every $\beta > \alpha$.*

Proof. We restrict ourselves only to the case of a nonatomic infinite measure. First, assume that φ vanishes outside 0 and define

$$u_o = \sup \{u > 0 : \varphi(u) = 0\}.$$

Given an arbitrary $\alpha > 0$, choose $u > 0$ in such a manner that $\alpha u = u_o$ and let $A \in \Sigma$ be an arbitrary set of positive measure. Define $x_n = u\chi_A$ for $n = 1, 2, \ldots$. Then $I_{\varphi}(\alpha x_n) = \varphi(u_o)\mu(A) = 0$ for every $n \in \mathbb{N}$. Moreover, for every $\beta > \alpha$, we have $I_{\varphi}(\beta x_n) = \varphi(\beta u_o/\alpha)\mu(A) > 0$; so $I_{\varphi}(\beta x_n) \not\to 0$.

Now, assume that φ vanishes only at zero and does not satisfy the condition Δ_2 for all $u \in \mathbb{R}$. Then there exists a sequence (u_n) of positive reals such that

$$\varphi((1+\tfrac{1}{n})u_n) > 2^n \varphi(u_n); \quad n = 1, 2, \ldots .$$

Take a set (A_n) of measurable sets such that $\mu(A_n) = 2^{-n}(\varphi(u_n))^{-1}$ and define $x_n = u_n\chi_A$ for any $n\in\mathbb{N}$. Then

(10) $$I_\varphi(x_n) = 2^{-n} \quad \text{and} \quad I_\varphi((1+\tfrac{1}{n})x_n) \geq 1 \qquad (\forall\ n\in\mathbb{N}).$$

For an arbitrary $\alpha>0$ put $y_n = \alpha^{-1}x_n$. Then $I_\varphi(\alpha y_n) \longrightarrow 0$ and, for every $\beta > \alpha$, $I_\varphi(\beta y_n) \not\longrightarrow 0$. The proof is finished.

For every sequence (x_n) in $L^{\Phi}(\mu)$ (or even in $E^{\Phi}(\mu)$) we can define *the characteristic of the modular convergence*

$$\alpha((x_n)) = \sup \{\alpha>0:\ I_\varphi(\alpha x_n) \longrightarrow 0 \text{ as } n \longrightarrow +\infty\}$$

($\sup \emptyset \overset{\mathrm{def}}{=} 0$). Obviously, $\|x_n\|_\varphi \longrightarrow 0$ if and only if $\alpha((x_n)) = +\infty$ and (x_n) is modularly convergent to zero if and only if $0 < \alpha((x_n)) \leq +\infty$.

REFERENCES

[1] A.Alexiewicz, *Functional Analysis*, PWN, vol.49, Warsaw 1969, (in Polish).

[2] H.Hudzik, *Intersections and algebraic sums of Musielak-Orlicz spaces*, Portugaliae Math. 40, 3(1981), 287-296.

[3] W.Matuszewska, *Spaces of φ-integrable functions I and II*, Commentationes Math. (Prace Mat.), 121-139 and 149-174 (in Polish).

[4] S.Mazur, W.Orlicz, *On some classes of linear spaces*, Studia Math. 17 (1958), 97-119.

[5] J.Musielak, *Orlicz spaces and modular spaces*, Lecture Notes in Math. 1034, Springer Verlag 1983.

[6] J.Musielak, W. Orlicz, *On modul;ar spaces*, Studia Math. 18 (1959), 49-65.

[7] S.Rolewicz, *Metric linear spaces*, Polish Scientific Publishers, Warsaw 1984.

THE MIXED TOPOLOGY ON L^p-SPACES AND ORLICZ SPACES

MARIAN NOWAK

Poznań, Poland

In [6] and [7] we examined the mixed topology (cf. definition below) on normed function spaces from the point of view of the theory of locally solid Riesz spaces.In these investigations the notion of a uniformly absolutely continuous seminorm is of importance. In [3] T.Andô has given a description of uniformly absolutely continuous seminorms on Orlicz spaces L^{φ} defined by N-functions φ for finite measures.

In this paper we use the Andô result to give an important description of the mixed topology on L^{φ} in terms of some family of norms defined by certain N-functions depending on φ. As an application we obtain various topological characterizations of a natural two-norm convergence in Orlicz spaces. In particular we obtain interesting characterizations of this convergence in L^p-spaces for $1<p<\infty$.

1. Preliminaries.

In this section we shall recall some notions and results concerning uniformly-Lebesgue topologies and the mixed topology on Orlicz spaces.

Throughout this paper (Ω,Σ,μ) will be a finite measure space. Let L° denote the set of equivalence classes of all real-valued μ-measurable functions defined and a.e. finite on Ω. Let $\tau_{\circ}$ denote the topology of the F-norm $\|\cdot\|_{\circ}$ defined on L° by

$$\|x\|_{\circ} = \int_{\Omega} \frac{|x(t)|}{1+|x(t)|}\, d\mu \ .$$

This topology is the topology of convergence in measure. We will write $x_n \to x$ (μ) whenever $x_n \to x$ for $\tau_{\circ}$. For $\varepsilon>0$ we will write

$$B_{\circ}(\varepsilon) = \{x\in L^{\circ}:\ \|x\|_{\circ} \le \varepsilon\}.$$

By *an N-function* we mean a convex function $\varphi:[0,\infty) \to [0,\infty)$ such that $\varphi(0) = 0$ and $\varphi(u)/u \to \infty$ as $u \to \infty$ (cf. [3]). We denote by L^φ the Orlicz space associated with an N-function φ (see [3],[4],[5] for further details). Let

$$E^\varphi = \left\{x\in L^\circ: \int_\Omega \varphi(\lambda\,|x(t)|)d\mu < \infty \text{ for all } \lambda>0\right\}.$$

Let τ_φ denote the topology of the Luxemburg norm $\|\cdot\|_\varphi$ defined on L^φ by

$$\|x\|_\varphi = \inf\left\{\lambda>0: \int_\Omega \varphi(|x(t)|/\lambda)d\mu \le 1\right\}$$

(see [4]). For $r>0$ we will write

$$B_\varphi(r) = \{x\in L^\varphi: \|x\|_\varphi \le r\}.$$

Given an N-function φ we shall denote by φ^* the function complementary to φ in the Young sense, i.e.,

$$\varphi^*(v) = \sup\{uv - \varphi(u): u \ge 0\} \quad \text{for } v \ge 0.$$

In a natural way, a kind of sequential convergence, called *γ-convergence* (or *two-norm convergence*), can be defined in L^φ: a sequence (x_n) in L^φ is γ-convergent to $x\in L^\varphi$ if $x_n \to x$ (μ) and $\sup_n \|x_n\|_\varphi < \infty$ (see [1],[8]).

We say that a locally solid topology τ on L^φ satisfies *the uniformly-Lebesgue property* (or that *τ is an uniformly-Lebesgue topology*) if $x_n \to 0$ (μ) with $\sup_n \|x_n\|_\varphi < \infty$ implies $x_n \to 0$ for τ (see [6]).

A Riesz seminorm ρ on L^φ is called *uniformly absolutely continuous* if $\sup\{\rho(x\chi_E): \|x\|_\varphi \le 1\} \to 0$ as $\mu(E) \to 0$, where χ_E denotes the characteristic function of the set $E \subset \Omega$ (see [3,p.171],[6]).

The following theorem characterizes uniformly absolutely continuous seminorms on L^φ ([6, Theorem 1.2]).

THEOREM 1.1. *For a Riesz seminorm ρ on L^φ the following statements are equivalent:*

(i) ρ is uniformly absolutely continuous,

(ii) $x_n \to 0$ (μ) with $\sup_n \|x_n\|_\varphi < \infty$ implies $\rho(x_n) \to 0$.

The theory of γ-convergence is closely related to the theory of mixed topologies ([8]). Thus, with γ-convergence in L^φ we can associate the mixed topology $\gamma[\tau_\varphi, \tau_\circ|_{L^\varphi}]$ (see [8] for details). It has a base of neighborhoods of zero consisting of all sets of the form:

$$\bigcup_{N=1}^{\infty}\Big(\sum_{n=1}^{N} B_o(\varepsilon_n) \cap n\cdot B_\varphi(r) \Big),$$

where (ε_n) is a sequence of positive numbers and $r>0$. Henceforth we shall write γ_φ instead of $\gamma[\tau_\varphi,\tau_o{}_{|L^\varphi}]$. If $\varphi(u) = u^p$ for $u\geq 0$, where $1 < p < \infty$, then we will write γ_p and $\|\cdot\|_p$ instead of γ_φ and $\|\cdot\|_\varphi$ respectively.

The basic properties of γ_φ are summarized in the following theorem.

Theorem 1.2. *The following statements hold:*

(i) γ_φ *is a locally convex-solid topology,*

(ii) γ_φ *is the finest uniformly-Lebesgue topology on* L^φ,

(iii) the following conditions are equivalent:

(1) $$x_n \longrightarrow 0 \text{ for } \gamma_\varphi,$$

(2) $$x_n \longrightarrow 0 \ (\mu) \text{ and } \sup \|x_n\|_\varphi < \infty,$$

(3) $$\int_\Omega |x_n(t)y(t)|\, d\mu \longrightarrow 0 \quad \text{for every } y \in E^{\varphi^*},$$

(4) *if* τ *is a Hausdorff uniformly-Lebesgue topology on* L^φ,

$$x_n \longrightarrow 0 \text{ for } \tau \text{ and } \sup_n \|x_n\|_\varphi < \infty.$$

Proof. (i) See [6,Theorem 4.1]. (ii) See [6,Theorem 2.1]. (iii) It follows from [7,Theorem 1.2] and [6,Theorem 1.4].

We say that an N-function φ increases essentially more rapidly than another N-function ψ, briefly $\psi \ll \varphi$, if, for every $c>0$, $\psi(cu)/\varphi(u) \longrightarrow 0$ as $u \longrightarrow \infty$ (see [4.p.114]).

T.Andô [3,Theorem 4] has presented the following description of uniformly absolutely continuous seminorms on L^φ.

Theorem 1.3. *For a Riesz seminorm* ρ *on* L^φ *the following statements are equivalent:*

(i) ρ *is uniformly absolutely continuous,*

(ii) there exists an N-function ψ *such that* $\psi \ll \varphi$ *and* $\rho(x) \leq \|x\|_\psi$ *for* $x \in L^\varphi$.

2. Results.

We are now ready to state our main result.

Theorem 2.1. *Let φ be an N-function and let $\Psi^{\ll\varphi}$ be the collection of all N-functions ψ such that $\psi \ll \varphi$. Then the mixed topology γ_φ on L^φ is generated by the family $\{\|\cdot\|_{\psi}|_{L^\varphi}: \psi \in \Psi^{\ll\varphi}\}$.*

Proof. Denote by τ_* the linear topology on L^φ generated by the family $\{\|\cdot\|_{\psi}|_{L^\varphi}: \psi \in \Psi^{\ll\varphi}\}$. According to Theorem 1.3, for each $\psi \in \Psi^{\ll\varphi}$, the norm $\|\cdot\|_\psi$ is uniformly absolutely continuous on L^φ, and, by Theorem 1.1, we obtain that τ_* satisfies the uniformly-Lebesgue property. Therefore, by Theorem 1.2.(ii), $\tau_* \subset \gamma_\varphi$.

Since γ_φ is a locally convex-solid topology there exists a family $\{\rho_\alpha\}$ of Riesz seminorms on L^φ that generates γ_φ (cf. [2, Theorem 6.1]). Then each seminorm ρ_α is uniformly absolutely continuous on L^φ. Indeed, let $x_n \to 0$ (μ) and $\sup_n \|x_n\|_\varphi \to \infty$. Then, by Theorem 1.2.(ii), $x_n \to 0$ for γ_φ, so $\rho_\alpha(x_n) \to 0$. In view of Theorem 1.1 this means that ρ_α is uniformly absolutely continuous. According to Theorem 1.3 we obtain $\gamma_\varphi \subset \tau_*$. Thus the proof is finished.

Combining Theorem 1.2 with Theorem 2.1 we obtain the following interesting characterizations of γ-convergence in L^φ.

Corollary 2.2. *Let φ be an N-function. Then for a sequence (x_n) in L^φ the following statements are equivalent:*

(i) $x_n \to 0$ (μ) *and* $\sup_n \|x_n\|_\varphi < \infty$,

(ii) $\|x_n\|_\psi \to 0$ *for some N-function ψ such that $\psi \ll \varphi$ and* $\sup_n \|x_n\|_\varphi < \infty$,

(iii) $\|x_n\|_\psi \to 0$ *for every N-function ψ such that $\psi \ll \varphi$,*

(iv) $\int_\Omega |x_n(t)y(t)|\,d\mu \to 0$ *for every* $y \in E^{\varphi^*}$.

In particular, we have:

Corollary 2.3. *Let $p>1$. Then the mixed topology γ_p on L^p is generated by the family $\{\|\cdot\|_{\psi}|_{L^p}\}$, where ψ is taken over all N-functions such that $\psi(u)/u^p \to 0$ as $u \to \infty$.*

Corollary 2.2. *Let $p>1$. Then for a sequence (x_n) in L^p the following statements are equivalent:*

(i) $x_n \to 0$ (μ) *and* $\sup_n \left(\int_\Omega |x_n(t)|^p d\mu\right) < \infty$,

(ii) $\|x_n\|_q \to 0$ *for some* $0 < q < p$ *and* $\sup_n \left(\int_\Omega |x_n(t)|^p d\mu\right) < \infty$,

(iii) $\|x_n\|_\psi \to 0$ *for every N-function* ψ *such that* $\psi(u)/u^p \to 0$ *as* $u \to \infty$,

(iv) $\int_\Omega |x_n(t)y(t)|\,d\mu \to 0$ *for every* $y \in L^q$, *where* $q = \frac{p}{p-1}$.

REFERENCES

[1] A.ALEXIEWICZ, *On two-norm convergence*, Studia Math. 14 (1954), 49–56.

[2] C.D.ALIPRANTIS, O. BURKINSHAW, *Locally solid Riesz spaces*, New York 1978.

[3] T.ANDÔ, *Weakly compact sets in Orlicz spaces*, Canad. J.Math. 14 (1962), 170–176.

[4] M.A.KRASNOSELSKI, YA. B. RUTICKII, *Convex functions and Orlicz spaces*, Groningen 1961.

[5] W.A.LUXEMBURG, *Banach function spaces*, Thesis, Delft 1955.

[6] M.NOWAK, *Mixed topology on normed function spaces I*, Bull. Pol. Ac.: Math. 36 (1988), 35–46.

[7] M.NOWAK, *Mixed topology on normed function spaces II*, Bull. Pol. Ac.: Math. 36 (1988), 47–51.

[8] A.WIWEGER, *Linear spaces with mixed topologies*, Studia Math. 20 (1961), 47–68.

MAXIMAL INEQUALITIES AND IMBEDDINGS IN WEIGHTED ORLICZ SPACES

LUBOŠ PICK

Praha, Czechoslovakia

1. Our aim is to study integral inequalities involving either Hardy-Littlewood maximal operator or identity operator in the context of weighted Orlicz spaces.

Throughout, U is a measurable subset of $\mathbb{R}^n$; ρ,σ are weights (measurable and almost everywhere positive functions); and the capitals F, P, H, N stand for Young functions. The complementary function to F is defined by the formula $\tilde{F}(t) = \sup_{\tau>0} |t\tau - F(\tau)|$, $t\geq 0$. For basic properties of Young functions and Orlicz spaces we refer to [7]. We denote with $f(E)$ the integral of f over E and with f_E the integral mean $f(E)/|E|$, $|E|$ is the Lebesgue measure of E. In what follows, Q will always mean nondegenerate cube in $\mathbb{R}^n$ with sides parallel to coordinate axes.

The Hardy-Littlewood maximal operator defined for $f \in L_{1,loc}$ by $Mf(x) = \sup_{Q\ni x} |f|_Q$ is known to be of weak type (1,1) and of type (p,p) whenever p>1. It is shown e.g. in recent Gallardo paper [3] that M acts boundedly from the Orlicz space L_F into L_F iff $\tilde{F} \in \Delta_2$. The question "When M maps boundedly $L_{p,\rho}$ into $L_{p,\rho}$?" had remained open for rather long period. In 1972, B.Muckenhoupt ([10]) solved this problem. More precisely, he proved the following two results:

a) the inequality

$$\int (Mf)^p \rho \leq C\int |f|^p \rho \tag{1}$$

holds with C independent of f iff $\rho \in A_p$, that is,

$$\sup_Q (\rho_Q)\,[(\rho^{-1/(p-1)})_Q]^{p-1} = C < \infty\ ;$$

b) the inequality

$$\rho(\{Mf>\lambda\}) \leq C\cdot\int(|f(x)|/\lambda)^p \sigma(x)dx \tag{2}$$

holds with C independent of f iff $(\rho,\sigma) \in A_p$, that is,

$$\sup_Q \langle\rho_Q\rangle\, [\langle\sigma^{-1/(p-1)}\rangle_Q]^{p-1} = C < \infty .$$

We call the estimates of type (1) or (2) the strong or weak type maximal inequality, respectively. The fundamental result of Muckenhoupt was continued by a large number of subsequent papers, and, actually, has been generalized in many directions.

2. There are at least two different ways how to write the analogue of the weak type maximal inequality (2) in Orlicz space. Indeed, one may consider either

(3) $$\rho(\{Mf>\lambda\}) \le C\cdot\int F(|f|/\lambda)\sigma,$$

or

(4) $$\rho(\{Mf>\lambda\})\, F(\lambda) \le C\cdot\int F(|f|)\sigma.$$

It is needless to say that if (4) is valid for all f and λ, then the same holds for (3), that is, (4) implies (3). Employing elementary but useful fact that

(5) $$p'/p = p'-1 = 1/(p-1),$$

we easily deduce that $(\rho,\sigma) \in A_p$ iff

$$\sup_Q \int_Q \left(\frac{\rho_Q}{\sigma(x)}\right)^{p'} \frac{\sigma(x)}{\rho(Q)}\, dx = C < \infty.$$

This observation offers the analogous statement for Orlicz spaces:

(6) $$\sup_Q \int_Q \tilde{F}\left(\varepsilon\cdot\frac{\rho_Q}{\sigma(x)}\right) \frac{\sigma(x)}{\rho(Q)}\, dx = C < \infty \qquad \text{for some } \varepsilon>0.$$

(A note on the role of ε: When dealing with general Young functions we cannot rely on the homogeneity or subhomogeneity arguments applicable if $F = t^p$ or $F \in \Delta_2$, respectively. We pay for that by introducing a constant into the argument of $\tilde{F}$.)

It is not hard to check that always (6) $\Longrightarrow$ (3). Considerably less trivial fact is that in special cases $F \sim t(1+\log t)^K$ the condition (6) becomes also necessary for (3) – see [2,8,11]. The method of proofs can be called "the limit approach" (cf. [9]); actually, it uses ideas about limit treatment of necessary and sufficient conditions in L_p cases when p tends to 1. However, the implication (3) $\Longrightarrow$ (6) for general F seems to be still open.

Turning attention to inequalities of type (4) we find rather better situation. Gallardo ([4]) characterized completely the pairs (ρ,σ) for which (4) holds under the assumption of reflexivity of Orlicz spaces, or, equivalently, $F \in \Delta_2$ and $\tilde{F} \in \Delta_2$. He proved that (ρ,σ) is such a pair iff

$$\sup_{\alpha,Q} \alpha\cdot\rho_Q\cdot\varphi([\varphi^{-1}(1/\alpha\sigma)]_Q) = C < \infty,$$

where $\alpha>0$ and $\varphi = F'$. The idea how to generalize Gallardo result for the general setting of Orlicz spaces (i.e., with no restriction on the growth of the Young function in question) is surprisingly simple. We define new auxiliary functions $R_F(t) := F(t)/t$, and $S_F(t) := \tilde{F}(t)/t$, $t>0$. The basic properties of these functions read as follows (in fact, they in some sense replace (5)):

(i) $R_F S_F(t) \le t$,

(ii) $\tilde{F}R_F(t) \le F(t) \le \tilde{F}(2R_F(t))$,

(iii) $R_F(t) \le F'(t)$,

(iv) $F'(t) \le C\, R_F(t)$ iff $F \in \Delta_2$.

With the help of the functions R_F and S_F we are able to characterize the maximal inequality of type (4) involving two Young functions.

Theorem 1. *The following statements are equivalent.*

(i) There exists $C,\varepsilon > 0$ such that for all functions f and $\lambda>0$

$$P(\varepsilon\lambda)\cdot\rho(\{Mf>\lambda\}) \le C \int_{\mathbb{R}^n} H(|f|)\sigma \ ;$$

(ii) there exist $B,\gamma > 0$ such that for all functions f and cubes Q

$$P(\gamma|f|_Q)\cdot\rho(Q) \le B\cdot\int_Q H(|f|)\sigma \ ;$$

(iii) there exist $K,\delta > 0$ such that for all cubes Q and $\alpha>0$

$$R_P(\delta\cdot(S_H(1/\alpha\sigma))_Q)\cdot\alpha\cdot\rho_Q \le K.$$

On taking $\rho = \sigma = 1$, it follows from Theorem 1 that M is of weak type (H,P) iff $P(t) \le H(Ct)$ – compare the well known easy fact that, for G increasing, $\int G(Mf) \le C\cdot\int G(|f|)$ for all f iff G is quasiconvex. Another simple corollary of Theorem 1 asserts that each of the statements (i) – (iii) guarantees the two weight doubling condition $\rho(2Q) \le C\sigma(Q)$, or, more precisely, $\rho(Q) \le C(E,Q)\sigma(E)$, $E \subset \mathbb{R}^n$, where $C(E,Q) = (P(\varepsilon\cdot|E\cap Q|/|Q|))^{-1}$. The proof of the theorem provides us also with the equivalence of the following two estimates:

(a) $P(\varepsilon\lambda)\cdot\rho(\{M\chi_E>\lambda\}) \le C\cdot\sigma(E)$, E measurable, $\lambda>0$,

(b) $P(\gamma\cdot|E|/|Q|) \le B\cdot\sigma(E)/\rho(Q)$, $E \subset Q$, Q arbitrary.

This observation gives a slight generalization of Proposition 1 in [6].

Similar results may be obtained for the one-sided maximal operator, maximal operator given by quasimetric in $\mathbb{R}^n$ etc. ([12,13]).

3. Now we shall deal with the one weight strong type maximal inequality

(7) $$\int F(Mf)\rho \le C\cdot\int F(|f|)\rho.$$

As it has been discussed above, the set of weights ρ for which (7) is valid for $F = t^p$, $p>1$, was described in [10]. This result was generalized to the more general setting of Orlicz spaces by Kerman and Torchinsky ([6]), namely, they showed that if F together with $\tilde{F}$ satisfies Δ_2, then (7) holds with C independent of f if and only if

$$\sup_{\alpha,Q} \alpha\cdot\rho_Q\cdot\varphi([\varphi^{-1}(1/\alpha\rho)]_Q) = C < \infty,$$

where $\varphi = F'$. Claiming to find necessary and sufficient conditions for (7) to hold with no apriori restrictions on the growth of F we arrive at the following particular result.

Theorem 2. *For $\delta>0$ define $\tilde{F}_\delta(t) = t\cdot S_F^{1+\delta}(t)$. Suppose FF_δ^{-1} is a Young function such that the complementary function satisfies Δ_2 for small δ. Then (7) is equivalent to the validity of the two conditions*

(i) $\tilde{F} \in \Delta_2$,

(ii) $\rho \in A_F^*$, *that is, for some $\varepsilon,K > 0$ and all $\alpha>0$, Q*

$$R_F(\varepsilon[S_F(1/\alpha\rho)]_Q)\cdot\alpha\cdot\rho_Q \le K.$$

The theorem may be found of some interest since $\tilde{F} \in \Delta_2$ is not presented as its assumption but as a statement. On the other hand, this is paid by another ugly looking assumption which is desired to be avoided.

4. Avantaggiati ([1]) and Kabaila ([5]) proved an interesting result on the imbeddings of weighted Lebesgue spaces: if $1 \le p < q < \infty$, then the imbedding $L_{q,\rho} \hookrightarrow L_{p,\sigma}$ holds iff

$$\int \sigma^{q/(q-p)}\rho^{p/(p-q)} < \infty.$$

Employing the above mentioned limit approach, again, we get that for $F_K \sim t\cdot\log^K t$ the imbedding $L_{F_K,\rho} \hookrightarrow L_{1,\sigma}$ holds iff there is $\varepsilon>0$ such that

$$\int \exp(\varepsilon\sigma/\rho)\cdot\rho < \infty.$$

However, before we come to general imbeddings of weighted Orlicz spaces, we are entailed to make some auxiliary notes. First, it is clear that the function H should grow "faster" than P in some sense. It turns out that the proper way of comparison of P and H is that

HP^{-1} is supposed to be a Young function, again. Then, on denoting with N the complementary function to HP^{-1}, the above limit result offers the condition

(8) $$\int_U N(\varepsilon\sigma/\rho)\cdot\rho < \infty \qquad \text{for some } \varepsilon>0,$$

for the imbedding

(9) $$L_{H,\rho}(U) \hookrightarrow L_{P,\sigma}(U).$$

Indeed, we obtain the following results.

Theorem 3. *The condition (8) is always sufficient for the imbedding (9).*

Theorem 4. *If at least one of the functions P, H obeys Δ_2, then (9) implies (8).*

The main tools of the proofs of Theorems 3, 4 are: Young inequality, indices of Young function, and the relations between modular and norm topologies on weighted Orlicz space. For details see [9].

It remains to consider the case when neither P nor H satisfies Δ_2. In such a situation the assumption of convexity of HP^{-1} must be replaced by some more convenient one. Namely, we have the concluding

Theorem 5. *Suppose that for C large $H(CP^{-1})$ is Young function and denote with N_C its complementary function. Then the imbedding (9) holds if and only if there exists $\varepsilon>0$ and $C \geq 1$ such that*

$$\int_U N_C(\varepsilon\sigma(x)/\rho(x))\cdot\rho(x)\, dx < \infty .$$

Finally, let us note that it is not hard to find an example showing that the assumption of the last theorem is not automatically implied by HP^{-1} being a Young function.

The results of the last section were obtained in cooperation with M.Krbec.

REFERENCES

[1] A. Avantaggiati, *On compact embedding theorems in weighted Sobolev spaces*, Czechoslovak Math. J. 29(104) (1979), 635–648.

[2] A. Carbery, S. Y. Chang, J. Garnett, *Weights and LlogL*, Pacific J. Math. 120, (1985), 33–45.

[3] D. Gallardo, *Orlicz spaces for which the Hardy-Littlewood maximal operator is bounded*, Publ. Mathematiques 32(1988), 261–266.

[4] D. Gallardo, *Weighted weak type integral inequalities for the Hardy-Littlewood maximal operator*, preprint.

[5] V.P. Kabaila, *On imbeddings of $L^p(\mu)$ into $L^r(\nu)$*, Lit. Mat. Sb. 21 (1981), 143–148 (in Russian).

[6] R.A. Kerman, A. Torchinsky, *Integral inequalities with weights for the hardy maximal function*, Studia Math. 71 (1982), 277–284.

[7] M.A. Krasnosel'skii, Ya. B. Rutitskii, *Convex functions and Orlicz spaces*, Nordhoff, Grøningen, 1961.

[8] M. Krbec, *Two weights type inequalities for the maximal function in the Zygmund class*, in: Function Spaces and Applications, Proceedings of the US-Swedish Seminar, Lund 1986, Springer Verlag, 1988.

[9] M. Krbec, L. Pick, *On imbeddings between weighted Orlicz spaces*, Math. Inst. Czech. Acad. Sci. 37(1988), 1–18 (preprint); to appear in Z. Anal. Anwend.

[10] B. Muckenhoupt, *Weighted norm inequalities for the Hardy maximal function*, Trans. Amer. Math. Soc. 165 (1972), 207–226.

[11] L. Pick, *Two weights type inequality for the maximal function in $L(1+\log^+ L)^K$*, in: Constructive Theory of Functions, Proceedings of the International Conference, Varna 1987, Bulg. Acad. Sci. 1988, 377–381.

[12] L. Pick, *Weighted inequalities for the Hardy-Littlewood maximal operator in Orlicz spaces*, Math. Inst. Czech. Acad. Sci. 46 (1989), 1–22 (preprint).

[13] L. Pick, *Weighted inequalities in Orlicz spaces*, Ph.D. thesis, Prague 1989 (in Czech).

ON SUBSPACES OF ORLICZ FUNCTION SPACES SPANNED BY SEQUENCES OF INDEPENDENT SYMMETRIC RANDOM VARIABLES

CÉSAR RUIZ

Madrid, Spain

The aim of this paper is to present a characterization of the subspaces spanned by sequences of independent symmetric random variables in Orlicz function spaces $L^{\varphi}(0,1)$. Namely, when these sequences are symmetric basic sequences, we show here that they span Orlicz sequences spaces ℓ^{ψ} (Theorem 6). Lindenstrauss and Tzafriri have solved in [8] these problems in the case of sequences of mutually disjoint functions in $L^{\varphi}(0,1)$. By the way, it is known the existence of Orlicz function spaces containing subspaces with symmetric basis which are not isomorphic to any Orlicz sequence space ℓ^{ψ} (see, f.i., [1]).

We use the disjointification process developed by Johnson and Schechtman in [6] which permits to connect independent random variables with disjoint functions. Some of the results presented here generalize some ones given earlier for L^p-spaces by Rosenthal in [13] and [14].

Let us start recalling some definitions. Let φ be an Orlicz function (i.e., a non-decreasing convex continuous function from $\mathbb{R}^+$ into $\mathbb{R}^+$ vanishing at 0, $\varphi(1) = 1$ and unbounded). Moreover we assume $\exists$ K>0 such that $\varphi(2x) < K\varphi(x)$ for every $x>0$ (in short $\varphi \in \Delta_2$). The Orlicz function space $L^{\varphi}(0,1)$ is the space of all measurable functions f on (0,1) such that $\int_0^1 \varphi(|f(t)|)dt < \infty$ equipped with the Luxemburg norm

$$\|f\|_{\varphi} = \inf \{u>0: \int_0^1 \varphi(u^{-1}|f(t)|)dt \le 1\}.$$

We say that two Orlicz functions φ and ψ are equivalent at ∞, we write $\varphi \overset{\infty}{\sim} \psi$, (resp. at 0, $\varphi \overset{0}{\sim} \psi$) if there exists K>1 and $x_0>0$ such

Supported in part by CAICYT grant 0338-84

that $K^{-1}\varphi(x) \le \psi(x) \le K\varphi(x)$ for every $x > x_0$ (resp. $x < x_0$). If φ and ψ are equivalent at ∞ and 0 we write $\varphi \approx \psi$.

By $\bar{\varphi}$ we mean an Orlicz function which is equivalent to φ at ∞ and with $\bar{\varphi}(x) = x^2$ if $x \in (0,1)$. The Orlicz function space $L^{\bar{\varphi}}(0,\infty)$ is defined as $L^{\varphi}(0,1)$ taking the interval $(0,\infty)$ instead of $(0,1)$. It is well known that $L^{\varphi}(0,1) = L^{\bar{\varphi}}(0,1)$ and the identity embedding is an isomorphism (see, f.i., [11]).

We will consider the weighted Orlicz sequence spaces $\ell^{\varphi}(w)$ defined by the sequences $x=(x_n)$ such that $\sum_1^{\infty}\varphi(|x_n|)w_n < \infty$, where $w=(w_n)$ is an arbitrary positive sequence, with the norm

$$\|x\|_{\varphi,w} = \inf\{u>0: \textstyle\sum_n \varphi(u^{-1}|x_n|)w_n \le 1\}.$$

The canonical unit vector basis, (e_n), is an unconditional basis sequence of $\ell^{\varphi}(w)$ (see [9] about basis sequences). When $w_n = 1$ for every $n \in \mathbb{N}$ we have the usual Orlicz sequence space ℓ^{φ} and then (e_n) is a symmetric basis.

An Orlicz function φ is called *distinctive* ([5]) if for every weight sequence $w = (w_n)$ the space $\ell^{\varphi}(w)$ is isomorphic to the Orlicz sequence space ℓ^{φ}. Among these functions we have the minimal functions introduced by Lindenstrauss and Tzafriri (see [9] and [4]).

The set $C_{\varphi}(0,\infty)$ has been defined in (12) as the closed convex hull of the compact set of continuous function on $[0,1]$

$$C_{\varphi}(0,\infty) = \overline{\{\psi \in C(0,1): \exists\ t>0 \text{ with } \psi(x) = \frac{\varphi(tx)}{\varphi(t)}\quad 0 \le x \le 1\}}.$$

We denote by $w \in \Lambda$ when the weight sequence $w = (w_n)$ has a subsequence (w_{n_k}) converging to 0 and $\sum_k w_{n_k} = \infty$. The following result is in [15].

Proposition 1. *A Banach space X with a symmetric basis is isomorphic to a subspace of $\ell^{\varphi}(w)$, with $w \in \Lambda$, if and only if there exists $\psi \in C_{\varphi}(0,\infty)$ such that the canonical basis of ℓ^{ψ} is equivalent to the symmetric basis of X.*

From Proposition 3.9 in [7] we get that the symmetric basis of X is equivalent to a block basic sequence of the canonical basis of $\ell^{\varphi}(w)$.

A sequence of measurable functions (f_n) in $L^{\varphi}(0,1)$ is called *a sequence of independent symmetric random variables* (since now on we will write in short i.s.r.v.) if

i) every f_n is symmetric, i.e., for every measurable set A of $\mathbb{R}^+$, $\lambda\{t: f_n(t) \in A\} = \lambda\{t: f_n(t) \in -A\}$ where λ is the Lebesgue measure;

ii) for every finite set $I \subset \mathbb{N}$ and for every family $(A_i)_{i\in I}$ of measurable sets of $\mathbb{R}$ it holds that:

$$\lambda\left(\bigcap_{i\in I} f_i^{-1}(A_i)\right) = \prod_{i\in I} \lambda(f_i^{-1}(A_i)).$$

Whenever we have an i.s.r.v. sequence (f_n) in $L^{\varphi}(0,1)$, we will denote by $(\bar{f}_n)$ the sequence of $L^{\bar{\varphi}}(0,\infty)$ defined by $\bar{f}_n(t) = f(t-n)\chi_{(n,n+1)}(t)$ for every $n \in \mathbb{N}$, where χ_A is the characteristic function of the measurable set A. Thus, the sequence $(\bar{f}_n)$ is a disjointification of the sequence (f_n). A recent inequality proved by Johnson and Schechtman ([6]) allows us to find a constant $C \geq 1$, which only depends on φ, such that if (f_n) is an i.s.r.v. sequence then:

$$(*) \qquad C^{-1}\|\textstyle\sum_n \bar{f}_n\|_{L^{\bar{\varphi}}(0,\infty)} \leq \|\sum_n f_n\|_{L^{\varphi}(0,1)} \leq C\,\|\sum_n \bar{f}_n\|_{L^{\bar{\varphi}}(0,\infty)}.$$

Proposition 2. *Suppose (s_n) is a sequence of independent symmetric simple functions in $L^{\varphi}(0,1)$ which take only 2J non-vanished values. Then the subspace spanned by (s_n) in $L^{\varphi}(0,1)$, $[s_n]$, has (s_n) as an unconditional basis which is equivalent to the canonical basis of the space $\ell^{\bar{\varphi}}(w)$ where $w_n = \lambda\{t:\ |s_n(t)|>0\}$.*

Proof. We may suppose $s_n = \sum_{i=1}^{2J} a_i\chi_{B_{n,i}}$, $n\in\mathbb{N}$, and $2^m > |a_i| > 1$ for $i = 1,2,\dots,2J$. From $(*)$ $\sum x_n s_n$ converges in $L^{\varphi}(0,1)$ if and only if $\sum x_n \bar{s}_n$ converges in $L^{\bar{\varphi}}(0,\infty)$. And we have

$$\sum_n \bar{\varphi}(|x_n|)\sum_{i=1}^{2J}\lambda(B_{n,i}) \leq \sum_n \sum_{i=1}^{2J}\bar{\varphi}(|a_i x_n|)\,\lambda(B_{n,i})$$

$$= \int_0^{\infty}\bar{\varphi}(|\sum_n x_n s_n(t)|)dt \leq K^m \sum_n \bar{\varphi}(x_n)\lambda(\bigcup_{1\leq i\leq 2J} B_{n,i})$$

where the last inequality is obtained from $\bar{\varphi} \in \Delta_2$. Thus $\sum x_n \bar{s}_n$ converges in $L^{\bar{\varphi}}(0,\infty)$ if and only if the sequence (x_n) belongs to $\ell^{\bar{\varphi}}(w)$.

Definition 3. X_{φ} is the subspace of $L^{\varphi}(0,1)$ spanned by a sequence of i.s. 3-valued random variables (s_n) with $\lambda\{t:\ |s_n(t)| > 0\} = w_n$ satisfying the condition Λ.

It follows from Proposition 2 that the space X_{φ} is isomorphic to the weighted sequence space $\ell^{\bar{\varphi}}(w)$. Moreover, since given two sequences w and w' in Λ then $\ell^{\varphi}(w)$ is isomorphic to $\ell^{\varphi}(w')$ (see

[15]), we have that Definition 3 is well declared.

Recall that an Orlicz function φ is p-convex, $p \geq 1$, if $\varphi(x)/x^p$ is a non-decreasing function.

Proposition 4. *Let φ be a p-convex Orlicz function for some $p>2$. Then X_φ is isomorphic to a subspace of $\ell^2 \oplus \ell^\varphi(w)$, for every $w \in \Lambda$. In particular when φ is distinctive then X_φ is isomorphic to a subspace of $\ell^2 \oplus \ell^\varphi$.*

Proof. Let $\ell^2(w) \oplus \ell^\varphi(w)$ be the Cartesian square with the maximum norm. Since φ is p-convex we have $\varphi(x) \leq x^2$ if $x \in (0,1)$ and $\varphi(x) \geq x^2$ if $x > 1$. Let K and 1 be the constants for the equivalence at ∞ between φ and $\bar{\varphi}$. By the above remark it is sufficient to prove that the operator $T : \ell^{\bar{\varphi}}(w) \longrightarrow \ell^2(w) \oplus \ell^\varphi(w)$, defined by $T((x_n)) = ((x_n),(x_n))$, is an isomorphism. Let (x_n) be a sequence of numbers, if

$$N_1 = \{n: |x_n| \leq 1\} \quad \text{and} \quad N_2 = \{n: |x_n| > 1\}$$

then

$$\sum_n |x_n|^2 w_n \leq \sum_{n \in N_1} |x_n|^2 w_n + K \sum_{n \in N_2} \bar{\varphi}(|x_n|) w_n \leq K \sum_n \bar{\varphi}(|x_n|) w_n$$

and

$$\sum_n \varphi(|x_n|) w_n \leq \sum_{n \in N_1} |x_n|^2 w_n + K \sum_{n \in N_2} \bar{\varphi}(|x_n|) w_n .$$

On the other hand

$$\sum_n \bar{\varphi}(|x_n|) w_n \leq \sum_{n \in N_1} |x_n|^2 w_n + K \sum_{n \in N_2} \varphi(|x_n|) w_n$$

$$\leq \sum_n |x_n|^2 w_n + K \sum_n \varphi(|x_n|) w_n .$$

Therefore

$$\frac{1}{2K} \|(x_n\|_{\ell^{\bar{\varphi}}(w)} \leq \|T((x_n))\|_\infty \leq K \|(x_n)\|_{\ell^{\bar{\varphi}}(w)} .$$

Finally note that ℓ^2 is isomorphic to $\ell^2(w)$ and the same happens taking φ instead of x^2 when φ is distinctive.

Remark. If $\varphi(x) = x^p$ then the space X_φ is the Rosenthal space X_p and, in addition, if $p>2$ then Proposition 4 has been proved in [13]. Rosenthal also showed that X_p, $p>1$, is isomorphic to a complemented subspace of $L^p(0,1)$ (i.e., there exists a continuous projection from $L^p(0,1)$ onto the subspace). Now, if $L^\varphi(0,1)$ is reflexive, we have that X_φ is isomorphic to a complemented subspace of $L^\varphi(0,1)$ too. Indeed, from Theorem 2.f.1 of [10] we have $L^\varphi(0,1)$ is isomorphic to $L^{\bar{\varphi}}(0,\infty)$ and by Proposition 2 of (4) we have that $\ell^{\bar{\varphi}}(w)$ is isomorphic to a complemented subspace of $L^{\bar{\varphi}}(0,\infty)$ for every weight sequence w.

Drewnowski proved in [3] that given $w \in \Lambda$ for every positive sequence $u = (u_n)$ the canonical basis of $\ell^{\varphi}(u)$ is equivalent to a block basic sequence with constant coefficients of the canonical basis of $\ell^{\varphi}(w)$. This fact allows us to generalize Corollary 4.3 of [14].

Proposition 5. *i) Every i.s.r.v. sequence (f_m) in $L^{\varphi}(0,1)$ is equivalent to a block basic sequence of $\ell^{\bar{\varphi}}(w)$ for every $w \in \Lambda$.*

ii) Suppose $u_m = \sum_{i \in \sigma_m} a_{i,m} e_i$ is a block basic sequence of $\ell^{\bar{\varphi}}(w)$ such that there exist a constant M with $\sum_{i \in \sigma_m} w_i < M$ for every $m = 1,2,3,\ldots$. Then there exists an i.s.r.v. sequence of $L^{\varphi}(0,1)$ equivalent to (u_m).

iii) If (u_m) is a symmetric block basic sequence of $\ell^{\bar{\varphi}}(w)$ with $w \in \Lambda$ then the thesis of ii) is verified.

Proof. i) First let us consider the case of a sequence of simple i.s. functions (s_n) in $L^{\varphi}(0,1)$, $s_m = \sum_{i \in \sigma_m} a_{i,m} \chi_{B_{i,m}}$. By (*) and Proposition 2 we get that (s_n) is equivalent to (u_m), $u_m = \sum_{i \in \sigma_m} a_{i,m} e_{i,m}$, a block basic sequence of $\ell^{\bar{\varphi}}(v)$ where v is the weight sequence defined by:

$$\lambda(B_{1,1}),\lambda(B_{2,1}),\ldots,\lambda(B_{\max\sigma_1,1}),\lambda(B_{1,2}),\ldots,\lambda(B_{\max\sigma_2,2}),\ldots$$

So i) is followed immediately from Drewnowski's result. Now let (f_m) be an i.s.r.v. sequence, with $\|f_m\|_{L^{\varphi}(0,1)} = 1$, $m = 1,2,\ldots$, and $\frac{1}{2^{p(m)}} < |f_m(t)| < 2^{p(m)}$ for every $t \in \{t : |f(t)| > 0\}$ and for some $p(m) \in \mathbb{N}$. Let D be the basic constant of (f_m). We define the simple functions:

$$s_m^l(x) = \begin{cases} k-1/2^l & \text{if} \quad k-1/2^l \le f_m(x) < k/2^l \\ -(k-1/2^l) & \text{if} \quad -(k/2^l) < f_m(x) \le -(k-1/2^l) \\ 0 & \text{if} \quad f_m(x) = 0 \end{cases}$$

$$\text{for} \quad k = 2^{1-p(m)}+1, 2^{1-p(m)}+2, \ldots, 2^{1+p(m)}$$

and take $l(m)$ such that $\|f_m - s_m^{l(m)}\|_{L^{\varphi}(0,1)} < \frac{1}{D}\cdot\frac{1}{2^{m+1}}$, $m = 1,2,3,\ldots$

Thus $(s_m^{l(m)})$ is an i.s.r.v. sequence, therefore by the first part and proposition 1.a.9 of (9), i) is verified. The general case follows now trivially.

ii) Since $\ell^{\bar{\varphi}}(w) = \ell^{\bar{\varphi}}(w/M)$ we may assume M = 1. Also we can suppose $a_{i,m} > 0$ for all $m,i \in \mathbb{N}$. Reasoning as if Rademacher functions are built, we can find a sequence of simple i.s.r.v. on (0,1), (s_m), with

$$s_m = \sum_{i\in\sigma_m} a_{i,m}\chi_{B_{i,m}} + \sum_{i\in\sigma_m} -a_{i,m}\chi_{C_{i,m}} \quad \text{and} \quad \lambda(B_{i,m}) = \lambda(C_{i,m}) = \frac{1}{2}\cdot w_i$$

for every $i \in \sigma_m$ and all $m \in \mathbb{N}$. By (*) the sequence (s_m) is equivalent to (u_m).

iii) From Proposition 1 we have that (u_m) is equivalent to the canonical basis of any ℓ^{ψ} where $\psi \in C_{\bar{\varphi}}(0,\infty)$ and ℓ^{ψ} is isomorphic to a subspace of $\ell^{\bar{\varphi}}(w')$ for every $w' \in \Lambda$. Using the remark after Proposition 1 we may suppose that $w_n < 1$ for every n. Now we get a sequence of 3-valued i.s.r.v. (f_n) in $L^{\varphi}(0,1)$ such that $\lambda\{t: |f_n(t)| > 0\} = w_n$, which is equivalent to the canonical basis of $\ell^{\bar{\varphi}}(w)$. Now the i.s.r.v. sequence defined by $g_m = \sum_{i\in\sigma_m} a_{i,m}f_i$ is equivalent to (u_m).

Theorem 6. *Let (f_n) be an i.s.r.v. sequence in $L^{\varphi}(0,1)$. Then (f_n) is equivalent to a symmetric basis of some Banach space X if and only if there exists a function $\psi \in C_{\bar{\varphi}}(0,\infty)$ such that ℓ^{ψ} is isomorphic to X.*

Proof. The necessary condition follows from the above Proposition, part i), and Proposition 1. Now Propositions 1 and 5 part iii) prove the sufficient condition.

Last result is more precise and general than 3.3.8 in [2].

Remark. Notice that by Proposition 5 all the subspaces of $L^{\varphi}(0,1)$ spanned by a sequence of i.s.r.v. are isomorphic to subspaces of X_{φ}. However the space X_{φ} is not isomorphic to $L^{\varphi}(0,1)$. Indeed, it holds that X_{φ} is isomorphic to a subspace of a suitable Orlicz sequence space ℓ^{M} (see Theorem 4.d.5 in [9]) and an Orlicz function space, different of L^2, is never isomorphic to a subspace of a separable Orlicz sequence space (see Theorem 3 in [8]).

Remark. We know from [13] that every subspace spanned by a sequence of i.s.r.v. in $L^p(0,1)$ for, p>2, is isomorphic to a complemented subspace of $L^p(0,1)$. This property can not be extended to Orlicz function spaces. Indeed, it is known ([4]) that there exists minimal Orlicz function spaces $L^{\varphi}(0,1)$ with φ p-convex for some p>2 such that for every $q \geq 1$ ($q \neq 2$) ℓ^q is not isomorphic to any complemented subspace of $L^{\varphi}(0,1)$. On the other hand, X_{φ} always has subspaces isomorphic to some ℓ^q, q>2, because $x^q \in C_{\bar{\varphi}}(0,\infty)$ (see [12]). Now by Proposition 5, iii) we can find an i.s.r.v. sequence (f_n) in $L^{\varphi}(0,1)$ such that the span $[f_n]$ is isomorphic to ℓ^q.

This work is a part of author's Doctoral Thesis prepared under the supervision of F.L.Hernández.

REFERENCES

[1] D.Dacunha-Castelle, *Variables aleatoires echangeables et espaces d'Orlicz*, Seminaire Maurey-Schwartz 1974-1975.

[2] D.Dacunha-Castelle, M.Shreiber, *Techniques probabilistes pour l'étude de problémes d'isomorphismes entre espaces de Banach*, Ann. Inst. Henri Poincaré 10 (1974), 229-277.

[3] L.Drewnowski, *F-spaces with a basis which is shrinking but not hiper-shrinking*, Studia Math. 64 (1979), 97-104.

[4] F.L.Hernandez, V. Peirats, *Orlicz function spaces without complemented copies of ℓ^p*, Isr. J. Math. 56 (1986), 355-360.

[5] F.L.Hernandez, B.Rodriguez-Salinas, *On ℓ^p-complemented copies in Orlicz spaces*, Isr. J. Math. 62 (1988), 37-55.

[6] W.Johnson, G. Schechtman, *Sums of independent random variables in rearrangement invariant function spaces*, (to appear).

[7] K.Lindberg, *On subspaces of Orlicz sequence spaces*, Studia Math. 45 (1973), 119-146.

[8] J.Lindenstrauss, L. Tzafriri, *On Orlicz sequence spaces III*, Isr. J. Math. 14 (1973), 368-389.

[9] J.Lindenstrauss, L. Tzafriri, *Classical Banach Spaces I*, Springer Verlag (1977).

[10] J.Lindenstrauss, L. Tzafriri, *Classical Banach Spaces II*, Springer Verlag (1979).

[11] J.Musielak, *Orlicz Spaces and Modular Spaces*, Lect. Notes in Math. 1034, Springer Verlag (1983).

[12] N.J.Nielsen, *On the Orlicz function spaces $L_M(0,\infty)$*, Isr. J. Math. 20 (1975), 237-259.

[13] H.Rosenthal, *On the subspaces of L^p (p>2) spanned by sequences of independent random variables*, Isr. J. Math. 8 (1970), 273-303.

[14] H.Rosenthal, *On the span in L^p of sequences of independent random variables (II)*, Proc. 6th Berkeley Symposium, Probability Theory, (1972).

[15] C.Ruiz, *Estructura de Espacios de Orlicz de funciones y de sucesiones con pesos. Subespacios distinguidos*, Doctoral Thesis, Universidad Complutense de Madrid, 1989.

SOME GEOMETRIC PROPERTIES OF ORLICZ SPACES

WANG TINGFU

Harbin, China

Since the first international conference on Function Spaces, we have discussed some new geometric properties of the classical Orlicz spaces and Orlicz sequence spaces, and obtained the following results.

1. Roughness.

The conception of roughness was introduced be Leach and Whitfield [15] in 1973. after that, strong roughness and pointwise roughness were introduce by John, Zizler [11] and Li Xiaojian [16] respectively.

X – Banach space, X^* – the dual of X, S(X), B(X) – the unit sphere and unit ball. For $x \in S(X)$, denote

$$\varepsilon(x) = \sup\{\varepsilon>0: \exists f_n, g_n \in S(X^*),\ \|f_n-g_n\| \geq \varepsilon,\ \lim_n f_n(x)=\lim_n g_n(x)=1\}.$$

X is said to be *pointwise rough* iff $\varepsilon(x) > 0$ for every $x \in S(X)$; X is said to be *rough* iff $\inf_{x\in S(X)} \varepsilon(x) > 0$; X is said to be *strong rough* iff $\inf\{\operatorname{diam} A(x): x \in S(X)\} > 0$, where $A(x) = \{f\in S(X^*): f(x)=\|x\|=1\}$.

Three roughnesses [15] characterize the levels of non-differentiability of the norms. For example, strong roughness is equivalent to uniformly Gateaux non-differential, pointwise roughness is equivalent to Frechet non-differentiability everywhere.

Cui Yunan and I [6] found all criteria of three roughness for both Orlicz norm and Luxemburg norm on Orlicz spaces.

For Orlicz norm, roughness is equivalent to pointwise roughness and is equivalent to "$M \notin \nabla_2$" – M(u) doesn't satisfy ∇_2 condition; the strong roughness is impossible. For Luxemburg norm, pointwise roughness is also equivalent to "$M \notin \nabla_2$" and strong roughness is also impossible. But the criterion of roughness is unexpected: $\|\cdot\|_M$ is rough iff $M \notin \nabla_2$ and $M \in \Delta_2$.

2. Normal structure.

It is known that normal structure, weakly normal structure and Lami-Dozo property are the fundamental tools in the fixed point theory of nonexpansive mappings. For example, Browder's theorem [4]: every non-expansive self map of A has a fixed point provided A has the normal structure and is weakly compact and convex. Edelstein's theorem [8]: If X has LD property, nonexpansive mapping T has a fixed point provided there exists a bounded sequence (x_n) satisfying $\|Tx_n-x_n\| \to 0$.

The set A is said *diametral* if A is bounded and

$$\sup\{\|x-y\|: y\in A\} = d \quad \text{for all } x\in A \text{ and some } d>0.$$

And X is said to have *normal structure* [3] if X has no convex diametral subset. X is said to have *weakly normal structure* if every weakly compact subset of A has normal structure. Lami-Dozo [13] introduced the following condition:

(LD) every bounded sequence has a subsequence which is pointwise and almost convergent.

We say a sequence (x_n) almost converges to x provided that

$$\overline{\lim_n}\ \|x_n-x\| < \underline{\lim_n}\ \|x_n-y\| \quad \text{for any } y \neq x.$$

Thomas Landes [14] proved that NS $\Leftrightarrow$ WNS $\Leftrightarrow$ LD $\Leftrightarrow$ $M \in \Delta_2$ in Orlicz sequence space $\ell_{(M)}$ endowed with Luxemburg norm. Recently, Wang Baoxiang and I [22] obtained the following: in Orlicz sequence space endowed with Orlicz norm,

ℓ_M has LD property $\Longleftrightarrow M \in \Delta_2$;

ℓ_M has WNS without any condition;

If $M \in \nabla_2$ or there is $r>0$ that $M(u)$ is strictly convex on $[0,r]$ then ℓ_M has the normal structure.

3. G and K property.

In 1958, the property (G) was introduced by Fan and Glicksberg [9]. X is said to have (G) if every point $x_o \in S(X)$ is denting point of $B(X)$, i.e., for all $\varepsilon>0$, $x_o \notin \overline{\text{conv}}\ \{x\in B(X): \|x-x_o\| \geq \varepsilon\}$. 28 years passed, Bor-Lublin, Pei-Keelin and Troyanski [2] discovered that (G) is equivalent to (K) + rotundity. X is said to have the property (K) if the norm topology and the weak topology coincide on S(X).

Clearly, (K) is stronger than (H). But I proved recently [20] that (K) $\Longleftrightarrow$ (H), therefore (G) $\Longleftrightarrow$ (H)+(R) for either Orlicz function space or Orlicz sequence space endowed with either Luxemburg norm or Orlicz norm.

4. NUC and UKK.

Partington and Istratescu [17,10] introduced nearly uniform convexity (NUC) and uniformly Kadec-Klee property (UKK) in 1983.

X is said to be *NUC* provided for any $\varepsilon>0$, there exists δ, $0<\delta<1$, such that $\|x_n\| \le 1$, $\|x_n-x_m\| \ge \varepsilon$ $(n,m=1,2,\ldots;\ n\ne m)$ implies

$$\operatorname{conv}\{(x_n)_1^\infty\} \cap \{x: \|x\| \le 1-\delta\} \ne \emptyset.$$

X is said to have *UKK* property provided for any $\varepsilon>0$, there exists δ, $0<\delta<1$, such that $\|x_n\| \le 1$, $\|x_n-x_m\| \ge \varepsilon$ $(n,m=1,2,\ldots;\ n\ne m)$ and $x_n \xrightarrow{w} x_o$ implies $\|x_o\| \le 1-\delta$.

It is easy to see UC $\Longrightarrow$ NUC $\Longrightarrow$ UKK $\Longrightarrow$ H. Shi Zhongrui and I [21] obtained all criteria of NUC and UKK for L_M, $L_{(M)}$, ℓ_M, $\ell_{(M)}$.

	NUC	UKK
L_M, $L_{(M)}$	=UC	=UC
ℓ_M, $\ell_{(M)}$	=Reflex	= H

(H means that norm convergence and weak convergence coincide on S(X)). This shows that Orlicz function space and Orlicz sequence space are quite different.

5. P, O, Q, H convexity and fully K-convexity.

Kottman (1970, [12]), Sastry and Naidu (1978, [18]), Amir and Franchett (1984, [1]) introduced some new convexities: P-,O-,Q-, H-convexity respectively.

P-convexity: there exist n and $\varepsilon>0$ such that for any $x_1,\ldots x_n \in S(X)$, $\min_{j\ne k} \|x_j-x_k\| < 2-\varepsilon$.

O-convexity: there exist n and $\varepsilon>0$ such that for any $x_1,\ldots x_n \in S(X)$, $\min_{j\ne k} (\|x_j+x_k\|, \|x_j-x_k\|) < 2-\varepsilon$.

Q-convexity: there exist n and $\varepsilon>0$ such that for any $x_1,\ldots x_n \in S(X)$, $\|\sum_{i=1}^{k-1} x_i - x_k\| < k-\varepsilon$, $(k=2,\ldots,n)$.

H-convexity: $\sup_{f\in S(X^*)} \inf_{z\in f^{-1}(1)} \sup_{x\in B(X)} \|x-f(x)z\| < 2$.

It is known that $P \Rightarrow O \Rightarrow Q \Rightarrow H \Rightarrow B$, where B-convexity means that there exists n such that X is uniform non-$l_n^{(1)}$.

Danker and Kombrink [7] have proved that in Orlicz spaces (B) $\Longleftrightarrow$ Refl.

Ye Yining, He Miaohong [24] and I [19] proved that in L_M, $L_{(M)}$, ℓ_M, $\ell_{(M)}$, (P) $\Longleftrightarrow$ Reflexivity. Therefore all the P-, O-, Q-, H-, B-convexity superreflexivity and reflexivity are equivalent in Orlicz spaces.

Fully K-convexity means that for any sequence (x_n),

$$\lim_{n_1,\ldots,n_k\to\infty} \|\frac{1}{k}\sum_{i=1}^{k} x_{n_i}\| = 1 \text{ implies } (x_n) \text{ is a Cauchy sequence.}$$

For Orlicz spaces $L_{(M)}$ and $\ell_{(M)}$, this year, Chen Shutao, Bor-Luh Lin and Yu Xintai proved [5] that FKC $\Longleftrightarrow$ Reflexivity + (C). Then, Wang Baoxiang, Zhang Yunfeng and I [23] proved that the same statement is true in L_M and ℓ_M.

REFERENCES

[1] P. Amir, C. Franchetti, Trans. Amer. Math. Soc. 282 (1984), 275–291.

[2] Bor-Luh Lin, Pei-Kee Lin, S. L. Troyanski, Math. Ann. 274 (1986), 613–616.

[3] M.S. Brodskii, D. P. Milman, Dokl. Akad. Nauk USSR, 59 (1948), 837–840.

[4] F. E. Browder, Bull. Amer. Math. Soc., 73 (1967), 875–881.

[5] Chen Shutao, Bao-Luh Lin, Yu Xintai, Contemporary Math. 85 (1989), 79–86.

[6] Cui Yunan, Wang Tingfu, *The roughness of the norms on Orlicz space*, (to appear).

[7] M. Danker, R. Kombrink, Lectures Notes in Math. 79 (1979), 87–95.

[8] M. Edelstein, Bull. Amer. Math. Soc., 78 (1972), 206–208.

[9] K. Fan, I. Glicksburg, Duke Math. J. 25 (1958), 553–568.

[10] V.I. Istratescu, J. R. Partington, Math. Proc. Comb. Phil. Soc. 95 (1984), 325–327.

[11] K.John, V.Zizler, Comment. Math. Univ. Carolinae, 19 (1978), 335–349.

[12] C.A.Kottman, Trans. Amer. Math. Soc. 150 (1970), 565–576.

[13] E.Lami-Dozo, Lecture Notes in Math., 686 (1981).

[14] T.Landes, Trans. Amer. Math. Soc. 285 (1984), 523–534.

[15] F.B.Leach, J. H. H. Whitfield, Proc. Amer. Math. Soc., 33 (1973), 120–136.

[16] Li Xiaojian, China Math. Ann., 8A (1987), 621–625.

[17] J. R. Partington, Math. Proc. Comb. Phil. Soc. 93 (1983), 127–129.

[18] K.P.R.Sastry, S. N. R. Naidu, J. Reine Ange Math. 297 (1978), 35–53.

[19] Wang Tingfu, *P-convexity of Orlicz spaces endowed with Orlicz norm*, Chinese Math. quarterly (to appear).

[20] Wang Tingfu, *Property (G) and (K) of Orlicz spaces*, (to appear).

[21] Wang Tingfu, Shi Zhongrui, China North-East Math. J. 3 (1987), 160–172.

[22] Wang Tingfu, Wang Baoxiang, *Normal structure and LD property of Orlicz sequence spaces*, (to appear).

[23] Wang Tingfu, Wang Baoxiang, Zhang YunFeng, *Fully K-convexity in Orlicz spaces*, J.Harbin Normal Univ., (to appear).

[24] Ye Yining, He Miaohong, *P-convexity of Orlicz spaces*, (to appear).

STABLE UNIT BALLS IN FINITE DIMENSIONAL GENERALIZED ORLICZ SPACES

MAREK WISŁA

Poznań, Poland

SUMMARY. Every unit ball of the finite dimensional generalized Orlicz space is stable. This result is an extension of theorems presented by R.Grząślewicz [2] and A.Suárez-Granero [8].

A convex subset C of a Hausdorff topological vector space X is said to be *stable* if the midpoint map $(x,y) \longrightarrow \frac{1}{2}(x+y)$ from $C\times C$ into C is open. The other characterizations of stability and its consequences has been studied in [7] and [1]. For instance, the set ExtC of all extreme points of C is closed. The above property is also useful in investigations of extreme points of unit balls of spaces C(K,X), where K is a compact set and X is a Banach space. Namely, using the Michael selection theorem [3], $f \in \mathrm{Ext}B_{C(K,X)}$ iff $f(k) \in \mathrm{Ext}B_X$ for every $k\in K$ provided the unit ball B_X of the space X is stable.

The descriptions of stable unit balls of Orlicz spaces have been presented in [2] and [8]. However, the assumptions which have been made there ($\varphi(t)=0$ iff $t=0$ and $I_\varphi(\frac{x}{\|x\|_\varphi})=1$ respectively) do not allow us to deduce from that results that ℓ_n^∞ has a stable unit ball. On the other hand, it is well known that the unit ball of ℓ_n^∞ is stable because ℓ_n^∞ has 3.2 intersection property (cf [1,p.195]). Further, ℓ_n^∞ can also be treated as an Orlicz space generated by the function φ defined as follows

$$\varphi(t)=\begin{cases} 0 & \text{for } |t|\le 1 \\ +\infty & \text{otherwise.}\end{cases}$$

In this paper we shall present a theorem on stability of finite dimensional Orlicz spaces which will include the case of Orlicz spaces generated by a function with values in $[0,+\infty]$.

Let $\varphi_i:\mathbb{R} \to [0,+\infty]$, $i=1,2,\dots,n$ be convex, even and non-identically equal to 0 functions with $\varphi_i(0)=0$ and $\varphi_i(t)<+\infty$ for some $t>0$ and each $i=1,2,\dots,n$. By a *generalized Orlicz space* (or *Musielak-Orlicz space*) ℓ_n^φ we mean the space $\mathbb{R}^n$ equipped with the Luxemburg norm ([6])

$$\|x\|_\varphi = \inf \{\lambda>0: I_\varphi(\tfrac{x}{\lambda})\le 1\},$$

where $I_\varphi(x)= \sum_{i=1}^{n}\varphi_i(x_i)$ (cf. [4]). The unit ball of ℓ_n^φ will be denoted by B_φ.

We shall say that a *point* $z\in B_X$ *is stable* if, for every $x,y\in B_X$ such that $\frac{1}{2}(x+y)=z$, the following condition holds

(1) for every $\varepsilon>0$ there exists $\delta>0$ such that, for every $w\in B_X$,

$$\|w-z\|<\delta \Longrightarrow \exists_{u,v\in B_X} \|u-x\|<\varepsilon,\ \|v-y\|<\varepsilon \text{ and } \tfrac{1}{2}(u+v)=w.$$

1. Lemma. *The unit ball B_X is stable iff every point $z\in B_X$ is stable.*

Proof. ($\Longrightarrow$) Let $z,x,y \in B_X$ be such that $\frac{1}{2}(x+y)=z$. Further, let $\varepsilon>0$ and $B(x,\varepsilon):=\{y\in X: \|x-y\|<\varepsilon\}$. Then $U=B(x,\varepsilon)\cap B_X$ and $V=B(y,\varepsilon)\cap B_X$ are open neighborhoods of x and y respectively. Thus $W = \frac{1}{2}(B(x,\varepsilon)\cap B_X+ B(y,\varepsilon)\cap B_X)$ is an open set in $(B_X,\|\cdot\|)$. Hence there is $\delta>0$ such that $B(z,\delta)\cap B_X\subset W$, where $z=\frac{1}{2}(x+y)$. Thus condition (1) holds.

($\Longleftarrow$) Let U,V be open sets in X. We claim that $\frac{1}{2}(U\cap B_X+V\cap B_X)$ is an open set in B_X. Let $z \in \frac{1}{2}(U\cap B_X+V\cap B_X)$, i.e. $z=\frac{1}{2}(x+y)$ for some $x\in U\cap B_X$ and $y\in V\cap B_X$. Since U,V are open, there is $\varepsilon>0$ such that $B(x,\varepsilon)\subset U$ and $B(y,\varepsilon)\subset V$. By (1), we can find $\delta>0$ such that

$$B(z,\delta)\cap B_X \subset \tfrac{1}{2}(B(x,\varepsilon)\cap B_X+ B(y,\varepsilon)\cap B_X) \subset \tfrac{1}{2}(U\cap B_X+V\cap B_X)$$

which ends the proof.

Evidently, every extreme point $z \in \operatorname{Ext}B_X$ is stable. Suppose that $z,x,y \in B_X$, $\|z\|<1$, $\frac{1}{2}(x+y)=z$ and fix $\varepsilon>0$. Choose $\lambda>0$ such that $\|x-\bar{x}\|<\varepsilon/2$ and $\|y-\bar{y}\|<\varepsilon/2$ where $\bar{x}= \lambda x+(1-\lambda)y$ and $\bar{y}= (1-\lambda)x+\lambda y$. Then both $\bar{x}$ and $\bar{y}$ are elements of $\operatorname{int}B_X$. Let $0<\eta<\varepsilon$ be such a number that $B(\bar{x},\eta) \subset \operatorname{Int}B_X$. Then, for any $y\in B_X$, the set

$$W = \tfrac{1}{2}(B(\bar{x},\eta)+B(y,\eta)\cap B_X) \subset \operatorname{Int}B_X$$

is an open neighborhood of $z=\frac{1}{2}(x+y)$ in $(B_X,\|\cdot\|)$. Hence $B(z,\delta) \subset W$ for some $\delta>0$, i.e. the point z is stable. Thus the thesis of the above Lemma can be reformulated as follows: *The unit ball B_X is stable iff every point $z\in B_X$ for which there are $x,y\in B_X$ such that $x\neq y$,*

$\frac{1}{2}(x+y) = z$ and $\|x\|=\|y\|=\|z\|=1$ *is stable*. The fact that every strictly convex Banach space has stable unit ball (cf [5]) is now evident.

2. Theorem. *Every finite dimensional generalized Orlicz space ℓ_n^φ has stable unit ball.*

Proof. Let $x,y\in B_\varphi$ be such that $\|x\|_\varphi=\|y\|_\varphi=\|z\|_\varphi=1$, where $z=\frac{1}{2}(x+y)$. Let $\varepsilon>0$ be fixed. We can find $\frac{1}{2}<\alpha<1$ such that $\|x-\bar{x}\|_\varphi<\varepsilon/2$ and $\|y-\bar{y}\|_\varphi<\varepsilon/2$, where $\bar{x} = \alpha x+(1-\alpha)y$ and $\bar{y} = (1-\alpha)x+\alpha y$. Let $D:=\{1\le i\le n: x_i\neq y_i\}$ and let σ be such a number that

$$0 < \sigma < (1-\alpha)\cdot\min_{i\in D}|x_i-y_i| = \min_{i\in D}|x_i-\bar{x}_i| = \min_{i\in D}|y_i-\bar{y}_i|.$$

Since ℓ_n^φ and ℓ_n^∞ are isomorphic, there is $0<\rho<\varepsilon/2$ such that

(2) $$\|z\|_\varphi<\rho \implies \max_{1\le i\le n}|z_i|<\sigma.$$

Note that

(3) $$\bar{x}_i+t,\ \bar{y}_i+t \in \{\beta x_i+(1-\beta)y_i:\ \beta\in[0,1]\}$$

for every $|t|<\sigma$ and $i\in D$.

Further, let $t_i = \sup\{t\ge 0: \varphi_i(t)\le 1\}$ for $i=1,2,\ldots,n$ and let $A:= \{1\le i\le n: |z_i|=t_i\}$. Note that

$$A = \{1\le i\le n: |\bar{x}_i| =t_i\} = \{1\le i\le n: |\bar{y}_i|=t_i\}.$$

The proof will be split into two parts.

1) $I_\varphi(z)=1$. Then, for every $\beta\in[0,1]$, $I_\varphi(\beta x+(1-\beta)y)=1$. Indeed, if $I_\varphi(\beta x+(1-\beta)y)<1$ for some $0\le\beta\le 1$, then

$$I_\varphi(z) = I_\varphi\left(\tfrac{1}{2}[\beta x+(1-\beta)y] + \tfrac{1}{2}[(1-\beta)x-\beta y] \right)$$
$$\le \tfrac{1}{2}\ I_\varphi(\beta x+(1-\beta)y) + I_\varphi((1-\beta)x+\beta y) < 1$$

and we get a contradiction. Similarly,

$$\varphi_i(\beta x_i+(1-\beta)y_i) = \beta\varphi_i(x_i)+ (1-\beta)\varphi_i(y_i)$$

for every $\beta\in[0,1]$ and $1\le i\le n$. Thus $\varphi_i(t) = a_it+b_i$ for some $a_i,b_i\in\mathbb{R}$ and every $t \in \{\beta x_i+(1-\beta)y_i:\ \beta\in[0,1]\}$, $i=1,2,\ldots,n$.

Let $\delta=\rho$ and $w\in B_\varphi$ be an element such that $\|w-z\|_\varphi<\delta$. Denote $u=\bar{x}+w-z$, $v=\bar{y}+w-z$. Then $\frac{1}{2}(u+v)=w$, $\|u-x\|_\varphi\le \|x-\bar{x}\|_\varphi+\|w-z\|_\varphi<\varepsilon$ and $\|v-y\|_\varphi<\varepsilon$. Further, by (2) and (3),

$$\bar{x}_i, w_i, z_i, \bar{x}_i+w_i-z_i \in \{\beta x_i+(1-\beta)y_i:\ \beta\in[0,1]\}$$

for every $i\in D$. Hence

$$I_\varphi(u) = I_\varphi(\bar{x}+w-z) = \sum_{i\in D}[a_i(\bar{x}_i+w_i-z_i)+b_i] + \sum_{i\in E}\varphi_i(w_i) =$$

$$= I_\varphi(\bar{x}) + I_\varphi(w) - I_\varphi(z) = I_\varphi(w) \le 1,$$

where $E = \{1,2,\dots,n\}\setminus D$. Therefore $\|u\|_\varphi \le 1$. Analogously, $\|v\|_\varphi \le 1$.

2). $I_\varphi(z)<1$. Then

$$a := \sum_{i\in A} \varphi_i(t_i) \le I_\varphi(z) < 1.$$

We claim that $I_\varphi(\beta x+(1-\beta)y)<1$ for every $\beta\in(0,1)$. On the contrary, suppose that $I_\varphi(\beta x+(1-\beta)y)=1$ for some $\beta\in(0,1)$. Without loss of generality we can assume that $\frac{1}{2}<\beta<1$. Let $\beta_m := 1-2^m(1-\beta)$ for $m=1, 2,\dots,m_o$, where $m_o := \min\{m\in\mathbb{N}: 1-2^m(1-\beta)<\frac{1}{2}\}$. Then

$$\beta_m x+(1-\beta_m)y = \frac{1}{2}([\beta_{m+1}x+(1-\beta_{m+1})y] + x)$$

for $m=1,2,\dots,m_o-1$ and arguing in the same manner as in the part 1) of this proof we obtain $I_\varphi(\beta x+(1-\beta)y)=1$ for every $\beta\in[\beta_{m_o},1]$. Since $\frac{1}{2}\in[\beta_{m_o},1]$, we have $I_\varphi(z)=1$ – a contradiction. Therefore $I_\varphi(\bar{x})<1$ and $I_\varphi(\bar{y})<1$. Now, we can find $\gamma>1$ such that

$$\sum_{i\notin A} \varphi_i(\gamma\bar{x}_i) \le 1-a \qquad \text{and} \qquad \sum_{i\notin A} \varphi_i(\gamma\bar{y}_i) \le 1-a.$$

Let $\delta = \min\{(1-\frac{1}{\gamma})(1-a),\rho\}$ and let $w\in B_\varphi$ be such that $\|w-z\|_\varphi<\delta$. Denote $u=\bar{x}+w-z$, $v=\bar{y}+w-z$. Then $w=\frac{1}{2}(u+w)$, $\|u-x\|_\varphi<\varepsilon$ and $\|v-y\|_\varphi<\varepsilon$. Since $\|w-z\|_\varphi< (1-\frac{1}{\gamma})(1-a)$, $I_\varphi(\frac{\gamma}{\gamma-1}(w-z))\le 1-a$. Therefore,

$$\begin{aligned}
I_\varphi(u) &= \sum_{i\in A} \varphi_i(\bar{x}_i+w_i-z_i)+\sum_{i\notin A} \varphi_i(\bar{x}_o+w_i-z_i) \le \\
&\le \sum_{i\in A}\varphi_i(t_i) + \sum_{i\notin A}\Big[\frac{1}{\gamma}\varphi_i(\gamma\bar{x}_i) + (1-\frac{1}{\gamma})\varphi_i(\frac{\gamma}{\gamma-1}(w_i-z_i))\Big] \\
&\le a + \frac{1}{\gamma}(1-a) + (1-\frac{1}{\gamma})I_\varphi(\frac{\gamma}{\gamma-1}(w-z)) \\
&\le (1-\frac{1}{\gamma})a + \frac{1}{\gamma} + (1-\frac{1}{\gamma})(1-a) = 1
\end{aligned}$$

i.e. $\|u\|_\varphi\le 1$. Analogously, $\|v\|_\varphi\le 1$ and the proof is finished.

Let x be a point of a convex compact set K. By a *face* F_x *generated* by x we mean the set

$$F_x := \{y\in K: x=\alpha y+(1-\alpha)z \text{ for some } z\in K \text{ and } \alpha\in(0,1]\}.$$

Further, let $K^{(m)}$ be the m–th skeleton of K, that is the set of all points $x\in K$ such that the face F_x has dimension less or equal to $m\in\{1,2,\dots,n\}$. Then, in virtue of the above Theorem 2 and [7,Theorem 2.3] we get the following

3. COROLLARY. *For a unit ball of a finite dimensional generalized Orlicz space the following assertions hold:*

(i) the correspondence $x \to F_x$ is lower semicontinuous,

(ii) all skeletons $K^{(m)}$, $m=0,1,2,\dots,n$ of B_φ are closed.

4. COROLLARY. *Let K be a compact set. Then a function $f \in C(K,\ell_n^\varphi)$ is an extreme point of $B_{C(K,\ell_n^\varphi)}$ iff $f(k) \in \operatorname{Ext}B_\varphi$ for every $k \in K$.*

REFERENCES

[1] A.CLAUSING, S.PAPADOPOULOU, *Stable convex sets and extremal operators*, Math. Ann. 231 (1978), 193-203.

[2] R.GRZĄŚLEWICZ, *Finite dimentional Orlicz spaces*, Bull. Pol. Ac.: Math., 33 (1985), 277-283.

[3] E.MICHAEL, *Continuous selections I*, Ann. of Math. 63 (1956), 361-382.

[4] J.MUSIELAK, *Orlicz spaces and modular spaces*, Lecture Notes in Math. 1034 (1983).

[5] A.J.LAZAR, *Affine functions on simplexes and extreme operators*, Israel J. Math. 5 (1967), 31-43.

[6] W.A.J.LUXEMBURG, *Banach functions spaces*, Thesis, Delft 1955.

[7] S.PAPADOPOULOU, *On the geometry of stable compact convex sets*, Math. Ann. 229 (1977), 193-200.

[8] A.SUÁREZ-GRANERO, *Stable unit balls in Orlicz spaces*, Proc. A.M.S., in print.

[9] J.VERSTERSTRØM, *On open maps compact convex sets and operator algebras*, J. London Math. Soc. 6 (1973), 289-297.

AUTOMATIC CONTINUITY RESULTS FOR LINEAR MAPS ON BANACH ALGEBRAS

EDWARD BECKENSTEIN, LAWRENCE NARICI

New York, USA

Introduction.

Let T and S be compact, Hausdorff spaces and let C(T) and C(S) be the algebras of continuous $\mathbb{F}$-valued functions on T and S, respectively, equipped with their supremum norm topologies. For most of our results the underlying field $\mathbb{F}$ in which the function take their values can be any complete field with valuation, including, of course, the real or complex numbers. If the valuation is nonarchimedean, we also assume S and T to be 0-dimensional.

By an *automatic continuity result* we mean one in which continuity of a linear map is deduced from other conditions as happens, for example, in the case of the Banach-Steinhaus theorem and the Closed Graph theorem; similarly, the class of bornological spaces is of interest because a bounded linear map must be continuous there. We have considered a property of linear maps $A:C(T) \to C(S)$ called *separating*, defined below, which implies continuity in many important cases. Sometimes a separating map A must even be a *weighted composition map:* there must exist a continuous map h of S into T and a function *(the weight)* $a \in C(S)$ such that, for any $x \in C(T)$ and any $s \in S$, $Ax(s) = a(s)x(h(s))$. A result which motivated much of our thinking about these things is the Banach-Stone theorem in which weighted compositions feature prominently:

If A is a linear isometry of C(T) onto C(S) then there is a homeomorphism h of S onto T and a function $a \in C(S)$ such that, for any $x \in C(T)$ and any $s \in S$, $Ax(s) = a(s)x(h(s))$.

It is time to introduce "separating".

Definition. The *cozero set*, coz x, of a function x is the set of points in the domain of x at which x is not 0. A linear map $A:C(T) \to C(S)$ is *separating* if, for any x,y in $C(T)$, $\text{coz } x \cap \text{coz } y = \emptyset$ implies $\text{coz } Ax \cap \text{coz } Ay = \emptyset$; equivalently, $xy \equiv 0$ implies $AxAy \equiv 0$.

Associated with any separating linear map $A:C(T) \to C(S)$ there is a continuous $h:S \to T$; If A is injective, then h is surjective [Prop. 1.7]. The map h also establishes a close connection between the cozero set of a function x and the cozero set of its image Ax, namely $h(\text{coz } Ax) \subset \text{cl}(\text{coz } x)$, where "cl" denotes closure.

Having seen that the existence of a separating linear isomorphism between $C(T)$ and $C(S)$ implies that T is the continuous image of S, it is natural to ask under what conditions are S and T homeomorphic? i.e., under what conditions must h be a homeomorphism? To get this sharper statement about h we must sharpen the conditions on A. Since we are dealing with maps A which have a certain sensitivity to cozero sets, we attempt to phrase things in terms of cozero sets. In view of the Banach-Stone theorem, it seems reasonable to expect that if $C(T)$ and $C(S)$ are very similar, then so are S and T. Therefore, for h to be a homeomorphism, we want $A(C(T))$ to closely resemble $C(S)$. In other words, images Ax of functions x in $C(T)$ should well-describe or well-approximate functions in $C(S)$ – from a dense subset of $C(S)$, perhaps, or something like that. But how to get cozero sets back into the picture? Here, after all, is where our leverage is.

To motivate our ultimate choice, let us look back at completely regular spaces for a moment. Completely regular spaces are topological spaces in which points and closed sets may be "separated" by continuous functions. This feature enables the space to be "described" by its continuous functions: the original topology is the weakest topology with respect to which those functions are continuous; furthermore, incidentably, the cozero sets of continuous real-valued functions form a base for any completely regular topology.

To achieve the "similarity" of $A(C(T))$ and $C(S)$ by means of cozero sets we settled on the map A having the following property. We say that A is *detaching* if for any two distinct points s and t of S there exist functions x and y in $C(T)$ whose cozero sets have disjoint closures with $s \in \text{coz } Ax$ and $t \in \text{coz } Ay$. Since $h(\text{coz } Ax) \subset \text{cl}(\text{coz } x)$, etc., it was hoped that this may permit the functions Ax to describe the topology on S. As it turns out [Prop. 1.9], if A is detaching then the continuous map $h:S \to T$ is injective (and conversely). Since a continuous injection of a compact space into a Hausdorff space is a homeomorphism, S and T must be homeomorphic. We achieve something of our goal in the form of a Banach-Stone type theorem, Theorem 1 below. Note that no continuity-type assumption about A is made.

Theorem 1. (Props 1.7, 1.9 and [4, Th. 2.1]) *If the separating isomorphism A is detaching then S and T are homeomorphic.*

What about that other feature of the Banach-Stone theorem, the part about linear isometries being necessarily weighted composition maps? Let e stand for the map which takes every point in T into 1 and let a be Ae. To avoid trivialities we make the standing assumption that Ae never vanishes. By assuming continuity but dropping detaching, we obtain:

Theorem 2. (Prop. 1.11 and [4, Theorem 2.2]) *If A is continuous and separating then, for any x in $C(T)$ and any s in S, $Ax(s) = a(s)x(h(s))$. In addition, h is surjective if and only if A is injective.*

Let us keep continuity but weaken detaching: We say that A is *extremely regular* if, for any $r>0$, and distinct points u and v of S, there exists an x in $C(T)$ such that $Ax(v) \neq 0$ and $|Ax(u)/Ax(v)| < r$. The points of S are still separated to some extent by the functions Ax, but not as sharply. Nevertheless:

Theorem 3. [4,Theorem 2.3] *If A is a continuous, separating, extremely regular isomorphism then*

(a) S and T must be homeomorphic (h being the homeomorphism) and

(b) A is a surjective weighted composition map.

Must a separating map be continuous? As shown by a fairly simple example [4, Example 3.5], differentiation is separating on a certain subspace of $C([0,1])$. As is well-known, differentiation is almost never continuous. In [6] Jarosz shows that if A is separating and onto then it is a continuous weighted homomorphism; he deduces that A is detaching.

As we have mentioned, separating alone does not imply continuity but often does in combination with other conditions. What about statements in the converse direction? Must a continuous map be separating? The following simple example shows that this is not so.

Example. [4, Example 3.5] Let $S=T=[0,1]$ and let A be integral from 0 to $s \in [0,1]$. A is obviously continuous; by considering two shifted triangular pulses, however, it is easy to see that A is not separating.

Indeed (Example 1.13, [4, Example 3.6] and [5, Example 5.1]) there are linear maps which are separating and even injective but discontinuous.

However, they are not detaching. Separating and detaching does imply continuous in a wide category of spaces: if T is metrizable, for example, or even an infinite product of metrizable spaces; or, dropping metrizability, but assuming that the isolated points of T form a dense subset. An important question remains with significant implications. The discontinuous separating map, differentiation, was defined on a carefully selected proper subspace of C(T). Does category play a role? In particular, we have the following conjecture, similar in form to the closed graph theorem.

Conjecture. A separating map defined on a Baire (= nonmeager = second category) subspace of C(T) must be continuous.

Many of the results just mentioned may be extended to Banach algebras ([5]). In order to discuss this, let X and Y be commutative, semisimple, Banach algebras which contain an identity e.

For most of the results about Banach algebras, the underlying field $\mathbb{F}$ can be any complete field with valuation; sometimes it must be a local field (see Sec.2). When the underlying field is not the complex numbers, we assume that one of two situations occurs: Either X and Y are "Gelfand" algebras in the sense that $X/M = \mathbb{F}$ for any maximal ideal M or the intersection of the kernels of the field-valued homomorphisms of the algebras is {0}. In either case both X and Y can be viewed as subalgebras of the algebra of continuous functions on a compact Hausdorff space. This permits the definition of straightforward analogs of separation and detaching in this context.

Let H(X) and H(Y) denote the $\mathbb{F}$-valued homomorphisms of X and Y, respectively, endowed with their Gelfand topologies. When A is separating, then by the same arguments as used prior to Def. 1.6, there exists an associated continuous function $h:H(Y) \rightarrow H(X)$. If the separating map A is detaching then h is a homeomorphism. In many cases, if A is detaching and separating then A is continuous [Theorems 2.5 and 2.7].

1. Algebras of functions.

Let T and S be at least compact Hausdorff spaces and let C(T) and C(S) be the algebras of continuous $\mathbb{F}$-valued functions on T and S, respectively, equipped with their supremum norm topologies. If the underlying field $\mathbb{F}$ is nonarchimedean-valued, we also assume that S and T are 0-dimensional spaces. In the version of the Banach stone Theorem developed in [2] when the underlying field $\mathbb{F}$ is nonarchimedean-valued, it was assumed that A was a **surjective** isometry, but the proof required only that A be continuous and that the range "separate points of S" in the sense that for any $r,s \in S$, there exists $x \in C(T)$ with $Ax(r) = 1$, and $Ax(s) = 0$. We state this strengthened version of the theorem (i.e. without assuming that A is surjective) as Theorem 1.1 here.

For results prior to Def. 1.3 we assume that S and T are 0-dimensional and that $\mathbb{F}$ is a complete nonarchimedean-valued field. Note that characteristic functions of "clopen" (= closed + open) sets are continuous in this case.

In proving the main result of [2] it was noted that the collections of clopen subsets of T and S are algebras of sets and that the subalgebras of continuous functions generated by the characteristic functions of those clopen sets are dense in the function algebras C(T) and C(S). This was instrumental in the proofs.

If f is a nontrivial continuous linear form from C(T) into $\mathbb{F}$, then f is called *separating* if it has the property that when k_U and k_V are the characteristic functions of disjoint clopen subsets U and V of T, then $f(k_U)f(k_V) = 0$. The collection $L = \{U \subset T: U \text{ is clopen and } f(k_U) \neq 0\}$ of clopen sets for separating linear forms f is shown to have the following properties ([2, Lemma 2]):

(a) *L is closed under finite intersections.*

(b) *Supersets of sets in L also belong to L.*

(c) *A clopen subset of T belongs to L if and only if it contains the intersection of all the sets in L.*

(d) *The intersection of all the sets in L is a singleton.*

An evaluation map $t^{\wedge}$ on a function space C(T) is the map sending $x \in C(T)$ into $x(t)$. As observed in [2] a nontrivial continuous linear form f on C(T) is a scalar multiple of an evaluation map (i.e., for some $a \in \mathbb{F}$, $f(x) = at\hat{}(x) = ax(t)$ for some $t \in T$ and all $x \in C(T)$) if and only if f is separating.

Using this we are able to prove:

Theorem 1.1. [2, Theorem] *If $A:C(T) \to C(S)$ is continuous and the range of A separates points of S, then the following are equivalent:*

(a) A is separating

(b) There exists a homeomorphism h from S onto T and a function $a \in C(S)$ such that for all $x \in C(T)$, $Ax(s) = a(s)x(h(s))$.

Unlike the classical situation (i.e., the real or complex case), isometric isomorphisms $A:C(T) \to C(S)$ need not be weighted homomorphisms as the following example illustrates.

Example 1.2. Let V and W be disjoint homeomorphic compact 0-dimensional Hausdorff spaces. Topologize $V \cup W$ by taking open sets to mean union of open sets from V and W and let $T = S = V \cup W$. Let $a, b \in \mathbb{F}$ with $a \neq b$, $0 < |a| < 1$, $0 < |b| < 1$. Let the homeomorphism from V to W be denoted by h, and define $A:C(T) \to C(S)$ by taking for any $x \in C(T)$,

$$Ax(t) = \begin{cases} ax(t) + x(h(t)), & t \in V \\ x(h^{-1}(t)) + bx(t), & t \in W \end{cases}$$

As a consequence of the ultrametric inequality, $|c+d| = |d|$ when $|d| > |c|$; this implies that A is an isometry. A is not separating because Ak_U and Ak_V are invertible functions which never vanish; hence $Ak_UAk_V \neq 0$ while $k_Vk_U = 0$.

Up to this point one could say that we had considered when a continuous, separating linear map is a weighted homomorphism. We then dropped the assumption that the linear map A be continuous and investigated the broader question of connection between separating and continuity. In [3] we did this in the same context as [2] for compact 0-dimensional spaces S and T and for continuous functions on them taking values in a nonarchimedean valued field. In [4] we consider the case of $\mathbb{R}$-valued and $\mathbb{C}$-valued functions. The proofs in [4] apply **mutatis mutandis** to any complete valued field; the examples remain true with some modification.

Assumptions: For the remainder of the results in this section we assume that

- the image Ae of the identity e of C(T) is invertible in C(S);

- A^ denotes the restriction of the algebraic adjoint A^* of A to the set S^ of evaluation maps s^ determined by the points s of S; for any $x \in C(T)$, note that A^s^(x) = Ax(s);

- $\mathbb{F}$ is a complete valued field;

- S and T are compact Hausdorff spaces; if $\mathbb{F}$ is nonarchimedean-valued, S and T are 0-dimensional.

- For a function x, the cozero set of x, the set of points in its domain where it is not 0, is denoted coz x.

Definition 1.3. Let $s \in S$. An open subset U of T is called a *vanishing set for A^s^* if when coz $x \subset U$, A^s^x = 0. The *support of A^s^*, denoted supp A^s^, is the complement of the union of all the vanishing sets of A^s^.

With the exception of Example 1.13, the remaining results of this section are all proved in [4].

Proposition 1.4. *supp A^s^ is not empty.*

Sketch of proof. The proof applies to any linear transformation and only uses the assumption that Ae is invertible and therefore A^s^ is a nonvanishing linear functional for all $s \in S$. Assume that the vanishing sets cover U_i, $i = 1,\dots,n$, such that $T = \sum_{i=1}^{n} U_i$. But C(T) is a normal algebra and therefore there exists a decomposition of the identity $e = \sum_{i=1}^{n} x_i$ where coz $x_i \subset U_i$. But then $x = \sum_{i=1}^{n} xx_i$ and from the definition of vanishing set Ax(s) = 0 for all x and this is a contradiction.

Proposition 1.5. *supp A^s^ is a singleton.*

Sketch of the proof. If $t,u \in$ supp A^s^, then there exist disjoint open sets V and W such that $t \in V$ and $u \in W$, and V and W are not vanishing sets. Then there exist z,w such that coz $z \subset V$, coz $w \subset W$, A^s^(z) $\neq$ 0 and A^s^(w) $\neq$ 0. This contradicts the separating property of A.

Definition 1.6. Let $h: S \to T$ where h(s) = supp A^s^ and by supp A^s^ we mean the single point in supp A^s^.

Proposition 1.7. *The map h of Def. 1.6 has the following properties.*

(a) If $x \equiv 0$ on an open set U, then $Ax \equiv 0$ on $h^{-1}(U)$. If $x \equiv y$ on U, then $Ax \equiv Ay$ on $h^{-1}(U)$. If, for some scalar b, $x \equiv b$ on U, then $Ax \equiv bAe = ba$ $(a = Ae)$ on $h^{-1}(U)$.

(b) h is continuous.

(c) If A is injective, then h is surjective.

(d) If $x \in C(T)$, then $h(\text{coz } Ax) \subset \text{cl}(\text{coz } x)$.

Definition 1.8. A is *detaching* if for $r,s \in S$ there exists $z,w \in C(T)$ such that cl(coz z) and cl(coz w) are disjoint while $Az(r) \neq 0$ and $Aw(s) \neq 0$.

Proposition 1.9. *A is detaching if and only if h is injective.*

Proof. The proof is direct application of Prop. 1.7.

Example 1.10. Consider the following map.

$$A: C[0,1] \to C[0,2]$$

$$Ax(s) = \begin{cases} x(s), & s \in [0,1] \\ x(1), & s \in [1,2]. \end{cases}$$

The map A is separating but not detaching. In fact, $h(s) = s$ for $0 \leq s \leq 1$ while $h(s) = 1$ for $s>1$.

The following result can be proved in a fairly direct manner.

Proposition 1.11. *If $A:C(T) \to C(S)$ is separating and continuous, then there exists $a \in C(S)$ such that $Ax(s) = a(s)x(h(s))$ for all $x \in C(T)$ and all $s \in S$.*

Sketch of the proof. Let $s \in S$ and $t = h(s)$. Suppose that $x(h(s)) = 0$. Let (U_n) be a sequence of neighborhoods of t with the property that $\sup(|x(t)|: t \in U_n) < 1/n$. Then let V_n be a neighborhood of t such that $t \in V_n$, $\text{cl}(V_n) \subset U_n$ with $x_n \equiv 1$ on $\text{cl}(V_n)$ while $x_n \equiv 0$ on CU_n. Since $xx_n \to 0$ and A is continuous, then $Axx_n(s) \to 0$. But $x \equiv xx_n$ on $h^{-1}(V_n)$ so $Axx_n(s) = Ax(s) = 0$. If $x(t) \neq 0$, apply the previous paragraph to $x - x(t)e$.

The closed graph theorem implies that A is continuous if and only if $A^\wedge s^\wedge$ is continuous for all $s \in S$. If the topological spaces S and T are "full" (Def. 2.3) as first countable spaces are, for example, the separated, detaching, injective maps $A:C(T) \to C(S)$ are continuous. As shown in Theorem 1.12 below in the case when T is first countable, the map A is a weighted homomorphism.

Theorem 1.12. *If A is separating, detaching, and injective then for $Ae = a \in C(S)$ and with h as in Def. 1.6,*

(a) $A^{\wedge}s^{\wedge}x = Ax(s) = a(s)x(h(s))$ for all s belonging to the closure of the set of isolated points of S and all $x \in C(T)$;

(b) if T is first countable then A is continuous and $A^{\wedge}s^{\wedge}x = Ax(s) = a(s)x(h(s))$ for all $s \in S$ and all $x \in C(T)$.

Sketch of the proof. By Props 1.7 and 1.9, h is a surjective homeomorphism.

(a) If s is an isolated point of S, then h(s) is an isolated point of T. Thus, by Prop. 1.7(a), $A(x-x(h(s))) = 0$ on $h^{-1}(h(s)) = \{s\}$. In other words $Ax(s) = a(s)x(h(s))$ for isolated points of S. The continuity of Ax completes the proof of (a).

(b) To prove (b), let $t \in T$ and (W_n) be a base of neighborhoods of t. Let h be as in Def. 1.6. Let s be such that $t = h(s)$ and assume that $A^{\wedge}s^{\wedge}$ is not continuous. This means that there exists $x_n \in C(T)$ such that $\|x_n\| < 1/2^n$ while $|Ax_n(s^*)| > 2r > 0$ for some $r>0$ and all $s^* \in W_n$. There exist open sets U_n, V_n, where U_n and V_n are contained in $W_n - W_{n+1}$, such that $cl(U_n)$ is disjoint from both $cl(\bigcup_{m\neq n} U_m)$ and $cl(\bigcup_n V_n)$. There exist $z_n \in C(T)$ such that $z_n \equiv 1$ on a closed subneighborhood of U_n and $z_n \equiv 0$ on CU_n. Let $w_n \in C(T)$ where $w_n \equiv 1$ on a closed subneighborhood of V_n and $w_n \equiv 0$ on CV_n.

As a result of the previous construction the functions $x = \sum_n x_n z_n$ and $y = \sum_n x_n w_n$ satisfy $xy = 0$ while $Ax(s) \neq 0$ and $Ay(s) \neq 0$. This contradicts the assumption that A is separating.

In the following example, we retain all the hypotheses of Theorem 1.12 except for A being detaching. The conclusions (a) and (b) both fail in this case.

Example 1.13. A separating, discontinuous map which is not detaching. [5, Example 5.1] Let $T = \{1/n: n>0\} \cup \{0\}$ with the topology inherited from the real line. Let $H = \{s_n: n\geq 0\}$ be any denumerable set with discrete topology. Let $\mathcal{F}$ be an ultrafilter of subsets of T with the property $\bigcap_{F\in\mathcal{F}} F = \emptyset$. Let $M = \{x\in C(T)$: $x\equiv bt$ for some scalar b on some $F\in\mathcal{F}\}$. Because $\mathcal{F}$ is a filter, M is a subspace of C(T). Let $v(t) = 1-t$ and N be an algebraic complement of M to which v belongs.

Let a linear functional on C(T) be defined as follows.

$$f: C(T) \to F, \quad f(x) = f(m+n) = b$$

where $x = m+n$, $m \in M$, $n \in N$, $m \equiv bt$ on $F \in \mathcal{F}$. Because $\mathcal{F}$ is an ultrafilter, it follows that f is a separating linear functional. Define $B: C(T) \to C(H)$ where

$$B(x(s_n)) = \begin{cases} x(s_n), & n>0 \\ f(x), & n=0 \end{cases}$$

Since f is separating it follows that B is separating; clearly B is injective. Since Bx is virtually a copy of the bounded function x, Bx is bounded and can be continuously extended to $(Bx)^{\beta}$ on the Čech-Stone compactification βH of H. Now the linear transformation

$$A: C(T) \to C(\beta H)$$
$$x \mapsto (Bx)^{\beta}$$

is separating and injective, but not detaching. In fact for all $s \in \beta H$, where $s \neq s_n$ for some $n>0$, $h(s) = 0$.

The fact that A is not continuous can be observed by considering $x \in C(T)$ where $x(t) = t$ on a set $F \in \mathcal{F}$; let x be 0 on $\complement F$. Choose F such that if $1/n \in F$ then $1/n$ is less than some fixed number ε. By the way f is constructed, $f(x)=1$. It follows that x can be made arbitrarily small while $f(x)=1$. Thus Ax has norm equal to 1 and A is discontinuous.

It can be readily seen that A is not surjective because the range of B does not include all the functions in C(H) which are continuous at s_0. Thus, this example does not contradict the result of Jarosz [6] in which he showed that if A is onto then it must be continuous.

The preceding example can be reconstructed to apply to nonarchimedean valued fields by constructing T with a sequence of field elements which converge to 0 rather than using the real numbers $1/n$. In order that the above argument be used, we assume that the field F is a local field and use the Banaschewski compactification ([1]) of H, an analog of the Čech-Stone compactification.

2. Banach agebras.

In [5] the work of the preceding section is broadened to include maps from a Banach algebra X into a Banach algebra Y, each with an identity e. The maps A are called *separating* if when $x, w \in X$ and $xw = 0$, then $AxAw = 0$. Continuous separating maps need not be weighted homomorphisms anymore. When the underlying field is not the complex numbers, it need not be true that $X/M = \mathbb{F}$ for each maximal ideal M of X. If $X/M = \mathbb{F}$ for each maximal ideal M of X we assume that our algebras X are semisimple so that our algebras X and Y can be viewed as algebras of continuous functions on the compact Hausdorff space of maximal ideals endowed with the Gelfand topology.

We also treat algebras X (and Y) in which there are maximal ideals M for which $X/M \neq \mathbb{F}$ but for which the intersection of the kernels of field-valued homomorphisms is $\{0\}$. These algebras are what could be called "homomorphically semisimple" and are isomorphic to a subalgebra of the algebra of continuous functions on the carrier space of all field-valued homomorphisms. In order that the carrier space be compact we assume that the field $\mathbb{F}$ is local.

Suppose that S and T are compact Hausdorff spaces and that X and Y are Banach subalgebras with identity of C(T) and C(S), respectively, and that X is a normal subalgebra of C(T). We assume that $A:X \longrightarrow Y$ is separating and that there exists $w \in X$ such that Aw is invertible in C(S). We assume further that all of the $\mathbb{F}$-valued homomorphisms of X and Y are given by evaluation maps at points of T and S, respectively. The sup-norm topologies inherited by X and Y from C(T) and C(S) (the spectral radius topologies) are of course weaker than the Banach algebra topologies on them. As a consequence of Gelfand theory, the assumption that Aw is invertible in C(S) actually implies that Aw^{-1} belongs to Y. The notions of separating and detaching as described in Sec. 1 can now be restated here in exactly the same manner as we done there (involving cozero sets).

A mapping h from S into T is defined exactly as in Def. 1.6. Because of the discussion of the preceding paragraphs, Props. 1.4, 1.5, 1.7, 1.9 and 1.11 of Sec. 1 remain true in this setting with virtually identical proofs. About the only difference which occurs is that we must use the function Aw in the way we use Ae earlier. Proposition 1.7(c) is replaced by the more general statement Prop. 2.1 which gives an equivalent condition for h to be onto.

Proposition 2.1. *Let S and T be compact Hausdorff spaces, let X and Y be Banach subalgebras of C(S) and C(T), respectively, whose homomorphisms are given by the evaluation maps on T and S. Let $A:X \longrightarrow Y$ be separating. Then the map h is surjective if and only if there does not exist an open subset U of T such that when $\operatorname{coz} x \subset U$, $Ax = 0$.*

We can now characterize separating maps which are weighted

homomorphisms.

Proposition 2.2. *Let A, X and Y be as in Prop. 2.1 and let a=Ae. If Aw is invertible for some w∈X, then A is continuous when X and Y carry their respective spectral radius topologies if and only if Ax(s) = a(s)x(h(s)) for all x∈X and all s∈S.*

The proof of Theorem 1.12 depends entirely on the existence of sequences (t_n) and (t_n^*) in T which are contained in open subsets of the complement of the closure of the set of isolated points and have the following properties:

- these sequences are disjoint from one another, have none of their own limit points (are discrete subspaces), and
- have a common limit point.

Because of this, $cl((t_n)\cup(t_n^*))$ is not homeomorphic to (is smaller than) $\beta\mathbb{N}$, the Čech-Stone compactification of the integers. A stronger requirement ("sequentially small") than this is considered by Kunen in [7].

The result of Theorem 1.12(a) depends on the fact that continuous functions are locally constant at isolated points. This happens on a possibly larger set than the isolated points called *P-points*. A point of T is a *P-point* if and only if each continuous function is constant in some neighborhood of the point. Equivalent to this is the statement that each denumerable intersection of neighborhoods of the point is a neighborhood. Thus the functions A^s^ are continuous at all points in the closure of the set of P-points of S and are in fact of form described in Theorem 1.12(a).

Definition 2.3. A compact Hausdorff space T is called *a full space* if any open subset W of the complement of the closure of the P-points contains a discrete denumerable set (t_n) with the property that two disjoint subsets of the set have a common limit point.

If the cardinality of T is less than that of $\beta\mathbb{N}$, then T is a full space ([5]). In [5] it is also shown that a finite product $\Pi_{i=1}^{n}T_i$ is a full space when a particular T_i is a full space. Also every infinite product of compact Hausdorff spaces is a full space. Thus, there are a large number of full spaces. The nature and number of these spaces is presently under investigation.

The closed graph theorem yields the following result:

Proposition 2.4. *Let X and Y be a Banach subalgebras of $C(T)$ and $C(S)$ (taking values in $\mathbb{R}$, $\mathbb{C}$, or any complete nonarchimedean-valued field) as mentioned earlier and let $A:X \to Y$ be linear. Then:*

(a) A is continuous if and only if $A^\wedge s^\wedge$ is continuous for all $s \in S$.

(b) If A is not continuous then there exists $r>0$, $s \in S$, an open neighborhood of s, and a sequence of vectors $x_n \in X$ such that $x_n \to 0$ and $\inf_{s^* \in V} |Ax_n(s^*)| \geq r$.

This result is then used to prove the following theorem.

Theorem 2.5. [5, Theorem 5.3] *With X and Y as in Prop. 2.4, let $A:X \to Y$ be separating and detaching and let Aw be invertible for some $w \in X$. Let X be a normal subalgebra of $C(T)$ (e.g., X could be a regular Banach algebra), let S be a full space and let $a=Ae$.*

(a) If S has no P-points then A is continuous.

(b) If S has P-points and h is onto (as happens, for example, when A in injective), then A is continuous and $Ax(s) = a(s)x(h(s))$ for all s in the closure of the P-points of S.

The proof of Theorem 2.5 is quite similar to that of Theorem 1.12. It is necessary to require h to be onto in (b) because then the P-points of S are mapped into P-points of T by h. Through this one obtains the form of $A^\wedge s^\wedge$ on the closure of the P-points of S. It is an open question as to whether or not continuity of A is lost if h is not assumed to be onto when S has P-points; Example 2.8 shows that the form Ax takes on the closure of the P-points of S is indeed lost.

The map of the next example satisfies the hypotheses of Theorem 2.5(a). Although it is not continuous, it is not a weighted homomorphism.

Example 2.6. (5, Example 3.1] The space $D_n[a,b]$ of functions with continuous derivatives up to order n carrying the norm given by

$$||x|| = \sum_{i=0}^{n} \frac{1}{i!} \sup_{t \in [a,b]} |x^{i}(t)|$$

is a regular Banach algebra and therefore a normal subalgebra of C[a,b]. The derivative operator from $D_n[a,b]$ to $D_{n-1}[a,b]$ is separating and detaching. there is no open subset U of [a,b] on which all functions whose cozero set is contained in U will have a derivative equal to 0. Thus the map $h:[a,b] \to [a,b]$ is onto and is in fact the identity. There are no P-points in S = T = [a,b].

This linear map is not injective because the derivative of the identity function is equal to 0. It is also not continuous when the domain and range algebras carry the spectral radius (sup norm) topologies. Consequently, it is continuous but not a weighted homomorphism.

The previous material applies to commutative semisimple Banach algebras X and Y. The semisimplicity implies that these algebras can be viewed as isomorphic to the subalgebras of the Gelfand functions X^ and Y^ of the algebra of continuous functions on their maximal ideals carrying the Gelfand topologies. If the topologies on X and Y are carried over to their images in these algebras via the injection maps, i_X and i_Y, then X and Y are isometric to X^ and Y^. A linear transformation A^* can then be defined making the diagram below commutative.

$$\begin{array}{rcccc} A: & X & \longrightarrow & Y \\ & \Big\downarrow i_X & & \Big\downarrow i_Y \\ A^*: & X\hat{} & \longrightarrow & Y\hat{} \end{array}$$

Thus, $i_Y A = A^* i_X$ and all the properties of A relevant to investigations of this paper are carried out to A^* via the isometries i_Y and i_X. Thus proof that A^* is a continuous weighted homomorphism implies that A is also such an operator. By Theorem 2.5 and the previous discussion we can now show:

Theorem 2.7. [5, Theorem 5.4] *Let X and Y be commutative semisimple Banach algebras with identity and X a regular algebra. Let $A:X \longrightarrow Y$ be separating and Aw be invertible for some $w \in X$.*

(a) Suppose that the maximal ideals M(Y) of Y with Gelfand topology is full and contains no P-points. Suppose also that for any pair $s,t \in M(Y)$ there exists $x,y,z \in X$ such that $Ax(s) \neq 0$, $Ay(t) \neq 0$, $xz=x$ and $yz=0$. Then A is continuous.

(b) Let M(Y) be full and contain P-points. Suppose that the remaining hypothesis of (a) is true. In addition, suppose that there exists no nonzero element $v \in X$ such that when $u \in X$ and $uv=v$ then $Au = 0$ (e.g., A is injective). Then A is continuous and if M(Y) is equal to the closure of its P-points it follows that A is a weighted homomorphism.

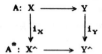

Sketch of proof. The hypotheses of (a) and (b) imply that A^* is detaching and in (b) that $h:M(Y) \longrightarrow M(X)$ is onto. Thus Theorem 2.5 can be applied to A^*. Part (b) in Theorem 2.7 was stated differently than (b) in Theorem 2.5 to avoid the awkward situation of trying to express the form of $A^*x^\wedge$ (where $x^\wedge$ denotes the Gelfand function associated with the algebra element x) on certain points of M(Y) into a corresponding property of A.

EXAMPLE 2.8. Let $X = D_1[a,b]$ and $Y = C(T)$ where $T = \{1/n: n>0\} \cup \{0\}$. The map A is defined by

$$A: X \longrightarrow Y$$
$$x \longmapsto dx/dt\,|_T$$

A is separating and detaching and $h:T \longrightarrow [a,b]$ is the identity. But T has P-points and h is not onto. Because of this, the form of A in Theorem 2.5(b) is not valid. In fact according to Theorem 2.7(b), if h were surjective then A would be a weighted homomorphism and of course differentiation is not such a map.

There are a number of open questions.

(1) Can the Banach algebra results be extended to topological algebras? The authors have thus far been able to show that if T and S are locally compact and connected completely regular spaces, then a whole family of separating and detaching linear transformations from C(T) into C(S) are weighted homomorphisms. The proof hinges on a result which says that if K is a compact subset of a completely regular topological space T and $z,w \in C(K)$ with $zw=0$, then z and w can be extended continuously to functions z^* and w^* defined on all of T satisfying $z^*w^* = 0$.

(2) Concerning Theorem 2.5(b) of this section, can a separating, detaching, linear map A be constructed for which h is not onto and A is discontinuous?

(3) Concerning Example 2.8, the value of Ax is not determined by the value of $x|_{h(S)}$ because A is not a weighted homomorphism. Is the value of Ax uniquely determined by $x|_{h(S)}$ if and only if A is a weighted homomorphism?

(4) Can the results Theorems 2.5 and 2.7 in which full space was hypothesized be broadened to exclude this restriction? Or is this restriction actually necessary? This would be shown by the existence of a separating, detaching, discontinuous linear map, for which h is onto.

(5) The work of question (1) as presently developed is limited to the real and complex numbers because the continuous function extension theorem mentioned in (1) is presently unexamined in the nonarchimedean case.

(6) Determine which spaces are full, which are not.

REFERENCES

[1] G.BACHMAN, E.BECKENSTEIN, L.NARICI, S.WARNER, *Rings of continuous functions with values in a topological field*, Trans. A.M.S., 204 (1975), 91-112.

[2] E.BECKENSTEIN, L.NARICI, *A nonarchimedean Stone-Banach theorem*, Proc. A.M.S., 100 (1987), 242-246.

[3] E.BECKENSTEIN, L.NARICI, *Automatic continuity of certain linear isomorphisms*, Acad. Royale de Belgique, Bull. Soc. R. Sci. Bruxelles 73 (1987), 191-200.

[4] E.BECKENSTEIN, L.NARICI, A.TODD, *Automatic continuity of linear maps on spaces of continuous functions*, Manuscripta Math., 62 (1988), 257-275.

[5] E.BECKENSTEIN, L.NARICI, A.TODD, *An automatic continuity result for linear maps on Banach algebras*, to appear.

[6] K.JAROSZ, *Automatic continuity of separating linear isomorphism*, Bull. Can. Math. Soc., to appear.

[7] K.KUNEN, *Large Homogeneous Compact Spaces*, Dept. Math., University of Wisconsin, Madison, WI 53706, USA, preprint.

INEQUALITIES WITH WEIGHTS FOR HARDY OPERATORS IN FUNCTION SPACES

E.I. BEREZNOI

Jaroslavl, USSR

Let (X,Y) be a pair of Banach function spaces on $\mathbb{R}_+$ with the Lebesgue measure. There are given conditions on (X,Y) and the nonnegative weight functions v, w which ensure that the well known Hardy's integral operator P a is bounded between weighted Banach function spaces X_w and Y_v. Also the notion of weak type (X_w,Y_v) of operator is given and it is shown that P is of weak type (X_w,Y_v) if and only if certain condition holds. Similar results are given for the dual operator of P.

1. Introduction.

Let S be the space of all equivalence classes of Lebesgue measurable real valued functions defined on $\mathbb{R}_+ = (0,\infty)$, equipped with the topology of convergence in measure on sets of finite measure.

A linear subset X of S is said to be *a Banach function space* (on $\mathbb{R}_+$) whenever

$$(|y| \le |x| \text{ a.e., } y\in S,\ x\in X) \Longrightarrow (y\in X,\ \|y\| \le \|x\|).$$

By X' we denote the Köthe dual space to X (which can be identified with the space of all functionals with integral representations), defined by

$$X' = \{x' \in S\colon \|x'\|_{X'} = \sup_{\|x\|_X \le 1} \int_{\mathbb{R}_+} |x' x|\, d\mu < \infty\}.$$

Throughout the paper we assume that all Banach function spaces have the Fatou property. Recall (cf. [10]) that a Banach function space X has *the Fatou property* provided the unit ball of X is a closed subset of S. It is well known (see [10]) that X has the Fatou property if and only if X = X'' isometrically.

We shall say that a Lebesgue measurable function w∈S is *a weight* if it is positive a.e. on $\mathbb{R}_+$. For any Banach function space X and weight w, we define *the weighted Banach function space* X_w which consists of all $x \in S$ such that $xw \in X$. The norm on X_w is defined in the usual way, i.e.,

$$\|x\|_{X_w} = \|xw\|_X.$$

It is clear that

$$(X_w)' = X_{w^{-1}} \tag{1}$$

holds isometrically.

Let $x \in S$, then the Hardy operator P and its dual Q are defined by

$$Px(t) = \int_0^t x(s)ds, \qquad Qx(t) = \int_t^\infty x(s)ds.$$

For these operators hold the well-known Hardy inequalities (see [5]). The necessary and sufficient conditions on weight w, which ensure that P is bounded operator in L_w^p were given by Talenti and Tomaselli (see [3]). In the papers [1,2,8,9] more simple proofs of the above results and the best estimates of the norm of this operator were given. Moreover it was shown that Hardy's inequality remains true for a larger class of Banach function spaces than L_w^p spaces.

In this paper we shall prove Hardy's weighted inequalities in the general case of Banach function spaces. The obtained results are new even in the case of Orlicz spaces.

2. Main results.

Throughout this section L(E,F) denote the Banach space of all linear and bounded operators mapping a Banach space E into a Banach space F. Further, let X and Y be a Banach function spaces on $\mathbb{R}_+$ with the Fatou property and let v, w be some weights.

Definition 1. We say that *a pair (w,v) belongs to H(X,Y)* if

$$\sup_{t>0} \|w^{-1}\chi_{(0,t)}\|_{X'}\|v\chi_{(t,\infty)}\|_Y < \infty. \tag{2}$$

Here and in the sequel χ_A denote the characteristic function of the set A.

Definition 2. We write $(X,Y) \in G$ if there exists $C>0$ such that for any increasing sequence of positive numbers and any $x \in X$, $y \in Y'$ the following condition holds

$$\sum_k \|x\chi_{(t_{k-1},t_k)}\|_X \|y\chi_{(t_k,t_{k+1})}\|_{Y'} \le C \cdot \|x\|_X \|y\|_{Y'} . \tag{3}$$

Theorem 1. *(a) If $P \in L(X_w,Y_v)$, then $(w,v) \in H(X,Y)$.*

(b) If $(X,Y) \in G$ and $(w,v) \in H(X,Y)$, then $P \in L(X_w,Y_v)$.

Proof. (a) Fix $t_o>0$ and suppose $P \in L(X_w,Y_v)$. Then for some $c>0$ and any $0 \le x \in X_w$ with $\|x\|_{X_w} \le 1$, we have

$$C \ge \|P(x\chi_{(0,t_o)})\|_{Y_v} \ge \|v\chi_{(t_o,\infty)} \int_{\mathbb{R}_+} x\chi_{(0,t_o)}ds\|_Y$$

$$= \left|\int_{\mathbb{R}_+} x\chi_{(0,t_o)}ds\right| \; \|v\chi_{(t_o,\infty)}\|_Y \,, \tag{4}$$

since

$$\sup\left\{\int_{\mathbb{R}_+} x\chi_{(0,t_o)}ds \;:\; \|x\|_{X_w} \le 1,\; x \ge 0\right\} = \|w^{-1}\chi_{(0,t_o)}\|_{X'}$$

by (1). Thus $(w,v) \in H(X,Y)$ by (4).

(b) Suppose $(w,v) \in H(X,Y)$ and $(X,Y) \in G$. Then, it follows by (2) that

$$\int_0^t |x(s)|ds < \infty$$

for any $x \in X_w$ and $t>0$. Fix $0 \le x \in X_w$, and assume, without loss of generality that $x > 0$ a.e. and $x \notin L^1$. Then we can choose a positive increasing sequence $\{t_k\}$ such that

$$\int_0^{t_k} x(s)ds = 2^k$$

for any $k \in \mathbb{Z}$.

Now put $A_k = (t_{k-1},t_k)$ for $k \in \mathbb{Z}$. Then for any $y \in (Y_v)'$, $y \ge 0$, we have

$$\int_{\mathbb{R}_+} P(x(s))y(s)ds = \sum_k \int_{A_k} y(s)\left(\int_0^s x(u)du\right)ds$$

$$\le \sum_k 2^k \int_{A_k} y(s)ds = 4\sum_k \left(\int_{A_{k-1}} x(s)ds\right)\left(\int_{A_k} y(s)ds\right)$$

$$\le 4\sum_k \|w^{-1}\chi_{A_{k-1}}\|_{X'} \; \|wx\chi_{A_{k-1}}\|_X \; \|v^{-1}y\chi_{A_k}\|_{Y'} \; \|v\chi_{A_k}\|_Y$$

$$\le 4 \left(\sup_{t>0} \|w^{-1}\chi_{(0,t)}\|_{X'} \; \|v\chi_{(t,\infty)}\|_Y\right) \sum_k \|wx\chi_{A_{k-1}}\|_X \; \|v^{-1}y\chi_{A_k}\|_{Y'}$$

$$\le C \; \|x\|_{X_w} \; \|y\|_{Y_v}.$$

Hence, by arbitrariness of $y \in (Y_v)'$, $y \ge 0$, it follows that $P \in L(X_w,Y_v)$.

Let us remark that $(L^p,L^q) \in G$ if and only if $q \le p$. In this case the condition (3) is well known (see [3]).

By using the geometric notion it is possible to give simple conditions on X and Y under which the condition (3) holds.

Recall (cf. [6],p.82) that a Banach function space X on $\mathbb{R}_+$ is said to satisfy *a lower p-estimate* (resp. *an upper p-estimate*), $1 < p < \infty$, if there exists a constant $C>0$ such that, for every choice of measurable and pairwise disjoint sets $\{A_k\} \subset \mathbb{R}_+$ and every $x \in X$

(5) $$\left[\sum_k \|x\chi_{A_k}\|_X^p \right]^{1/p} \leq C \; \|x\|_X,$$

$$\left[\text{resp. } \|x\|_X \leq \left[\sum_k \|x\chi_{A_k}\|_X^p \right]^{1/p} \right].$$

By the duality between lower and upper estimate (see [6],p.83) and the Hölder inequality, we easily obtain the following fact:

If $1 < p \leq q < \infty$, X satisfies a lower p-estimate and Y satisfies an upper q-estimate, then $(X,Y) \in G$.

From Corollary 1.f.9 in [6] and results of [4] (resp. [7]), we easily obtain values of p for which (5) holds for Orlicz spaces (resp. for classical Lorentz space). For the Lorentz spaces $L_{p,q}$ suitable result easily follows from Lemma 2.5 in [10].

Now we give estimates of weak type for the operator P.

Definition 3. We say that an operator $T: X_w \longrightarrow S$ is *of weak type* (X_w, Y_v) if for any $x \in X_w$ and $\lambda > 0$

(6) $$\|\chi_{D(Tx,\lambda)}\|_{Y_v} \leq C \; \lambda^{-1} \; \|x\|_{X_w} ,$$

where $D(Tx,\lambda) = \{t \in \mathbb{R}_+ : \; |Tx(t)| > \lambda\}$.

Let us remark that, in the particular case $X = L^p$ and $Y = L^q$, we get the classical inequality of weak type.

Theorem 2. *The operator P is of weak type (X_w, Y_v) if and only if the condition (2) holds.*

Proof. Let $x \geq 0$ with $\|x\|_{X_w} \leq 1$ and $t_o > 0$ be fixed. Then $\{t \in \mathbb{R}_+ : |Px(t)| > \lambda\} = [t_o, \infty)$, where $\lambda = \int_0^{t_o} x(s)ds$. Thus the inequality

$$\lambda \; \|\chi_{D(Px,\lambda)}\|_{Y_v} \leq C$$

is equivalent to the following one

$$\int_0^{t_o} x(s)ds \; \|v\chi_{(t_o,\infty)}\|_Y \leq C.$$

Hence, by arbitrariness of x and t_o, the proof is finished.

In the same way, we obtain similar results for the operator Q.

Theorem 3. *Let $(X,Y) \in G$. Then $Q \in L(X_w, Y_v)$ if and only if*

$$\sup_{t>0} \|w^{-1}\chi_{(t,\infty)}\|_{X'} \cdot \|v\chi_{(0,t)}\|_Y < \infty. \tag{8}$$

Theorem 4. *The operator Q is of weak type (X_w, Y_v) if and only if the condition (8) holds.*

REFERENCES

[1] K.F.Anderson, H. P. Heinig, *Weighted norm inequalities for certain integral operators*, SIAM J. Math. Anal. 14 (4) (1983), 834–844.

[2] E.N.Batuiev, V. D. Stiepanov, *On weighted inequalities of the Hardy type*, Sibirsk. Mat. Z. 30 (1989), 13–22.

[3] E.M.Dynkin, B. P. Osilenker, *Weighted estimates of singular operators and applications*, Itogi Nauki i Tehniki, T. Mat. Analiz, 21, Moscow 1983, 42–130 (in Russian).

[4] Z.G.Gorgadze, V. I. Tarieladze, *On Geometry of Orlicz Spaces*, Lecture Notes in Math. 928 (1980), 47–51.

[5] G. Hardy, *Notes on some points in the integral calculus*, Messenger of Math. 57 (1922), 12–16.

[6] J.Lindenstrauss, L. Tzafriri, *Classical Banach Spaces II, Function Spaces*, Springer Verlag, Berlin–Heidelberg–New York 1979.

[7] S.Ja. Novikov,*Cotype and type of Lorentz function spaces*, Mat. Zametki 32 (1982), 213–221.

[8] E.Savyer, *Weighted Lebesgue and Lorentz norm inequalities for the Hardy operator*, Trans. Amer. Math. Soc. 281 (1984), 329–337.

[9] V.D.Stiepanov, *On weighted Hardy inequality*, Sibirsk. Mat. Z. 28 (1987), 205–207.

[10] P.P.Zabrejko, *Ideal spaces of functions*, Vestnik Jaroslavl Univ. 8 (1974), 12–52.

ON (V) AND (V*) SETS IN VECTOR VALUED FUNCTION SPACES

FERNANDO BOMBAL

Madrid, Spain

INTRODUCTION. The concept of (V) and (V*) sets were introduced by Pełczyński in his important paper [13]. In the same paper, the so called properties (V) and (V*) are defined by the coincidence of the (V) or (V*) sets with the weakly relatively compact sets. Pełczyński proved in the same paper that if a Banach space E (respectively, the topological dual E^*) has property (V), then E^* (resp., E) has property (V*), and asked whether the converse implications are true. We know now that the answer to this question is negative (see for instance [3] or [14]). Hence, it seems to be a natural problem the characterization of the Banach spaces E for which the answer to the Pełczyński's question is positive. To this general study is devoted Section I. In Section II we specialize to the Banach space $C(\Omega,E)$.

We shall try to follow the standard notations in Banach space theory, as in [7], [8] and [12]. In order to prevent any doubt, we shall fix some terminology: If E is a Banach space, B(E) will be its closed unit ball and E^* its topological dual. The word operator will always mean linear bounded operator, and $\mathcal{L}(E,F)$ will stand for the Banach space of all operators from E into F. A subset B of a Banach space is called *weakly conditionally compact* if every sequence in B has a weakly Cauchy subsequence. A series $\sum x_n$ in E is said to be *weakly unconditionally Cauchy* (w.u.c. in short) if $\sum |x^*(x_n)|<\infty$ for every $x^* \in E^*$ (equivalently, if $\{ \sum_\sigma x_n: \sigma \subset \mathbb{N} \text{ finite} \}$ is a bounded subset). If A is a subset of the normed space E, [A] will be the closed linear span of A. We shall denote by $\mathcal{B}(E)$, $\mathcal{WC}(E)$ and $\mathcal{W}(E)$ the classes of bounded, weakly conditionally compact and weakly relatively compact subsets of E, respectively.

Research partialy supported by DGICYT grant PB 88-0141

For a compact Hausdorff space Ω and a Banach space E, we shall denote by $C(\Omega,E)$ the Banach space of all the continuous E-valued functions on Ω, with the supremum norm ($C(\Omega) = C(\Omega,\mathbb{K})$). It is well known that the dual of $C(\Omega,E)$ can be identified with the Banach space $M(\Omega,E^*)$ of all the regular , E^*-valued, countably additive measures of bounded variation on the Borel σ-field, $Bo(\Omega)$, of Ω, with the variation norm.

I. (V) and (V*) sets.

Let us recall the following definitions of Pełczyński ([13]): Let E be a Banach space

-A subset $K \subseteq E$ is called a *(V^*) set*, and we shall write $K \in \mathcal{V}^*(E)$, if for every w.u.c. series $\sum x_n^*$ in E^*, the following holds:

$$\lim_{n\to\infty} \sup \{ |x_n^*(x)| : x \in K \} = 0$$

-A subset $K \subseteq E^*$ is called a *(V) set*, and we shall write $K \in \mathcal{V}(E^*)$, if for every w.u.c. series $\sum x_n$ in E, the following holds:

$$\lim_{n\to\infty} \sup \{ |x^*(x_n)| : x^* \in K \} = 0.$$

It is easy to see that $\mathcal{V}^*(E)$ and $\mathcal{V}(E^*)$ are formed by bounded sets and they are preserved by linear continuous images, linear combinations, closed absolutely convex hulls and passing to subsets. All these properties are consequences of the definition, or can be proved immediately by using the next useful characterization, that follows immediately from the bijection existing between the operators from E into ℓ_1 (resp., from c_0 into E) and series w.u.c. in E^* (resp. in E); see f.i. [7], Ch. VII, Ex. 2 and 5.):

Proposition 1.1. *(See [1] and [10]) Let E be a Banach space.*

a) $K \in \mathcal{V}^*(E)$ if and only if for every operator $T: E \longrightarrow \ell_1$, $T(K)$ is (weakly) relatively compact.

b) $K \in \mathcal{V}(E^*)$ if and only if for every operator $T: c_0 \longrightarrow E$, $T^*(K)$ is (weakly) relatively compact in $\ell_1 \approx (c_0)^*$.

The above result and the well known property of (weakly) relatively compact subsets of a Banach space (see [7], Ch. XIII, Lemma 2.), yields the following stability result: (see the argument of [1], Cor. 1.7.)

Corollary 1.2. *Let E be a Banach space.*

a) If $K \subseteq E$ is bounded and for every $\varepsilon>0$ there exists a (V^*) set $K_\varepsilon \subseteq E$ such that $K \subseteq K_\varepsilon + \varepsilon B(E)$, then K is a (V^*) set.

b) If $K \subseteq E^*$ is bounded and for every $\varepsilon>0$ a (V) set $K_\varepsilon \subseteq E^*$ exists such that $K \subseteq K_\varepsilon + \varepsilon B(E^*)$, then K is a (V) set.

With Proposition 1.1 at hand, it is obvious that the following relationships hold:

$$(\dagger) \qquad \mathcal{W}(E) \subseteq \mathcal{WC}(E) \subseteq \mathcal{V}^*(E) \subseteq \mathcal{B}(E)$$

and

$$(\ddagger) \qquad \mathcal{W}(E^*) \subseteq \mathcal{WC}(E^*) \subseteq \mathcal{V}^*(E^*) \subseteq \mathcal{V}(E^*) \subseteq \mathcal{B}(E^*).$$

Also, we have

Corollary 1.3. *a) For a Banach space E, the following assertions are equivalent:*

i) $\mathcal{B}(E^*) \subseteq \mathcal{V}(E^*)$.

ii) Every operator T from c_0 into E is weakly compact.

iii) E does not contain a copy of c_0.

b) For a Banach space E, the following assertions are equivalent:

i) $\mathcal{B}(E) \subseteq \mathcal{V}^*(E)$.

ii) Every operator T from E into ℓ_1 is (weakly) compact.

iii) E does not contain a complemented copy of ℓ_1.

Proof. (a) The equivalence of (i) and (ii) follows from Proposition 1.1 and Gantmacher's theorem. (ii) ⇒ (iii) is obvious, and (iii) ⇒ (ii) follows from [13], Th. 1 and Prop. 3.

(b) is proved with similar arguments. (See [1], Cor. 1.5.)

Many important Banach space properties are (or can be) defined by the coincidence of two classes of bounded sets. Following this direction, Pełczyński introduced in the aforementioned paper [13] the following definitions:

–E *has property (V)* (in short, $E \in (V)$) if $\mathcal{W}(E^*) = \mathcal{V}(E^*)$.

–E *has property (V^*)* (in short, $E \in (V^*)$) if $\mathcal{W}(E) = \mathcal{V}^*(E)$.

In view of (†) and (‡), it is obvious that reflexive spaces have both properties (V) and (V^*), and every $E \in (V^*)$ is weakly sequentially complete. Also, if E (resp., E^*) has property (V), then E^* (resp., E) has property (V^*). Pełczyński asked whether the converse is true. In general, the answer is negative and counter examples can be found in [3], [10] and [14]. Hence, the natural problem of characterizing the Banach spaces for which the answer to Pełczyński's question is positive, arises. Next proposition gives some insight on this problem:

Proposition 1.4. *([2], Th. 3) For a Banach space E the following assertions are equivalent:*

a) No basic sequence spanning a complemented subspace isomorphic to ℓ_1 in E^*, is a (V)-set.

b) $\mathcal{V}^*(E^*) = \mathcal{V}(E^*)$.

c) $E \in (V)$ if and only if $E^* \in (V^*)$.

For reasons of brevity, let us say that a Banach space E has *property (P)* if it satisfies any of the equivalent conditions of Proposition 1.4. Of course, if $E \in (V)$, obviously E has (P). But there are spaces with property (P) that does not have property (V). For example, any non reflexive space E such that E^{**} does not contain copies of c_0 has property (P) (because E^* does not contain a complemented copy of ℓ_1, and Prop. 1.4 (a) applies), but does not have property (V), because every non reflexive space with property (V) contains a copy of c_0 ([13], Prop. 8; it also follows immediately from Cor. 1.3.)

Pełczyński proved that every space $C(\Omega)$ has property (V), ([13], Th. 1) and so every (AL)-space has property (V^*). It is easy to see that every (AM) space has property (P) ([2], Prop. 7), what gives a quick proof of the fact that every (AM) space has property (V).

II. (V) sets in $C(\Omega,E)^*$.

Let Ω be a compact Hausdorff space and E a Banach space. We shall identify $C(\Omega,E)^*$ with $M(\Omega,E^*)$. When $E = \mathbb{K}$, the (V) sets of $M(\Omega)$ are just the weakly relatively compact (i.e., the uniformly σ-additive) subsets, because $C(\Omega)$ has property (V). But in general there is no complete characterization of weakly relatively compacts and (V)-sets in $M(\Omega,E^*)$. The first result we get is the following:

Proposition 2.1. *Let $K \subseteq M(\Omega,E^*)$ be a (V)-set. Then K is uniformly countably additive.*

Proof. If K were not uniformly countably additive, by [8] I.1.17 and VI.2.13, there would be an $\varepsilon>0$, a disjoint sequence (O_n) of open sets of Ω and a pair of sequences $(x_n) \subseteq B(E)$, $(m_n) \subseteq K$, such that $|\langle x_n, m_n(O_n)\rangle| > \varepsilon$ for every $n \in \mathbb{N}$. By the regularity of the Radon measures $x_n \circ m_n$, we can find functions $\varphi_n \in C(\Omega)$, $0 \leq \varphi \leq 1$, Supp $\varphi_n \subseteq O_n$ and such that

$$|\int \varphi_n \, d(x_n \circ m_n)| = |\int \varphi_n \otimes x_n \, dm_n| > \varepsilon \quad \text{for every } n \in \mathbb{N}.$$

As the series $\sum (\varphi_n \otimes x_n)$ is clearly w.u.c. in $C(\Omega, E)$, this shows that (m_n), and hence K, is not a (V)-set, contradicting the assumption.

We shall need the following well known result ([9], 13, Th. 5):

Theorem 2.2. *Let λ be a positive Radon measure on Ω and ρ a lifting of $\mathcal{L}^{\infty}(\lambda)$. If $m \in M(\Omega, E^*)$ is λ-continuous, there is a function $g: \Omega \longrightarrow E^*$, determined uniquely λ-almost everywhere, such that*

a) $\langle x, g\rangle$ is a λ-integrable function for every $x \in E$.

b) For every $x \in \Omega$ and $A \in Bo(\Omega)$,

$$\langle x, m(A)\rangle = \int_A \langle x, g\rangle d\lambda.$$

c) $\rho[g] = g$.

Furthermore, the function $\omega \longmapsto |g|(\omega) = \|g(\omega)\|$ is λ-intergrable and

$$|m|(A) = \int_A |g| \, d\lambda, \text{ for } A \in Bo(\Omega).$$

Given a positive Radon measure λ and a fixed lifting ρ of $\mathcal{L}^{\infty}(\lambda)$, if all the elements of a bounded set $K \subseteq M(\Omega, E^*)$ are absolutely continuous with respect to λ, we shall denote by $\varphi(m, \lambda, \rho) = \varphi(m)$ the function corresponding to m by Theorem 2.2. It is well known that K is uniformly countably additive if and only if so is $|K| = \{|m|: m \in K\}$ (see [8], Th. I.2.1 and I.2.4). Then it is clear that, with the previous notations, K is uniformly countably additive if and only if the set $\{|\varphi(m)|: m \in K\} \subseteq L_1(\lambda)$ is uniformly integrable, i.e., weakly relatively compact ([8] Th. III.2.15).

Theorem 2.3. *Let $K \subseteq M(\Omega, E^*)$ be a bounded, uniformly countably additive subset, λ a common control measure for K and ρ a lifting for $\mathcal{L}^{\infty}(\lambda)$. Suppose the following property holds:*

(Δ) *For every $\delta > 0$ there is a Borel set A with $\lambda(A) < \delta$, so that for each $\omega \notin A$, a (V)-set $V(\omega) \subseteq E^*$ exists satisfying $\varphi(m)(\omega) \in V(\omega)$, for all $m \in K$.*

Then, K is a (V) set.

Proof. We shall argue by contradiction. Suppose K is not a (V) set. Then, there is an $\varepsilon>0$, a w.u.c. serie $\sum f_n$ in $C(\Omega,E)$ and a sequence (m_n) in K such that

$$(*) \qquad |\langle f_n,m_n\rangle| = \left|\int \langle f_n(\omega),g_n(\omega)\rangle\, d\lambda\right| > \varepsilon \quad \text{for every } n\in\mathbb{N},$$

where we have put $g_n = \varphi(m_n)$. According to the observations made before the theorem, the sequence $\{|g_n|: n\in\mathbb{N}\} \subseteq L_1(\lambda)$ is uniformly integrable. Then, we can choose a $\delta>0$ such that for every Borel set $B\subseteq\Omega$ with $\lambda(B)<\delta$, we have

$$\int_B |g_n|\,d\lambda < \varepsilon/2M, \quad \text{for every } n\in\mathbb{N},$$

where M is a bound for $\{\|f_n\|: n\in\mathbb{N}\}$. Choose now a Borel set $A\subseteq\Omega$ with $\lambda(A)<\delta$ and satisfying the condition (Δ). Let us define

$$h_n(\omega) = |\langle f_n(\omega),g_n(\omega)\rangle|.$$

Since $\sum f_n(\omega)$ is w.u.c. in E for every $\omega\in\Omega$, we have

$$\lim_{n\to\infty} h_n(\omega) = 0, \quad \forall\ \omega \notin A,$$

and

$$|h_n(\omega)| \le M\ |g_n|(\omega), \quad \forall\ n\in\mathbb{N} \text{ and } \forall\ \omega\in\Omega.$$

In consequence, (h_n) is uniformly integrable, and Vitali's convergence theorem proves that $\lim\limits_{n\to\infty} \int_{\Omega\setminus A} h_n d\lambda = 0$. Choose a sufficiently large n so that $\int_{\Omega\setminus A} h_n d\lambda < \varepsilon/2$. Then

$$\left|\int\langle f_n,g_n\rangle d\lambda\right| \le \int_A h_n d\lambda + \int_{\Omega\setminus A} h_n d\lambda < \varepsilon/2 + \varepsilon/2 = \varepsilon,$$

which contradicts $(*)$. This contradiction proves that K is a (V) set.

Next result gives a characterization of some classes of (V) sets in $M(\Omega,E^*)$ that does not depend on the choice of a lifting.

Corollary 2.4. *Let $K \subseteq M(\Omega,E^*)$ be a bounded, uniformly countably additive subset, and λ a common control measure for K. Suppose the following condition holds:*

For every $\delta>0$ there is a $V(\delta) \in \mathcal{V}(E^*)$ such that for each $m\in K$ a Borel set $A(m,\delta) \subseteq \Omega$ exists satisfying

i) $|m|(\Omega\setminus A(m,\delta)) < \delta$

and

ii) $Av(m) = \left\{ \dfrac{m(D)}{\lambda(D)} : D\in Bo(\Omega),\ \lambda(D)>0,\ D \subseteq A(m,\delta)\right\} \subseteq V(\delta)$.

Then, K is a (V) set.

Proof. Given a Borel set $A \subseteq \Omega$ and $m \in M(\Omega, E^*)$, let as denote by $\chi_A m$ the member of $M(\Omega, E^*)$ defined by $(\chi_A m)(B) = m(A \cap B)$, for every $B \in Bo(\Omega)$. Clearly, $\|m - \chi_A m\| \leq |m|(A)$.

Let $\delta > 0$. Consider the set

$$K_\delta = \{ \chi_{A(m,\delta)} m : m \in K \}.$$

By Proposition 1.1.(b), we can suppose, and so we do, that every set $V(\delta)$ is absolutely convex and *weak* closed*. Also, it will be enough to prove that every *sequence* in K is a (V)-set. So we can suppose K countable. Then, if ρ is a lifting of $\mathcal{L}^\infty(\lambda)$ and g_m is the function associated to $\chi_{A(m,\delta)} m$ by Theorem 2.2, the Hahn-Banach theorem proves that $g_m(\omega) \in V(\delta) \cup \{0\}$ for every $m \in K$ and almost all ω. Theorem 2.3 assures that K_δ is a (V)-set. Since

$$K \subseteq K_\delta + \delta B(M(\Omega, E^*)),$$

Corollary 1.2 yields the result.

By analogy with the notation used in case of the spaces of Bochner integrable functions (see, f.i. [1], Section III), we shall call the uniformly countably additive subsets of $M(\Omega, E^*)$ sets satisfying the conditions of Corollary 2.4, *δV sets*. It is worth noting that when E has property V, every δV set $K \subseteq M(\Omega, E^*)$ can be isometrically imbedded in the space $L_1(\lambda, E^*)$ of all the E^*-valued, Bochner integrable functions. In fact, Theorem III.2.18 of [8] shows that every $m \in K$ has a Bochner density $g_m \in L_1(\lambda, E^*)$, i.e.,

$$m(A) = \int_A g_m(\omega) d\lambda(\omega), \text{ for every } A \in Bo(\Omega).$$

This fact can be used to prove that, in general, δV sets do not exhaust all the (V)-sets. Indeed, suppose E has property V and E^* does not have the Radon-Nikodym property (for example, $E = C([0,1])$). Take $\Omega = [0,1]$ and $m \in M(\Omega, E^*)$ without Radon-Nikodym derivative with respect to its variation λ. Then the set $K = \{m\}$ is obviously a (V)-set that, what we have said, is not a δV set.

In the next proposition some conditions under which δV sets coincide with (*V*)-sets are given.

Proposition 2.5. *Let E be a Banach space and Ω a compact Hausdorff space. Every (V)-set of $M(\Omega, E^*)$ is a δV set in the following cases:*

a) Ω is dispersed.

b) E does not contain copies of c_0.

Proof. a) Let $K \subseteq M(\Omega, E^*)$ be a (V)-set. By Proposition 2.1, K admits a common control measure λ. Since every Radon measure on a compact dispersed space is discrete ([11], Ch. 2, Sect. 8), λ is discrete; in particular, it is concentrated on the countable set $\{\omega_n: n\in\mathbb{N}\}$ of its atoms. If we put $A_k = \{\omega_n: n>k\}$, we have $\lim_{n\to\infty} \lambda(A_n) = 0$. So, given $\delta>0$ we can choose a $k\in\mathbb{N}$ such that $|m|(A_k)<\delta$ for every $m\in K$. Since the set

$$\left\{ \frac{m(A)}{\lambda(A)} : A \subseteq \Omega\setminus A_k,\ \lambda(A)>0\right\}$$

is finite, K is a $\delta\mathcal{V}$ set.

b) If $K \subseteq M(\Omega,E^*)$ is a (V)-set and λ a common control measure, the set function $n(A) = (m(A))_{m\in K}$ is a λ-continuous vector measure of bounded variation defined on $Bo(\Omega)$, with values in $\ell_\infty(K;E^*)$, the Banach space of all E^*-valued, bounded families, indexed by K, with the sup norm. Hence, reasoning as in [8], Th. III.1.5, there exists a sequence (A_n) of disjoint members of $Bo(\Omega)$ such that $\Omega = \cup A_n$ and with the property that

$$(n-1)\lambda(A) \le |n|(A) \le n\lambda(A),$$

for every Borel set $A \subseteq A_n$ $(n = 1,2,\ldots)$. Then, given $\delta>0$ there is a $k\in\mathbb{N}$ such that $|n|(\Omega\setminus \bigcup_{n>k} A_n) < \delta$ and so $|n|(A) \le k\lambda(A)$, for every Borel set $A \subset A_1\cup\ldots A_k$. In consequence, for every $m\in K$ we have $|m|(\Omega\setminus \bigcup_{n>k} A_n) < \delta$ and

$$\left\{ \frac{m(B)}{\lambda(B)} : B\in Bo(\Omega),\lambda(B)>0, B\subseteq A_1\cup\ldots\cup A_k\right\} \subseteq kB(E^*).$$

From Corollary 1.3(a) and the hypothesis, it follows that K is a $\delta\mathcal{V}$ set.

Remark 2.6. The proof of part (a) of the above proposition shows that every bounded, uniformly countably additive subset $K\subseteq M(\Omega,E^*)$, admitting a discrete control measure, is a $\delta\mathcal{V}$ set.

There is no complete characterization of weakly compact subsets in $M(\Omega,E^*)$, and this is one of the main difficulties to solve the question whether $C(\Omega,E)$ has properties (V) or (P) when E has. However, we have the following result:

Proposition 2.7. *Let E be a Banach space.*

a) If E has property (V), every $\delta\mathcal{V}$ subset of $M(\Omega,E^*)$ is weakly relatively compact.

b) If E has property (P) and E^* has the Radon-Nikodym property, every $\delta\mathcal{V}$ subset of $M(\Omega,E^*)$ is a (V^*) set.

Proof. a) By the commentaries following Corollary 2.4, there is a positive Radon measure λ on Ω such that every $m \in K$ has a Bochner integrable density $g_m \in L_1(\lambda,E^*)$, and the mapping $m \mapsto g_m$ is an isometry. The set $K_1 = \{g_m: m \in K\}$ satisfies the assumptions of [6], Cor. 3 (i.e., K_1 is a $\delta\mathcal{V}$-set, in the notations of Sect. III of [1]) and, consequently, it is relatively weakly compact.

b) By the hypothesis on E^*, there is again a positive Radon measure λ on Ω such that K is isometric to a set $K_1 = \{g_m: m \in K\} \subseteq L_1(\lambda,E^*)$ (where m is the indefinite integral of g_m). Mean value theorem and the definition of a $\delta\mathcal{V}$-set prove that K_1 is bounded, uniformly integrable and satisfies the following condition:

> For every $\delta>0$ there exists a set $V(\delta) \in \mathcal{V}(E^*)$ and for each $g \in K_1$ a Borel set $A(g,\delta)$ such that $\lambda(\Omega \setminus A(g,\delta))<\delta$ and $g(A(g,\delta)) \subseteq V(\delta)$.

Since E has property (P), each $V(\delta)$ is a (V^*) set. Hence, Corollary 3.3 of [1] show that K_1 is a (V^*) set in $L_1(\lambda,E^*)$ and, consequently, K is a (V^*) set in $M(\Omega,E^*)$.

Corollary 2.8. *Let E be a Banach space and Ω a compact Hausdorff space so that every (V)-set in $M(\Omega,E^*)$ is a $\delta\mathcal{V}$ set. Then*

a) If E has property (V), $C(\Omega,E)$ also has property (V).

b) If E has property (P) and E^* has the Radon-Nikodym property, $C(\Omega,E)$ also has property (P).

Proof. Follows immediately from Proposition 2.7.

Corollary 2.9. *a) (See [4]) If Ω is a dispersed compact space, $C(\Omega,E)$ has property (V) whenever E has.*

b) If E^* has the Radon-Nikodym property and does not contain complemented copies of ℓ_1, then $C(\Omega,E)$ has property (P), for every compact Hausdorff space Ω.

Proof. (a) follows from Corollary 2.8(a) and Proposition 2.5(a). As for (b), it follows from Corollary 1.3(b) that $\mathcal{B}(E^*) = \mathcal{V}(E^*) = \mathcal{V}^*(E^*)$. Corollary 2.8(b) and Proposition 2.5(b) yield now the result.

Remark 2.9. As far as we know, the best general condition under which can be assured that $C(\Omega,E) \in (V)$ if $E \in (V)$, was given in [5]; namely, E has property (u) and contains no copy of ℓ_1. However, this condition is far from being necessary. In fact, if $E = C(\Omega_1)$ or, more generally, E is an (AM) space, then $C(\Omega,E)$ is an (AM) space (with the pointwise order) and, consequently, it has property (V) (see the end of Section I).

REFERENCES

[1] F.Bombal, *On* (V^*) *sets and Pelczynski's property* (V^*), to appear in Glasgow Math. J.

[2] F. Bombal, *Sobre las propiedades (V) y* (V^*) *de Pelczynski*, to appear in the Proceedings of the XIV Jornadas Matemáticas Hispano-Lusas, la Laguna 1989.

[3] P.Cembranos, *Algunas propiedades del espacio de Banach C(K,E)*, Thesis. Univ. Comp. de Madrid, 1982.

[4] P.Cembranos, *On Banach spaces of vector valued continuous functions*, Bull. Austral. Math. Soc. **28** (1983), 175–186.

[5] P.Cembranos, N.Kalton, E. Saab and P.Saab, *Pelczynski's property V on* $C(\Omega,E)$ *spaces*, Math. Ann., **271** (1985), 91–97.

[6] J.Distel, *Remarks on weak compactness in* $L_1(\mu,X)$, Glasgow Math. J., **18** (1977), 87–91.

[7] J.Distel, *Sequences and series in Banach spaces*, Graduate Texts in Math., No. **92**, Springer, 1984.

[8] J.Distel, J.J.Uhl, *Vector measures*, Math. Surveys, No. 15, Amer. Math. Soc., 1977.

[9] N.Dinculeanu, *Vector measures*, Pergamon Press, 1967.

[10] G.Emmanuele, *On the Banach spaces with the property* (V^*) *of Pelczynski*, Annali Mat. Pura e Applicata, **152** (1988), 171–181.

[11] H.E.Lacey, *The Isometric Theory of Classical Banach Spaces*, Springer-Verlag, 1974.

[12] J.Lindensrauss, L. Tzafriri, *Classical Banach spaces I*, Springer 1977.

[13] A.Pelczynski, *On Banach spaces on which every unconditionally converging operator is weakly compact*, Bull. Acad. Pol. Sci., **10** (1962), 641–648.

[14] E.Saab, P.Saab, *On Pelczynski's properties (V) and* (V^*), Pac. J. of Math., **125** (1986), 205–210.

A GENERALIZED DYSON SERIES

KUN SOO CHANG, KUN SIK RYU

Seoul, Hannam, South Korea

ABSTRACT. The existence of an analytic operator-valued function space integral as an operator on L_2 has been established for certain functionals involving any Borel measure. Recently, the authors established the existence of the integral as an operator from L_p to $L_{p'}$ $(1<p<2)$ for certain functionals involving some Borel measures. In particular, they discussed the Dyson series for certain functionals involving a single Borel measure and a single potential. In this paper, we deal with the Dyson series for certain functionals involving many Borel measures and many potentials.

1. Notations and preliminaries.

In this section we present some necessary notations and lemmas which are needed in our subsequent section. Insofar as possible, we adopt the definitions and notations of [4] and [6].

A. Let $\mathbb{N}$ be the set of all natural numbers and let $\mathbb{R}$ be the set of all real numbers. Let $\mathbb{C}$, $\mathbb{C}_+$, $\mathbb{C}_+^\sim$ be the set of all complex numbers, all complex numbers with positive real part and all non-zero complex numbers with non-negative real part, respectively. Let ρ be a function on the set of all non-negative integers such that $\rho(0) = 0$

1980 Mathematics Subject Classification: Primary 28C20, Secondary 28A33, 47D45, 81C30, 81C35.

Key wards and phrases: Bochner integral, Feynman integral, Dyson series, Wiener integral, Lebesgue decomposition, Function space integral.

* Research partially supported by the grants from the Yonsei University and the Ministry of Education.

and $\rho(1) = 1$ for $n \geq 1$.

B. Given a number d such that $1 \leq d \leq \infty$. d and d' will be always related by $1/d + 1/d' = 1$. If $1 < p < 2$ is given, let α in $(1,\infty)$ be such that

$$\alpha = \frac{p}{2 - p} .$$

In our theorems, N will be a positive integer restricted so that

$$N < 2\alpha.$$

For $1 < p < 2$, let r be a real number such that

$$\frac{2\alpha}{2\alpha - N} < r < \infty.$$

The number $\frac{N}{2\alpha}$ will occur often so it is worthwhile introducing a symbol for it:

$$\delta \equiv \frac{N}{2\alpha} .$$

Note that $0 < r'\delta < 1$, where r and r' are conjugate indices.

C. For $1 \leq p < \infty$, $L_p(\mathbb{R}^N)$ is the space of C-valued Borel measurable functions ψ on $\mathbb{R}^N$ such that $|\psi|^p$ is integrable with respect to Lebesgue measure m_L on $\mathbb{R}^N$. $L_\infty(\mathbb{R}^N)$ is the space of C-valued Borel measurable functions ψ on $\mathbb{R}^N$ such that ψ is essentially bounded with respect to m_L. Let $L(L_p,L_{p'})$ be the space of bounded linear operators from $L_p(\mathbb{R}^N)$ into $L_{p'}(\mathbb{R}^N)$.

The notation $\|\cdot\|$ will be used both for the norm of vectors and for the norm of operators; the meaning will be clear from context.

D. Let $1 \leq p \leq 2$ be given. For λ in $\mathbb{C}_+^\sim$, ψ in $L_p(\mathbb{R}^N)$, ξ in $\mathbb{R}^N$, and a positive real numbers s, let

$$(C_{\lambda/s}\psi)(\xi) = \left(\frac{\lambda}{2\pi s}\right)^{N/2} \int_{\mathbb{R}^N} \psi(u)\exp\left(- \frac{\lambda\|u-\xi\|^2}{2s}\right)dm_L(u),$$

where if N is odd we always choose $\lambda^{-1/2}$ with non-negative real part and if $\operatorname{Re}\lambda = 0$ the integral in the above should be interpreted in the mean just as in the theory of the L_p Fourier transform. If $p = 1$, from [3] $C_{\lambda/s}$ is in $L(L_1,L_\infty)$ and $\|C_{\lambda/s}\| \leq \left(\frac{|\lambda|}{2\pi s}\right)^{N/2}$. And as a function of λ, $C_{\lambda/s}$ is analytic in $\mathbb{C}_+$ and weakly continuous in $\mathbb{C}_+^\sim$. If $1 < p \leq 2$ from [1] and [8] $C_{\lambda/s}$ is in $L(L_p,L_{p'})$ and $\|C_{\lambda/s}\| \leq \left(\frac{|\lambda|}{2\pi s}\right)^{\delta}$. And as a function of λ, $C_{\lambda/s}$ is analytic in $\mathbb{C}_+$ and strongly continuous in $\mathbb{C}_+^\sim$.

E. Let $t>0$ be given. $M(0,t)$ will denote the space of complex Borel measures η on the interval $(0,t)$. Every measure η in $M(0,t)$ has a unique decomposition, $\eta = \mu+\nu$ into a continuous part μ and a discrete part $\nu \equiv \sum_{p=1}^{\infty} w_p\delta_{r_p}$, where $\langle w_p\rangle$ is a summable sequence in C and δ_{r_p} is the Dirac measure [9]. In fact, this is the Lebesgue decomposition of η. And $M(0,t)^*$ will denote the subset of $M(0,t)$ which satisfies the following conditions:

(a) If μ is the continuous part of η in $M(0,t)^*$, then the Radon-Nikodym derivative $d|\mu|/dm$ exists and is essentially bounded, where m is the Lebesgue measure on $(0,t)$.

(b) If $\nu = \sum_{p=1}^{\infty} w_p\delta_{r_p}$ is the discrete part of η in $M(0,t)^*$, then either $\inf\{r_p : p\in\mathbb{N}\}$ is positive or $\sum_{p=1}^{\infty}|w_p|r_p^{-r'\delta}$ converges.

F. Let $C_0[0,t] \equiv C_0$ be the space of $\mathbb{R}^N$-valued continuous functions x on $[0,t]$ such that $x(0) = 0$. We consider C_0 as equipped with N-dimensional Wiener measure m_w. Let $C[0,t] \equiv C$ be the space of $\mathbb{R}^N$-valued continuous functions y on $[0,t]$.

G. For $1 < p \le 2$ and η in $M(0,t)$, let $L_{\alpha r:\eta}([0,t]\times\mathbb{R}^N) \equiv L_{\alpha r:\eta}$ be the space of all $\mathbb{C}$-valued Borel measurable functionals θ on $[0,t]\times\mathbb{R}^N$ such that

$$\|\theta\|_{\alpha r:\eta} \equiv \left\{ \int_{(0,t)} \|\theta(s,\cdot)\|_\alpha^r d|\eta|(s) \right\}^{1/r} < \infty.$$

Note that $L_{\alpha r:\eta} \subset L_{\alpha s:\eta}$ if $1 \le s \le r \le \infty$. If θ is in $L_{\alpha r:\eta}$ and if $\eta = \mu+\nu$ is the Lebesgue decomposition, it is not difficult to show that θ is in $L_{\alpha r:\mu} \cap L_{\alpha r:\nu}$ and $\|\theta\|_{\alpha r:\eta} = \|\theta\|_{\alpha r:\mu} + \|\theta\|_{\alpha r:\nu}$.

H. Let $1 < p \le 2$ be given and θ be in $L_\alpha(\mathbb{R}^N)$. From Lemma 1.3 in [8], a function $M_\theta : L_{p'}(\mathbb{R}^N) \longrightarrow L_p(\mathbb{R}^N)$ defined by $M_\theta(f) = f\theta$, is in $L(L_{p'},L_p)$ and $\|M_\theta\| \le \|\theta\|_\alpha$. It will be convenient to let $\theta(s)$ denote $M_\theta(s,\cdot)$ for θ in $L_{\alpha r:\eta}$. Let $\theta_1,\theta_2,\ldots,\theta_{m-1}$ be in $L_\alpha(\mathbb{R}^N)$, ψ in $L_p(\mathbb{R}^N)$ and $0 < s_1 < s_2 < \ldots < s_m < t$. From the Wiener integral formula [12],

$$\int_{C_0} \theta_1(x(s_1))\theta_2(x(s_2))\ldots\theta_{m-1}(x(s_{m-1}))\psi(x(s_m))dm_w(x)$$

$$= [(C_{1/s}\circ\theta_1(s_1)\circ\ldots C_{1/(s_{m-1}-s_{m-2})}\circ\theta_{m-1}(s_{m-1})\circ C_{1/(s_m-s_{m-1})})\psi](0).$$

I. Let $0 < k < 1$ be given and m be in $\mathbb{N}$. For $a < s_1 < s_2 < \ldots < s_m < b$,

$$\int_a^b\int_a^{s_m}\ldots\int_a^{s_1}\{(s_1-a)(s_2-s_1)\ldots(b-s_m)\}^{-k}ds_1ds_2\ldots ds_m$$

$$= \frac{(b-a)^{m-(m+1)k}\{\Gamma(1-k)\}^{m+1}}{\Gamma((m+1)(1-k))},$$

where Γ is a gamma function. Throughout this paper, this value is denoted by $E(a,b;m;k)$.

And let $0 < p \le 2$ be given and let $a_1,a_2,\ldots a_n$ be non-negative real numbers. From the Hölders inequality, we have

$$\sum_{i=1}^{n} a_i^p \le n^{(2-p)/2} \left[\sum_{i=1}^{n} a_i^2 \right]^{p/2}.$$

J. Let $1 \le p \le 2$ be given. Let F be a functional on C. Given $\lambda>0$, ψ in $L_p(\mathbb{R}^N)$ and ξ in $\mathbb{R}^N$, let

$$[I_\lambda(F)\psi](\xi) = \int_{C_0} F(\lambda^{-1/2}x+\xi)\cdot\psi(\lambda^{-1/2}x(t)+\xi)\, dm_v(x).$$

If for m_L-a.e. ξ in $\mathbb{R}^N$, $[I_\lambda(F)\psi](\xi)$ exists in $L_{p'}(\mathbb{R}^N)$ and if the correspondence $\psi \to [I_\lambda(F)]\psi$ gives an element of $L(L_p,L_{p'})$, we say that the operator-valued function space integral $I_\lambda(F)$ exists for λ. Suppose there exists λ_0 $(0 < \lambda_0 \le \infty)$ such that $I_\lambda(F)$ exists for all $0 < \lambda < \lambda_0$ and there exists an $L(L_p,L_{p'})$-valued function which is analytic in $\mathbb{C}_{+,\lambda_0} \equiv \mathbb{C}_+ \cap \{z\in\mathbb{C}: |z| < \lambda_0\}$ and agrees with $I_\lambda(F)$ on $(0,\lambda_0)$, then this $L(L_p,L_{p'})$-valued function is called *the operator-valued function space integral of F associated with λ* and in this case, we say that $I_\lambda(F)$ exists for λ in $\mathbb{C}_{+,\lambda_0}$. If $I_\lambda(F)$ exists for λ in $\mathbb{C}_{+,\lambda_0}$ and $I_\lambda(F)$ is strongly continuous in $\tilde{\mathbb{C}}_{+,\lambda_0}$, we say that *$I_\lambda(F)$ exists for λ in $\tilde{\mathbb{C}}_{+,\lambda_0} \equiv \tilde{\mathbb{C}}_+ \cap \{z\in\mathbb{C}: |z| < \lambda_0\}$*. When λ is purely imaginary, $I_\lambda(F)$ is called *the (analytic) operator-valued Feynman integral of F*.

K. Let X,Y be two Banach spaces, $L(X,Y)$ a space of bounded linear operators from X into Y and (Ω,m) be a measure space. Let $G:\Omega \to L(X,Y)$ be a function such that for each x in X, $(G(s))(x)$ is Bochner integrable with respect to m. Then there exists a linear operator J from X into Y such that

$$J(x) \equiv (B) - \int_\Omega (G(s))(x)dm(s) \qquad \text{for } x \text{ in } X,$$

where $(B)\int_\Omega (G(s))(x)dm(s)$ refers to the Bochner integral. Here, this linear operator J is denoted by $(BS)\int_\Omega G(s)dm(s)$ and it is called *the Bochner integral in the strong operator sense*. When $X = Y$, J is called *the strong integral of G*.

2. The generalized Dyson series.

In this section we study the Dyson series for certain functionals which involve many Borel measures and potentials. The new complications are mainly combinatorial in nature; otherwise, the proofs proceed much as in [4]. For this reason, we will state the results without proof.

Let $1 < p < 2$ be given, η_u be in $M(0,t)^*$ for $u = 1,2,\dots,m$. Let $\eta_u = \mu_u+\nu_u$ be the decompositions of η_u into its continuous and discrete parts. For u = 1,2,...,m, we will write $\nu_u = \sum_{p=1}^{\infty} \omega_{p:u}\delta_{\tau_{p:u}}$, where $\{\tau_{p:u}: p\in\mathbb{N}\} \subset (0,t)$ and $\{\omega_{p:u}: p\in\mathbb{N}\} \subset \mathbb{C}$ such that $\|\nu_u\| = \sum_{p=1}^{\infty}|\omega_{p:u}| < \infty$. Let θ_u be in $L_{\gamma r:\eta_u}$, where $\theta_u(\tau_{p:u},\cdot)$ is essentially bounded for $p\in\mathbb{N}$ and u = 1,2,...,m. Let

(1) $$F(y) = \Pi_{u=1}^{m} \int_{(0,t)} \theta_u(s,y(s))d\eta_u \quad \text{for } y \text{ in } C.$$

Note that by Lemma 1.1 in [4] for every $\lambda>0$, $F(\lambda^{-1/2}x+\xi)$ is defined for $m_v\times m_L$-a.e. (x,ξ) in $\mathbb{C}_0\times\mathbb{R}^N$.

We will write each of the factor in (1) as an absolutely convergent infinite series. We will then find it advantageous to multiply these series together. With this in mind, we introduce the following notation:

Given k between 0 and m, [k:m] will denote the collection of all subsets of size k of the set of integers {1,2,...,m}. If $\{\alpha_1,\alpha_2,\ldots,\alpha_k\}$ is in [k:m] we shall always write

$$\{\alpha_{k+1},\alpha_{k+2},\ldots,\alpha_m\} = \{1,2,\ldots,m\}\setminus\{\alpha_1,\alpha_2,\ldots,\alpha_m\}.$$

Then

(2) $$F(y) = \Pi_{u=1}^{m}\left[\int_{(0,t)}\theta_u(s,y(s))d\mu_u(s) + \sum_{p=1}^{\infty}\omega_{p:u}(\tau_{p:u},y(\tau_{p:u}))\right]$$

$$= \sum_{k=0}^{m}\ \sum_{\{\alpha_1,\ldots,\alpha_k\}\in[k:m]}\ \sum_{p_{k+1}=1}^{\infty}\ \sum_{p_{k+2}=1}^{\infty}\cdots\sum_{p_m=1}^{\infty}\Big[(\Pi_{i=1}^{k}$$

$$\int_{(0,t)}\theta_{\alpha_u}(s_u,y(s_u))d\mu_{\alpha_u}(s_u)) \times (\Pi_{u=k+1}^{m}\omega_{p_u:\alpha_u}\theta_{\alpha_u}(\tau_{p_u:\alpha_u},y(\tau_{p_u:\alpha_u})))\Big]$$

$$= \sum_{k=0}^{m}\ \sum_{\{\alpha_1,\ldots,\alpha_k\}\in[k:m]}\ \sum_{p_{k+1}=1}^{\infty}\ \sum_{p_{k+2}=1}^{\infty}\cdots\sum_{p_m=1}^{\infty}$$

$$\Big[\Big\{\int_{(0,t)}\Pi_{u=1}^{k}\theta_{\alpha_u}(s_u,y(s_u))\times_{u=1}^{k}d\mu_{\alpha_u}(s_u)\Big\}$$

$$(\Pi_{u=k+1}^{m}\omega_{p_u:\alpha_u}\theta_{\alpha_u}(\tau_{p_u:\alpha_u},y(\tau_{p_u:\alpha_u})))\Big].$$

Soon we will wish to calculate the Wiener integral defining $I_\lambda(F)$. For this purpose, we will order the time variables. We begin by ordering the τ's that appear within a given term of series in (2). For fixed k, $\{\alpha_1,\ldots\alpha_k\}$ in [k:m] and $p_{k+1},\ldots,p_m$, let σ be a permutation of $\{k+1,\ldots,m\}$ such that

(3) $$\tau_{p_{\sigma(k+1)}:\alpha_{\sigma(k+1)}} \le \tau_{p_{\sigma(k+2)}:\alpha_{\sigma(k+2)}} \ldots \le \tau_{p_{\sigma(m)}:\alpha_{\sigma(m)}}$$

If the τ's involved in (3) are distinct, the permutation σ is unique.

Theorem 2.1. *Let F be defined by (1). Then $I_\lambda(F)$ exists for λ in $\mathbb{C}_+^\sim$. Moreover, for all λ in $\mathbb{C}_+^\sim$,*

$$(4)\qquad I_\lambda(F) = \sum_{k=0}^{m} \sum_{(\alpha_1,\ldots,\alpha_k)\in(k:m)} \sum_{p_{k+1}=1}^{\infty} \sum_{p_{k+2}=1}^{\infty} \cdots \sum_{p_m=1}^{\infty}$$

$$\sum_{\rho\in S_k} \sum_{j_1+\ldots+j_{m-k+1}=k} (\Pi_{u=k+1}^{m} \omega_{p_{\sigma(u)}:\alpha_{\sigma(u)}})$$

$$\left\{(BS)\int_{\Delta_{k:j_1,\ldots,j_{m-k+1}}(\rho)} L_k\circ L_{k+1}\circ\ldots\circ L_m \times_{u=1}^{k} d\mu_{\alpha_{\rho(u)}}(S_{\rho(u)})\right\},$$

where ρ ranges through the group S_k of permutations of $(1,2,\ldots k)$ and

$$(5)\qquad \Delta_{k:j_1,j_2,\ldots,j_{m-k+1}}(\rho) = \{(s_1,s_2,\ldots,s_k)\in(0,t)^k:$$

$$0 < s_{\rho(1)} <\ldots< s_{\rho(j_1)} < \tau_{p_{\sigma(k+1)}:\alpha_{\sigma(k+1)}} < s_{\rho(j_1+1)} <\ldots s_{\rho(j_1+j_2)}$$

$$< \tau_{p_{\sigma(k+2)}:\alpha_{\sigma(k+2)}} < s_{\rho(j_1+j_2+1)} <\ldots< s_{\rho(k)} < t\}.$$

Beside, for $(s_1,s_2,\ldots,s_k)\in\Delta_{k:j_1,\ldots,j_{m-k+1}}(\rho)$ and $n = k,k+1,\ldots,m$,

$$L_n(\lambda:s_1,s_2,\ldots,s_k) = \theta_{\alpha_{\sigma(n)}}(\tau_{p_{\sigma(n)}:\alpha_{\sigma(n)}})\circ$$

$$C_{\lambda/(s_{\rho(j_1+\ldots+j_{n-k}+1)}-\tau_{p_{\alpha(n)}:\alpha_{\sigma(n)}})}\circ$$

$$\theta_{\alpha_{\rho(j_1+\ldots+j_{n-k}+1)}}(s_{\rho(j_1+\ldots+j_{n-k}+1)})$$

$$C_{\lambda/(s_{\rho(j_1+\ldots+j_{n-k}+2)}-(s_{\rho(j_1+\ldots+j_{n-k}+1)})}\circ\theta_{\alpha_{\rho(j_1+\ldots+j_{n-k}+2)}}$$

$$s_{\rho(j_1+\ldots+j_{n-k}+2)})\circ\ldots\circ\theta_{\alpha_{\rho(j_1+\ldots+j_{n-k+1})}}(s_{\rho(j_1+\ldots+j_{n-k+1})})\circ$$

$$C_{\lambda/(\tau_{p_{\alpha(n+1)}:\alpha_{\sigma(n+1)}})-s_{\rho(j_1+\ldots+j_{n-k+1})}}.$$

Here, σ is a permutation of $(k+1,k+2,\ldots,m)$ as defined in (3); in addition, we adopt the conventions $\tau_{p_{\sigma(k)}:\alpha_{\sigma(k)}} = 0$, $\tau_{\tau_{p_{\sigma(n+1)}:\alpha_{\sigma(n+1)}}} = t$ and $\theta(\tau_{p_{\sigma(k)}:\alpha_{\sigma(k)}}) = 1$, an identity map on $L_{p'}(\mathbb{R}^N)$. Further, we take $j_0=0$; then, when $n=k$, it is reasonable to interpret $j_1+\ldots+j_{n-k}+1$ as 1 and we also get $j_{n-k} = j_0 = 0$. The s' between two equal τ's are omitted in (5).

The series in (4) converges in operator norm. Further, for all λ in $\mathbb{C}_+^\sim$,

$$\|I_\lambda(F)\| \le \sum_{k=0}^{m}\left(\frac{|\lambda|}{2\pi}\right)^{(m+k+1)\delta}\left\{\frac{(m+1)!}{(m-k+1)!}\right\}^{1/2r'}(k!)^{1/2}$$

$$\times \sum_{(\alpha_1,\ldots,\alpha_k)\in(k:m)} (\Pi_{u=1}^{k}(\text{ess sup}\ \ d|\mu_{\alpha_u}|/dm))^{1/r'}$$

$$\times\ (\Pi_{q=1}^{k}\|\theta_{\alpha_q}\|_{\gamma r;\mu_{\alpha_q}}) \sum_{p_{k+1}=1}^{\infty} \sum_{p_{k+2}=1}^{\infty} \cdots \sum_{p_m=1}^{\infty}$$

$$(\Pi_{u=k+1}^{m} |w_{p_{\alpha(u)}:\alpha_{\sigma(u)}}|)(\Pi_{u=k+1}^{m} \|\theta_{\alpha_{\sigma(u)}}(\tau_{p_{\sigma(u)}:\alpha_{\sigma(u)}}, \cdot)\|_{\gamma})$$

$$\times \Bigg\{ \sum_{j_1+\cdots+j_{m-k+1}=k} \Pi_{l=1}^{m-k+1} E(\tau_{p_{\sigma(l-1)}:\alpha_{\sigma(l-1)}}, \tau_{p_{\sigma(l)}:\alpha_{\sigma(l)}}; j_l : r'\delta)^{2/r'} \Bigg\}^{1/2}.$$

Denote this norm estimate $b_F(|\lambda|)$.

Theorem 2.2. *(Generalized Dyson Series).*

Let $\langle F_n \rangle$ be a sequence of functionals each given by an expression of the type (1):

$$F_n(y) = \Pi_{u=1}^{m_n} \int_{(0,t)} \theta_{n,u}(s, y(s)) d\eta_{n,u}(s) \tag{6}$$

for y in C where every $\eta_{n,u}$ are in in $M(0,t)^*$ and every $\theta_{n,u}$, are in $L_{\gamma r:\eta_{n,u}}$. (Note that if $m_u=0$, we take $F_n \equiv 1$. For a discrete point τ of $\eta_{n,u}$, we assume that $\theta_{n,u}(\tau, \cdot)$ is essentially bounded for all u and n. Let $\lambda_0 > 0$ be given.

Assume that $\sum_{n=0}^{\infty} b_{F_n}(|\lambda|) < \infty$ for λ in $\mathbb{C}_{+,\lambda_0}^{\sim}$. Then for all $\lambda > 0$, the individual terms of the series $\sum_{n=0}^{\infty} F_n(\lambda^{-1/2}x+\xi)$ are defined and the series converges absolutely for $m_v \times m_L$-a.e. (x,ξ) in $\mathbb{C}_0 \times \mathbb{R}^N$. Let

$$F(y) = \sum_{n=0}^{\infty} F_n(y) \qquad \text{for } y \text{ in } \mathbb{C}. \tag{7}$$

Then for all λ in $\mathbb{C}_{+,\lambda_0}^{\sim}$, $I_\lambda(F)$ exists and is given by following pertubation series:

$$I_\lambda(F) = \sum_{n=0}^{\infty} I_\lambda(F_n) \qquad \text{for } \lambda \text{ in } \mathbb{C}_{+,\lambda_0}^{\sim}, \tag{8}$$

where $I_\lambda(F_n)$ is defined by (4) with the functional F from Theorem 2.1 replaced by F_n as in (6).

The series (8) converges in operator norm; furthermore,

$$\|I_\lambda(F)\| \le \sum_{n=0}^{\infty} b_{F_n}(|\lambda|) \qquad \text{for } \lambda \text{ in } \mathbb{C}_{+,\lambda_0}^{\sim}.$$

REFERENCES

[1] R.H.Cameron, D. A. Storvick, *An operator valued function space integral and a related integral equation*, J. Math. and Mech., 18 (1968), 517–552.

[2] R.H.Cameron, D. A. Storvick, *An operator valued function space integral applied to integrals of functions of class L_2*, J. Math. Anal. Appl., 42 (1973), 330–372.

[3] R.H.Cameron, D. A. Storvick, *An operator valued function space integral applied to integrals of functions of class L_1*, Proc. of London Math. Soc., 27 (1973), 345–360.

[4] K.S.Chang, K. S. Ryu, *Analytic operator valued function space integrals as an $(L_p, L_{p'})$ theory*, in preparation.

[5] E.Hille, R. S. Phillips, *Functional Analysis and Semi-groups*, A.M.S. Colloq. Publ., 31 (1957).

[6] G.W.Johnson, M. L. Lapidus, *Generalized Dyson series, Generalized Feynman Diagrams, the feynman integral and Feynman's operational calculus*, Mem. Amer. Math. Soc., 62, No 351 (1968).

[7] G.W.Johnson, D. L. Skoug, *The Carmeron-Storvick function space integral; The L_1 Theory*, J. Math. Anal. and Appl. 50 (1975), 647–667.

[8] G.W.Johnson, D. L. Skoug, *The Carmeron-Storvick function space integral; An $L(L_p, L_{p'})$ theory*, Nagoya Math. J., (1976), 93–137.

[9] M.Reed, B. Siman, *Methods of mathematical physics, Vol. I,II,III*, Rev. and enl. ed., New York, Acadèmic Press (1980).

[10] B.Siman,*Functional integration and quantum physics*, New York, Academic Press (1979).

[11] E.M.Stein, G. Weiss, *Introduction to Fourier Analysis on Euclidean Space*, Princeton U. Press, Princeton (1971).

[12] J.Yeh, *Stochastic process and the Wiener integral*, New York, Marcel Dekker (1973).

RESOLVENTS AND SELECTIONS OF MULTIVALUED MAPPINGS AND GEOMETRY OF BANACH SPACES

JOSEF KOLOMÝ

Prague, Czechoslovakia

The theory of monotone and accretive mappings, intensively studied in the last period, has fruitful applications in the theory of nonlinear partial, ordinary differential and integral equations. This paper contains some results concerning the resolvents, selections and singlevaluedness and continuity properties of accretive mappings. For the basic properties of accretive mappings, we refer the reader to Barbu [2], Browder [4], Cioránescu [5], Kato [11].

Definitions and Notation.

Let X be a real normed linear space, X^*, X^{**} its dual and bidual, $\langle , \rangle$ the pairing between X and X^*, $\tau : X \to X^{**}$ a canonical mapping of X into X^*, $\hat{x}$ the image of $x \in X$ in X^{**} under the mapping τ, $S_1(0)$, $S_1^*(0)$ the unit sphere in X and X^*, respectively. By $\mathbb{R}$, $\mathbb{R}_+$ we denote the set of all real and nonnegative numbers, respectively. We shall use the notions of Giles [8] for rotund (i.e. strictly convex) and uniformly rotund spaces, convex functions, Gâteaux and Fréchet derivatives. Recall that X is said to be:

(i) *smooth (Fréchet smooth)* if the norm of X is Gâteaux (Fréchet) differentiable on $S_1(0)$,

(ii) *uniformly (uniformly Fréchet) smooth*, if the norm of X is uniformly Gâteaux (uniformly Fréchet) differentiable on $S_1(0)$,

(iii) *an (H)-space*, if for each $(u_n) \subset X$, $u_n \to u$ weakly, $u \in X$, $\|u_n\| \to \|u\|$ we have $u_n \to u$ in the norm of X.

Let E, G be topological spaces, $F : E \to 2^G$ (where 2^G denotes a system of all subsets of G) a multivalued mapping, $D(F) = \{u \in E: F(u) \neq \emptyset\}$ its domain, $G(F) = \{(u,v) \in E \times G: u \in D(F), v \in F(u)\}$ its graph in the space $E \times G$. A mapping $F:E \to 2^G$ is said to be *upper-semicontinuous at* $u_o \in D(F)$ if for each open subset W of G such that $F(u_o) \subset W$ there exists an open neighborhood V of u_o such that $F(u) \subset W$ for each $u \in V \cap D(F)$.

Let X be a real normed linear space. Recall that a mapping $T:X \to 2^{X^*}$ is aid to be:

(i) *monotone on* $D(T)$ if for each $u,v \in D(T)$ and each $u^* \in T(u)$, $v^* \in T(v)$ we have $\langle u^*-v^*, u-v\rangle \geq 0$;

(ii) *maximal monotone on* $D(T)$ if T is monotone on $D(T)$ and its graph $G(T)$ is not properly contained in the graph of any other monotone map.

In particular, a subdifferential mapping $M \ni u \to \partial f(u) = \{u^* \in X^*: \langle u^*, v-u\rangle \leq f(v) - f(u)$ for all $v \in M\}$, where $f:M \to \mathbb{R}$ is a convex continuous functional on a convex open subset $M \subset X$, is maximal monotone. Another example of maximal monotone operator is a duality mapping $J:X \to 2^{X^*}$ defined by $J(u) = \{u^* \in X^*: \langle u^*, u\rangle = \|u\|^2, \|u^*\| = \|u\|\}$.

Note that a multivalued mapping $A:X \to 2^X$ is said to be:

(i) *accretive on* $D(A)$ if $I + \lambda A$, where I is an identity mapping in X, is expansive for each $\lambda > 0$, i.e. if for each $u,v \in D(A)$ and each $x \in A(u)$, $y \in A(v)$ there is $\|(u-v) + \lambda(x-y)\| \geq \|u-v\|$ for each $\lambda > 0$ (equivalently if for each $u,v \in D(A)$ and each $x \in A(u)$, $y \in A(v)$ there exists an element $x^* \in J(u-v)$ such that $\langle x-y, x^*\rangle \geq 0$);

(ii) *maximal accretive on* $D(A)$ if $(u,x) \in X \times X$ is a given element such that for each $v \in D(A)$ and $y \in A(v)$ there exists a point $x^* \in J(u-v)$ such that $\langle x-y, x^*\rangle \geq 0$, then $u \in D(A)$ and $x \in A(u)$;

(iii) *m-accretive* if A is accretive and the range $R(I + \lambda A)$ is equal to X for some $\lambda > 0$.

Results.

THEOREM 1. *Let X be a dual Banach space, $M \subset X^*$ a convex open subset, $\varphi:M \to \mathbb{R}$ a convex weakly* lower semicontinuous function on M. Assume that φ has the Gâteaux derivative $\varphi'(u_o^*)$ at some point $u_o^* \in M$. Then there exist a point $x_o \in X$, the sequences $(u_n^*) \subset M$, $(x_n) \subset X$ such that $\varphi'(u_o^*) = (\hat{x}_o)$, $\hat{x}_n \to \hat{x}_o$ weakly* in X^* (i.e. $x_n \to x_o$ weakly in X), $\|\hat{x}_n\| \to \|\hat{x}_o\|$ and $\hat{x}_n \in \partial\varphi(u_n^*)$. If X is an (H)-space, then $x_n \to x_o$ in the norm of X.*

Note that under the assumptions of Theorem 1 φ is continuous on M and hence there exists the subdifferential mapping $M \ni u^* \to \partial\varphi(u^*)$ on M. Recall that φ has the Gâteaux derivative $\varphi'(u_o^*)$ at $u_o^* \in M$ if the subdifferential map $\partial\varphi$ is singlevalued at u_o^*. This is the case for instance, when $\partial\varphi$ is continuous at u_o^* from the line-segments of X^* into the weak* topology of X^{**}.

Let X, M be the same as in Theorem 1, $\varphi: M \to \mathbb{R}$ a convex weakly* lower semicontinuous function on M. Consider the Kendorov function $f: M \to \mathbb{R}_+$ defined by $f(u^*) = \inf\{\|u^{**}\|: u^{**} \in \partial\varphi(u^*)\}$, $u^* \in M$. Since this function is finite and lower semicontinuous on M (see [12]), then the set C(f) of all points of M, where f is continuous, is dense G_δ subset of M. Note that each Orlicz space L_Φ with an arbitrary N-function Φ is a dual space.

Corollary 1. *Let X be a dual Banach space which admits an equivalent norm such that its second dual norm on X^{**} is rotund, $M \subset X^*$ a convex open subset, $\varphi: M \to \mathbb{R}$ a convex weakly* lower semicontinuous function on M. If $u_o^* \in C(f)$, then φ has the Gâteaux derivative $\varphi'(u_o^*)$ at u_o^* and moreover, the assertions of Theorem 1 hold. In particular, $\{\varphi'(u^*): u^* \in C(f)\} \subset X$.*

Asplund [1] proved that if X is a Banach space, $f: X \to (-\infty, +\infty]$ is a lower semicontinuous function, $f \not\equiv +\infty$ such that its dual function f^* has the Fréchet derivative at $u^* \in X^*$, then $(f^*)'(u^*) \in X$.

Let $A: X \to 2^X$ be an accretive mapping with $D_o = \operatorname{int} D(A) \neq \emptyset$. Denote by C(A) the set of all points of D_o, where A is simultaneously singlevalued and norm-to-norm upper semicontinuous. Let C(f) be a set of all points of D_o where the Kendorov function $f: X \to \mathbb{R}_+ \cup \{+\infty\}$, defined by $f(u) = \inf\{\|x\|: x \in A(u)\}$, $u \in X$, is continuous.

Theorem 2. *Let X be a uniformly Fréchet smooth Banach space such that X^* is Fréchet smooth, $A: X \to 2^X$ an accretive mapping with $D_o \neq \emptyset$. Then $C(A) = C(f)$ and C(f) is a dense G_δ subset of D_o.*

The proof of this theorem relies on the modifications of the Kendorov arguments [12]. For monotone operators, the relations between C(A) and C(f) and the properties of selections have been studied by Fabian [7], for the further results see also [13],[14], [20].

Let X be a real normed linear space, $A: X \to 2^X$ an accretive mapping with $D(A) \subseteq X$. From the definition of the accretive mapping A it follows that the singlevalued operator $J_\lambda = (I + \lambda A)^{-1}$, where I is an identity mapping X, exists for each $\lambda > 0$ with the domain $D(J_\lambda) = R(I + \lambda A)$ and the range $R(J_\lambda) = D(A)$. Moreover, J_λ is a nonexpansive mapping on $R(I + \lambda A)$ for each $\lambda > 0$. For $\lambda > 0$ we put $A_\lambda = \lambda^{-1}(I-J_\lambda)$. Recall that I_λ is a resolvent of $I + \lambda A$ and A_λ the Yoshida approximation of A. The basic properties of J_λ and A_λ are derived in [2],[5],[11].

Theorem 3. *Let X be a smooth Banach space such that X^* is Fréchet smooth, $A:X \to 2^X$ a maximal accretive mapping with $D(A) \subseteq X$. If $\overline{R(A)}$ is convex and $\cap\{R(I + \lambda A): \lambda > 0\} \supset D(A)$, then the strong limit $\lim_{\lambda\to+\infty} \frac{1}{\lambda}J_\lambda(u)$ exists for each $u \in D(A)$ and $\lim_{\lambda\to+\infty} \frac{1}{\lambda}J_\lambda(u) = -a^\circ$, where a° is a unique element of $\overline{R(A)}$ with the minimum norm.*

If X is a real Hilbert space, $A:X \to 2^X$ a maximal monotone operator with $D(A) \subseteq X$, then the conclusion of Theorem 3 holds (Morosanu [15]). The asymptotic properties of resolvents of accretive operators were intensively studied for instance by Gobbo [9], Reich [16-17] and Takahashi and Ueda [19]. Recall ([6]) that if X is a reflexive Banach space, then X^* is Fréchet smooth if and only if X is a rotund (H)-space. Let $X = L_\Phi(G)$ be an Orlicz space provided with the Orlicz norm, where $G \subset \mathbb{R}^n$ and mes $G < +\infty$ ($\mathbb{R}^n$ denotes an n-dimensional Euclidean space). If N-function Φ is strictly convex on $[0,\infty)$ and both Φ and its dual function Φ^* satisfy Δ_2-condition for large arguments, then X satisfies the assumptions of Theorem 3 (see [18]). A similar conclusions is valid if X is provided with the Luxemburg norm (see Hudzik [10]).

Definition 1 ([3]). Let X be a real normed linear space, $A:X \to 2^X$ an accretive mapping with $D(A) \subseteq X$. Then a singlevalued mapping A' is called *a main selection of A* if:

(i) A' is a selection of A;

(ii) a point $(u_o,x_o) \in \overline{D(A)}\times X$ is such that for each $u \in D(A)$ there exists $x^* \in J(u_o-u)$ such that $\langle x_o-A'(u),x^*\rangle \geq 0$, then $(u_o,x_o) \in G(A)$.

According to [11], for a set $G \subset X$, we define $|G|$ as follows:

$$|G| = \begin{cases} \inf\{\|u\|: u \in G\}, & \text{if } G \neq \emptyset \\ +\infty, & \text{if } G = \emptyset. \end{cases}$$

If $A:X \to 2^X$ is a mapping, we set

$$A^\circ u = \{z\in A(u): \|z\| = |A(u)|\},$$

$$D(A^\circ) = \{u \in D(A): A^\circ u \neq \emptyset\}$$

for each $u \in D(A)$. Then $A^\circ : D(A^\circ) \to 2^X$ is (in general) a multivalued mapping. If $D(A^\circ) = D(A)$ and $A^\circ u$ is exactly one point for each $u \in D(A)$, then $A^\circ : D(A^\circ) \to X$ is a selection of A with the property that $\|A^\circ u\| \leq \|x\|$ for each $u \in D(A)$ and for each $x \in A(u)$. In this case A° is called *a canonical restriction* (see [5]) or *a lower selection of A* ([7]). If X is a reflexive smooth and rotund Banach space and $A: X \to 2^X$ a maximal accretive mapping with $D(A) \subseteq X$, then there exists a unique canonical restriction A° of A. The following result is an extension of the Barbu theorem ([2]), where we do not assume that X, X^* are both uniformly rotund and A is m-accretive.

Theorem 4. *Let X be a Banach space such that X, X^* are both Fréchet smooth, $A: X \to 2^X$ a maximal accretive mapping with $D(A) \subseteq X$ and that $\cap\{R(I + \lambda A): \lambda > 0\} \supset D(A)$. Let $\tilde{A}$ be a selection of A such that there exists a nondecreasing function $\alpha: [0,\infty) \to \mathbb{R}_+$ with the property that $\|\tilde{A}(u)\| \leq \alpha(\|A^\circ u\|)$ for each $u \in D(A)$. If $(u_o, x_o) \in D(A) \times X$ is such that $\langle x_o - \tilde{A}(u), J(u_o - u)\rangle \geq 0$ for each $u \in D(A)$, then $x_o \in A(u_o)$ (i.e. $\tilde{A}$ is "almost the main selection" of A).*

Proposition 1. *Let X be a reflexive (H)-Banach space, $A: X \to 2^X$ an accretive mapping with $D(A) \subseteq X$, assume that there exists a main selection A' of A and let $D \subset X$. Suppose that the graph G(J) of the duality mapping $J: X \to 2^{X^*}$ is sequentially closed in $(X, \sigma(X,X^*)) \times (X^*, \sigma(X^*,X))$. Let $A_\alpha : X \to 2^X$ be accretive mappings with $D(A_\alpha) \subset X$ for each $\alpha > 0$ and assume that A_α satisfies the following two conditions:*

(i) $J_\lambda^\alpha \equiv (I + \lambda A_\alpha)^{-1} : D \to \overline{D(A)}$ for each $\lambda > 0$ and $\alpha > 0$;

(ii) for each $u \in D(A)$ there exists $y_\alpha \in A_\alpha u$ such that $y_\alpha \to A'(u)$ as $\alpha \downarrow 0$.

Then $J_\lambda^\alpha x \to J_\lambda x$, $A_\lambda^\alpha x \to Ax$ for each fixed $x \in D$ and fixed $\lambda > 0$, as $\alpha \downarrow 0$, where $A_\lambda^\alpha \equiv \lambda^{-1}(I - J_\lambda^\alpha)$.

Proposition 2. *Let X be a Banach space such that X^* is Fréchet smooth, $A: X \to 2^X$ an accretive mapping with $D(A) \subset X$. If $\cap\{R(I + \lambda A): \lambda > 0\} \supset \operatorname{conv} D(A)$, then $\overline{D(A)}$ is convex.*

Proposition 1 is an extension of the corresponding result of Brézis and Pazy [3], while Proposition 2 extends the known assertion (see Barbu [2], Cioranescu [5]). In the proof of Proposition 1 we do not use the result concerning the convexity of $\overline{D(A)}$. I thank M.Fabian and L.Veselý for pointing out the Asplund result.

REFERENCES

[1] E.Asplund, *Fréchet differentiability of convex functions*, Acta Math. 121 (1968), 31–47.

[2] V.Barbu, *Semigrupuri de contractii nelineare in spatii Banach*, Ed. Acad. Rep. Soc. Romania, Bucuresti, 1974.

[3] H.Brézis, A.Pazy, *Semigroups of nonlinear contractions on convex sets*, J.Functional Anal. 6 (1970), 237–281.

[4] E.F.Browder, *Nonlinear operators and nonlinear equations of evolution in Banach spaces*, Proc. Symp. Pure Math., 18, Part 2, Amer. Math. Soc., Providence, R.I., 1976.

[5] I.Cioranescu, *Aplicatii da dualitate in analiza funktionala neliniara*, Ed. Acad. Rep. Soc. Romania, Bucuresti, 1974.

[6] D.F.Cudia, *The geometry of Banach spaces: Smoothness*, Trans. Amer. Math. Soc. 110 (1964), 284–314.

[7] M.Fabian, *On single-valuedness (and strong continuity) of maximal monotone mappings*, Comment. Math. Univ. Carolinae 18 (1977), 19–39.

[8] J.R.Giles, *Convex Analysis with Application in the Differentiation of Convex Functions*, Pitman, Melbourne, 1982.

[9] L.Gobbo, *On the asymptotic behavior of the inverse of an m-accretive operator in a Banach space*, Rend. Sem. Mat. Univ. Politechn. Torino 42 (1984), 47–64.

[10] H.Hudzik, *Strict convexity of Musielak-Orlicz spaces with Luxemburg's norm*, Bull. Acad. Pol. Sci. Math. 29 (1981), 137–144.

[11] T.Kato, *Accretive operators and nonlinear equations in Banach spaces*, Proc. Symp. Pure Math. 18, Amer. Math. Soc., Providence, R.I., 1970, 138–161.

[12] P.S.Kenderov, *Multivalued monotone mappings are almost everywhere single-valued*, Studia Math. 56 (1976), 199–203.

[13] J.Kolomý, *On mapping and invariance of domain theorem and accretive operators*, Boll. U.M.I. 1-B (1987), 235–246.

[14] J.Kolomý, *Maximal monotone and accretive multivalued mappings and structure of Banach spaces*, Function Spaces, Proc. Int. Conf. Poznań, 1986, Teubner Texte zur Math. 103 (1988), 170–177.

[15] G.Morosanu, *Asymptotic behavior of resolvents for a monotone mapping in a Hilbert space*, Acad. Atti dei Lincei, Ser. 8 59 (1976), 565–570.

[16] S.Reich, *Approximating zeroes of accretive operators*, Proc. Amer. Math. Soc. 51 (1975), 381–384.

[17] S.Reich, *Strong convergence theorems for resolvents of accretive operators in Banach spaces*, J. Math. Anal. Appl. 75 (1980), 287–292.

[18] Chen Shutao, *Convexity and smoothness of Orlicz spaces*, *Geometry of Orlicz spaces I*, Function Spaces, Proc. Int. Conf. Poznań, 1986, Teubner Texte zur Math. 103 (1988), 12–19.

[19] W.Takahashi, Y.Ueda, *On Reich's strong convergence theorems for resolvents of accretive operators*, J. Math. Anal. Appl. 194 (1984), 546–553.

[20] L.Veselý, *Some new results on accretive multivalued operators*, Comment. Math. Univ. Carolinae 30 (1989), 45–55.

SPECTRAL PROPERTIES OF CANONICAL ISOMETRY IN WEIGHTED BERGMAN SPACE

Anna Krok, Tomasz Mazur

Radom, Poland

0. Introduction.

The Hilbert space methods in the theory of biholomorphic mappings were introduced and developed by S.Bergman [1,2]. In this approach, the space $L^2H(D)$ (sometimes denoted by $L^2H(D,m)$) of all functions, which are holomorphic and square integrable with respect to the Lebesgue measure on a domain $D \subset \mathbb{C}^N$, plays the central role. An operator of canonical isometry U_ϕ: $L^2H(D_2) \longrightarrow L^2H(D_1)$, generated by biholomorphic mapping ϕ: $D_1 \longrightarrow D_2$, is defined as follows

$$(0.1) \qquad (U_\phi f)(z) = f(\phi(z)) \cdot \phi'(z), \qquad z \in D$$

(here ϕ' is a complex Jacobian of ϕ).

The spectral theory of operators of H.Stone and J. von Neuman [13,8] is applied in a study of U_ϕ, when ϕ is biholomorphic automorphism of a domain $D \subset \mathbb{C}^N$, in [3,7]. In [5], this context is considered with group representations point of view.

On the other side every $\phi \in \mathrm{Aut}(D)$ ($\mathrm{Aut}(D)$ is a group of all biholomorphic automorphisms of a domain D) may be treated as a map, which leaves Lebesgue measure m on $D \subset \mathbb{C}^N$ invariant modulo holomorphic change of gauge ϕ', [4,6], i.e.,

$$(0.2) \qquad m(\phi(D')) = \int_{D'} |\phi'|^2 dm \quad \text{for each } D' \subset D$$

($|\phi'|^2$ is a real Jacobian of ϕ).

In the present paper we consider more general situation, when a domain $D \subseteq \mathbb{C}^N$ is equipped with an admissible measure μ.

Definition 0.1. ([10,11]). A positive measure μ on $D \subseteq \mathbb{C}^N$, absolutely continuous w.r.t. the Lebesgue measure m is called *admissible* if $L^2H(D,\mu)$ is a Hilbert space (is closed in the space of all μ-square integrable functions on D) and the evaluation functional

(0.3) $$\chi_z^*: L^2H(D,\mu) \longrightarrow \mathbb{C}, \qquad \chi_z^*(f) := f(z)$$

is continuous for each $z \in D$.

The Radon-Nikodym derivative of μ w.r.t. m is called *a weight*.

(0.4) $$(f,g)_\mu = \int_D f(z)\overline{g(z)}d\mu(z)$$

is a scalar product in $L^2H(D,\mu)$ ([6]).

The family G_μ of all maps μ-invariant modulo holomorphic change of gauge, (i.e., such $\phi \in Aut(D)$ for which the mapping $\psi: D \longrightarrow \mathbb{C}$ in the following expression:

(0.5) $$\mu(\phi(D')) = \int_{D'} |\psi|^2 d\mu \qquad \text{for each } D' \subseteq D$$

is non-zero and holomorphic), is a subgroup of $Aut(D)$ ([4,6]).

We consider a unitary representation of G_μ given by ([6])

$$V_\phi: L^2H(D,\mu) \longrightarrow L^2H(D,\mu),$$

(0.6) $$(V_\phi f)(z) = f(\phi(z))\cdot\psi(z) \qquad z \in D.$$

The purpose of the present paper is to characterize the spectrum of V_ϕ. It turns out that V_ϕ (similarly as U_ϕ [3]) has either discrete or purely continuous spectrum.

1. Preliminaries.

Fourier-Stieltjes coefficients and continuous measure.

Let F be a function of bounded variation on the unit circle $T \subset \mathbb{C}$. We will write

(1.1) $$dF \cong \sum_{-\infty}^{+\infty} c_k t^k \quad \text{if} \quad c_k = \int_T t^k dF(t)$$

in the sense of Riemman-Stieltjes integral.

Theorem 1.1. ([14]). *Let F be a function of bounded variation on T and $d_1, d_2, \ldots$ be all the jumps of F. Then, for c_k as in (1.1)*

(1.2) $$\lim_{n\to\infty} \frac{1}{2n+1} \sum_{k=-n}^{n} |c_k|^2 = \frac{1}{(2\pi)^2} \sum_j |d_j|^2.$$

Suppose Π is a Borel measure on the unit circle T. Call $B_p := \{t \in T: \Pi(\{t\}) \neq 0\}$ the set of pure points of Π. Define

(1.3) $$\Pi_{pp}(B) = \sum_{t \in B_p \cap B} \Pi(\{t\}) = \Pi(B_p \cap B)$$

for a Borel set B on T. Π_{pp} is a measure on T and $\Pi_{cont} := \Pi - \Pi_{pp}$

has a property: $\Pi_{cont}(\{t\}) = 0$ for all $t\in T$.

A Borel measure Π is called:

a) *continuous*, if it has no pure point

b) *pure point*, if $\Pi(B) = \sum_{t\in B} \Pi(\{t\})$ for each Borel set B.

Some ergodic theorem.

Let $(\Omega,\mathcal{A},P)$ be a propability space and $\zeta:\Omega \longrightarrow \mathbb{R}$ a random variable. Suppose that $\mathcal{F}$ is a σ-algebra in $\mathcal{A}$ and ζ is integrable (i.e., $\zeta \in L^1(\Omega,\mathcal{A},P)$). A Radon-Nikodym derivative of a measure $Q(A) := \int_A \zeta(\omega)dP(\omega)$, $A\in\mathcal{F}$, w.r.t. P restricted to $\mathcal{F}$ is called *a conditional expectation of* ζ *related to* $\mathcal{F}$ and denoted by $\mathbb{E}(\zeta\,|\mathcal{F})$. If ζ is a square integrable random variable, then $\mathbb{E}(\zeta\,|\mathcal{F}) = \mathrm{Proj}_{\mathcal{M}(\mathcal{F})}\zeta$ – orthogonal projection of ζ onto $\mathcal{M}(\mathcal{F}) := \{f\in L^2(\Omega,\mathcal{A},P):\ f - \mathcal{F}\text{-measurable}\}$.

Theorem 1.2. ([9]). *Assume that* $(\Omega,\mathcal{A},P)$ *is a propability space. If* $T:\Omega \longrightarrow \Omega$ *is P-invariant (i.e.,* $P(T^{-1}(A)) = P(A)$ *for all* $A \subset \Omega$*), then for every* $f \in L^1(\Omega,\mathcal{A},P)$ *and for P-a.e.* $\omega\in\Omega$

$$\lim_{n\to\infty} \frac{f(T^{-n}(\omega))+\ldots+f(T^{-1}(\omega))+f(\omega)+\ldots+f(T^{n}(\omega))}{2n+1} = \mathbb{E}(f\,|\mathcal{M})(\omega) \tag{1.4}$$

where $\mathcal{M}$ *is a* σ*-algebra of subsets of* Ω *such that:* $A\in\mathcal{M}$ *if and only if* $A = T^{-1}(A)$ *P-a.e.*

One of the basic properties of conditional expectation is:

$$P(B) = \mathbb{E}(\mathbb{E}(1_B\,|\mathcal{M})) \tag{1.5}$$

($\mathbb{E}$ denotes an expectation operator) for each Borel set $B \subset \Omega$. Here 1_B denotes the characteristic function of B.

The Bergman function.

On the base of Riesz representation theorem the evaluation functional $\chi_z^*:L^2H(D,\mu) \longrightarrow \mathbb{C}$ can be written in the form: $\chi_z^*(f) = f(z) = (f,\chi_z)_\mu$ for unique $\chi_z \in L^2H(D,\mu)$. The Bergman function on $L^2H(D,\mu)$ is defined as follows ([6,10]):

$$K_\mu(z,w) := (\chi_w,\chi_z) = \chi_w(z) \qquad z,w \in D. \tag{1.6}$$

Theorem 1.3. ([10,11]). *The measure* μ *on* $D \subset \mathbb{C}^N$ *is admissible if and only if for any compact set* $K \subset D$*, there exists* $c_K>0$ *such that*

$$\sup\,\{\|\chi_z\|:\ z\in K\} \le c_K < \infty\,. \tag{1.7}$$

2. The character of spectrum

Since V_ϕ is the unitary operator, there exists the unique complex measure $\pi_{f,g}$, $f,g \in L^2H(D,\mu)$, on the circle $T \subset \mathbb{C}$ and projection-valued measure P on T onto a Boolean algebra of projections in $L^2H(D,\mu)$ such that

(2.1) $$\pi_{f,g}(B) = (P(B)f,g)_\mu$$

for each Borel set B on T and

(2.2) $$(V_\phi^k f,g)_\mu = \int_T t^k d\pi_{f,g}(\{t\}).$$

$F_{f,g} := |\pi_{f,g}(\{t\})|$ is a function of bounded variation and $V_\phi^k = V_{\phi^k}$ for all $k \in \mathbb{Z}$, so we have the following expression for k-th Fourier-Stieltjes coefficient of $F_{f,g}$

(2.3) $$\hat{\pi}_{f,g}(k) = (V_\phi^k f,g)_\mu.$$

Main Theorem. *Let $\phi \in G_\mu$. Suppose that there exists a ϕ-invariant propability measure $\mathbb{P}$ on D ($\mathbb{P}(\phi^{-1}(B)) = \mathbb{P}(B)$), absolutely continuous w.r.t. μ. Then V_ϕ has discrite spectrum.*

Proof. Let H_{pp} be the closure of linear span of all eigenvectors of V_ϕ

(2.4) $$H_{pp} = \overline{\mathrm{lin}\,\{f \in L^2H(D,\mu) : \pi_{ff} \text{ is pure point measure}\}}$$

and H_c be a set of those $f \in L^2H(D,\mu)$ for which π_{ff} is a continuous measure. We have the following orthogonal decomposition:

(2.5) $$L^2H(D,\mu) = H_{pp} \oplus H_c$$

Suppose that the spectrum of V_ϕ is not discrete. This means that there exists a non-zero f in H_c. It is obvious that for each $g \in L^2H(D,\mu)$, $g = g'+g''$, $g' \in H_{pp}$, $g'' \in H_c$, we have $\pi_{f,g}(B) = \pi_{f,g''}(B)$ for all Borel sets $B \subset T$. From polarization formula

(2.6) $$\pi_{f,g''} = \frac{1}{4} \sum_{k=0}^{3} i^k \cdot \pi_{f+i^k g'', f+i^k g''}$$

it follows that $\pi_{f,g}$ is a continuous measure. Hence the function $F_{f,g}$ has no jumps on T. ($F_{f,g}(t) = 0$ for $t \in T$.) Theorem 1.1 in view of (2.3) yields

(2.7) $$\lim_{n\to\infty} \frac{1}{2n+1} \sum_{k=-n}^{n} |(V_{\phi^k} f,g)_\mu|^2 = \lim_{n\to\infty} \frac{1}{2n+1} \sum_{k=-n}^{n} |\hat{\pi}_{f,g}(k)|^2$$
$$\sum_{t\in T} |\pi_{f,g}(\{t\})|^2 = 0.$$

Let K be a compact subset of D, such that $\mathbb{P}(K) > 0$. For

(2.8) $$\frac{1}{n+1}\sum_{k=-n}^{n}|(V_{\phi^k}f,\chi_z)_\mu|^2 \le \|f\|_\mu\circ\|\chi_z\|_\mu$$

the following functions are uniformly bounded on K:

(2.9) $$z \longrightarrow \frac{1}{2n+1}\sum_{k=-n}^{n}|(V_{\phi^k}f,\chi_z)_\mu| \qquad n = 1,2,\dots .$$

Lebesgue denomination theorem applied to those functions via (2.7) yields:

(2.10) $$\lim_{n\to\infty}\frac{1}{2n+1}\sum_{k=-n}^{n}\int_K |(V_{\phi^k}f,\chi_z)_\mu|^2 d\mu(z) = 0.$$

In view of (0.6), for each $k\in\mathbb{Z}$

(2.11) $$\int_K |(V_{\phi^k}f,\chi_z)_\mu|^2 d\mu(z) = \int_{\phi^k}|f(z)|^2 d\mu(z).$$

Now, we can write (2.10) in the form

(2.12) $$\lim_{n\to\infty}\int_D\left[\frac{1}{2n+1}\sum_{k=-n}^{n}1_{\phi^k(K)}\right]|f|^2 d\mu = 0.$$

From Fatou's lemma we get:

(2.13) $$\int_D \liminf_{n\to\infty}\frac{1}{2n+1}\sum_{k=-n}^{n}1_{\phi^k(K)}|f|^2\, d\mu$$
$$\le \liminf_{n\to\infty}\int_D \frac{1}{2n+1}\sum_{k=-n}^{n}1_{\phi^k(K)}|f|^2\, d\mu.$$

Because f, as an holomorphic function, is not zero μ-a.e. in D so (2.12) implies

(2.14) $$\liminf_{n\to\infty}\frac{1}{2n+1}\sum_{k=-n}^{n}1_{\phi^k(K)} = 0.$$

Let $\mathcal{M}_\phi$ be a given σ-algebra of those Borel sets B in D such that $1_B = 1_{\phi(B)}$ $\mathbb{P}$-a.e. and $\mathbb{E}(\cdot\,|\mathcal{M}_\phi)$ – the corresponding conditional expectation operator. We may apply Theorem 1.2 to $T = \phi{:}D \longrightarrow D$ and $f = 1_K$ and we get

(2.15) $$\mathbb{E}(1_K|\mathcal{M}_\phi)(z) = 0 \ \mathbb{P}\text{-a.e.}.$$

This means, via (1.5) that

(2.16) $$\mathbb{P}(K) = \mathbb{E}(\mathbb{E}(1_K|\mathcal{M}_\phi)) = 0.$$

This contradicts to the assumption $\mathbb{P}(K) > 0$.

Theorem 2.1. *If V_ϕ has any eigenvector, then there exists propability measure on D, which is ϕ-invariant.*

Proof. Let $V_\phi h = \alpha\circ h$, $|\alpha| = 1$. Consider

$$\mathbb{P}(B) := \frac{1}{\|h\|_\mu}\int_B |h|^2 d\mu.$$

Notice that

$$\mathbb{P}(B) = \frac{1}{\|h\|_\mu} \int_B |h(\phi)|^2 |\psi|^2 d\mu = \frac{1}{\|h\|_\mu} \int_{\phi(B)} |h|^2 d\mu = \mathbb{P}(\phi(B))$$

and

$$\mathbb{P}(D) = \frac{1}{\|h\|_\mu} \int_D |h|^2 d\mu = 1.$$

Since $h \neq 0$ μ-a.e. in D, $\mathbb{P}$ is the searched measure.

Corollary 2.2. *The following conditions are equivalent:*

1) V_ϕ has discrete spectrum.

2) V_ϕ has any non-zero eigenvector.

3) there exists a probability measure on D which is absolutely continuous w.r.t. μ and ϕ-invariant.

REFERENCES

[1] S. Bergman, *The kernel function and conformal mapping*, Math. Surveys 5, second ed., Amer. Math. Soc. 1970.

[2] S.Bergman, *Über die Kernfunktion eines Bereiches und ihr Verhalten am Rande*, I.Reine Angew. Math. 169 (1933), 1–42; 172 (1934), 89–123.

[3] W.Chojnacki, On *some holomorphic dynamical systems*, Quart. J. Math. (2) 39 (1988), 159–172.

[4] S.Janson, J. Peetre, R. Rochberg, *Hankel forms and the Fock space*, Uppsala University, Dept. of Math. Report 6 (1986).

[5] S.Kobayashi, *On automorphism group of homogeneous complex manifold*, Proc. of AMS 12 (1961), 359–361.

[6] T.Mazur, *Canonical isometry on weighted Bergman spaces*, Pacific J. of Math. 136 (1989).

[7] T.Mazur, M. Skwarczyński, *Spectral properties of holomorphic with fixed point*, Glasgow Math. J. 28 (1986), 25–30.

[8] J.von Neumann, *Eine Spectraltheorie für allgemeine Operatoren eines unitares Räumes*, Math. Nacht. 4 (1951).

[9] W.Parry, *Topics in Ergodic Theory*, Cambridge University Press 1981.

[10] Z.Pasternak-Winiarski, *On the dependence of the reproducing kernel on the weight of integration*, to appear in J. Func. Analysis.

[11] Z.Pasternak-Winiarski, *On weight which admits the reproducing kernel of Bergman type*, preprint.

[12] M.Skwarczyński, *Biholomorphic invariants related to the Bergman functions*, Dissertationes Math. PWN Warsaw 1980.

[13] H.Stone, *Linear transformation in Hilbert space and their applications to analysis*, AMS Providence 9th ed. 1979.

[14] A.Zygmund, *Trygonometric Series, vol 1*, Cambridge Univ. Press 1959.

IDEAL SPACES OF VECTOR FUNCTIONS AND THEIR APPLICATIONS

NGUYỄN HỒNG THÁI, P. P. ZABREJKO

Minsk, USSR

This report deals with results on new functions spaces concept – ideal spaces of measurable vector functions defined on an arbitrary measure and taking values in a finite-dimensional linear space. These new spaces were introduced in [1] in 1984–1987 by Zabrejko: they are modulus over L_∞ with respect to usual multiplication of vector functions by scalar functions. It was noted by Nguyễn Hồng Thái in [2,3] that these new spaces are ordered in some new special (nonclassical) sense. More exactly, each ideal space of vector functions can be extended to a corresponding ideal semimodule of so-called infra-semi-units and semi-units, which are special measurable multifunctions in the sense of C.Castaing; this semimodule become an order complete algebraic system with respect to the order by means of usual inclusion relation of sets. This semimodule is closed to a semiring in the sense of V.P.Maslov that was introduced by him in 1979 and studied further by him and his colleagues. These new spaces contain all variants of Orlicz spaces of vector functions (see for example [4–6] and various references therein) and they are not and can not be reduced to vector lattices or K-spaces as in the scalar case. The presented theory allows to obtain the comprehensive general theory, the duality theory and the theory of linear and non-linear operators for these new spaces [5–8]. These new spaces have been applied [2,3,9] to systems of Hammerstein equations and boundary values problems for quasilinear eliptic differential equations with arbitrary nonlinearities, and also to the interpolation theory of linear and nonlinear operators.

1. Let Ω be an arbitrary set, $\mathcal{A}$ some σ-algebra of subsets of Ω and μ a complete countably additive and σ-finite measure on $\mathcal{A}$, μ_* a normalized and equivalent to μ measure. The algebra $\mathcal{A}$ is a complete metric space equipped with the metric $\rho(A,B) = \mu_*(A\Delta B)$ and it is an order complete algebra of countable type (Theorem of L.V. Kantorowicz).

Let $\mathbb{R}^m$ be the m-dimensional Euclidean space equipped with the scalar product $(\cdot,\cdot)$ and the norm $\|\cdot\|$ and T be the unit ball of $\mathbb{R}^m$. Let $O(\mathbb{R}^m)$ $[\mathcal{B}(\mathbb{R}^m)]$ be the family of all [resp. all bounded] closed convex and symmetric subsets of $\mathbb{R}^m$. The space $O(\mathbb{R}^m)$ equipped with the Hausdorff distance $d(\cdot,\cdot)$ and the space $\mathcal{B}(\mathbb{R}^m)$ equipped with the distance

$$\tilde{d}(A,B) = \sum_{n=1}^{\infty} \frac{1}{2^n} \frac{d(nT\cap A, nT\cap B)}{1+d(nT\cap A, nT\cap B)} \qquad (A,B \in \mathcal{B}(\mathbb{R}^m)),$$

are complete separable metric spaces.

Further, let $S(\mathbb{R}^m) = S(\Omega,\mathbb{R}^m)$ be the complete metric space of all measurable on Ω vector valued functions taking values in $\mathbb{R}^m$. A multifunction $B:\Omega \to 2^{\mathbb{R}^m}$ is called measurable in the sense of C.Castaing if there exists its Castaing representation, i.e., a countable family of measurable selectors $\{x_n\} \in S(\mathbb{R}^m)$ such that the family of vectors $\{x_n(s): n\in\mathbb{N}\}$ is dense in $B(s)$ for almost all $s\in\Omega$. We denote by $S(B)$ the family of all measurable selectors of a corresponding multifunction B.

Denote by $SO(\mathbb{R}^m)$ and $S\mathcal{B}(\mathbb{R}^m)$ the complete metric spaces of all measurable on Ω functions taking values in $O(\mathbb{R}^m)$ and $\mathcal{B}(\mathbb{R}^m)$, respectively. It is obvious that $B \in SO(\mathbb{R}^m)$ or $B \in S\mathcal{B}(\mathbb{R}^m)$ iff B is measurable in Castaing's sense. Analogously, we denote by $S\mathcal{L}(\mathbb{R}^m)$ the space of all multifunctions that are measurable in Castaing sense and take values in the space $\mathcal{L}(\mathbb{R}^m)$ of all linear subspaces of $\mathbb{R}^m$. We have the inclusions

$$SO(\mathbb{R}^m) \supset S\mathcal{B}(\mathbb{R}^m),\ S\mathcal{L}(\mathbb{R}^m).$$

Let $B \in SO(\mathbb{R}^m)$. The set $\operatorname{supp} B = \{s\in\Omega: B(s) \neq \{0\}\}$ is called *the support of B*. One can see that the multifunction $\mathcal{L}_B(s)$ defined by the formula

$$\mathcal{L}_B(s) = \bigcup_{\lambda>0} \lambda B(s) \tag{1}$$

belongs to $S\mathcal{L}(\mathbb{R}^m)$. Denote by M_B the complete seminormed space of all $x \in S(\mathcal{L}_B)$ equipped with the seminorm $\|x|M_B\| = \inf\{\lambda>0: x \in S(B)\}$. The seminorm $\|x|M_B\|$ is a norm iff $B \in S\mathcal{B}(\mathbb{R}^m)$.

The multifunctions B_x defined by

$$B_x(s) = \operatorname{conv}\{-x(s),x(s)\} \qquad (x \in S(\mathbb{R}^m)), \tag{2}$$

are simple examples of elements of $S\mathcal{B}(\mathbb{R}^m)$.

Denote by B° the usual polar to $B \in SO(\mathbb{R}^m)$.

Theorem 1. *The spaces $SO(\mathbb{R}^m)$ and $S\mathcal{B}(\mathbb{R}^m)$ are order complete semimodulus of countable type with respect to the usual addition, the multiplication by scalars and the order generated by means of usual inclusion relation of sets; in particular, every subset $\mathcal{N}$ that is order upper bounded has the supremum; moreover there exists a countable subset $\mathcal{N}_o = \{B_n: n\in\mathbb{N}\} \subset \mathcal{N}$ such that*

$$\sup \mathcal{N} = \sup \mathcal{N}_o,$$

$$\sup \mathcal{N}_o(s) = \overline{\text{conv}}\ \{B_n(s): n\in\mathbb{N}\} \qquad (s\in\Omega). \tag{3}$$

Further, the following statemens hold

1) every subset $\mathcal{N}$ of $SO(\mathbb{R}^m)$ has the supremum and the equality

$$(\sup \mathcal{N})^\circ = \inf \mathcal{N}^\circ \overset{\text{def}}{=} \inf \{B^\circ: B\in\mathcal{N}\} \tag{4}$$

is true;

2) the supremum of $\mathcal{N} \subset SO(\mathbb{R}^m)$ in $SO(\mathbb{R}^m)$ belongs to $S\mathcal{B}(\mathbb{R}^m)$ iff $\mathcal{N}$ is order upper bounded in $S\mathcal{B}(\mathbb{R}^m)$.

2. A vector space $X \subset S(\mathbb{R}^m)$ is called *ideal space* (cf. [1]) if the relations $x \in X$ and $\theta \in L_\infty$ imply that $\theta x \in X$, where L_∞ denotes the Banach space of all measurable on Ω and essentially bounded scalar functions equipped with the usual norm $\|\cdot\|_{L_\infty}$.

Let X be an ideal space. As usual we define the support of X by the relation

$$\text{supp } X = \sup \{\text{supp } \|x(\cdot)\|: x \in X\}.$$

We introduce the $\mathcal{L}$-support $\mathcal{L}_X \in S\mathcal{L}(\mathbb{R}^m)$ by the equality

$$\mathcal{L}_X = \sup \{\mathcal{L}_{B_x}: x \in X\}. \tag{5}$$

Let $\tilde{X} = \{B_x: x \in X\}$,

$\hat{E}_v(X) = \{B \in SO(\mathbb{R}^m): S(B) \subset X\}$, $\qquad E_v(X) = \hat{E}_v(X) \cap S\mathcal{B}(\mathbb{R}^m)$,

(6) $\quad \hat{E}(X) = \{B \in \hat{E}_v(X): P_{\text{supp } B}\, x \in S(\mathcal{L}_B) \quad (x\in X)\}$,

$$E(X) = \hat{E}(X) \cap S\mathcal{B}(\mathbb{R}^m),$$

$\hat{E}_o(X) = \{B \in \hat{E}_v(X): x \in S(\mathcal{L}_B) \quad (x\in X)\}$, $\qquad E_o(X) = \hat{E}_o(X) \cap S\mathcal{B}(\mathbb{R}^m)$,

where P_D denotes the operator of multiplication by a characteristic function of a set $D \in \mathcal{A}$. We call an element $B \in E_v(X)$, or $B \in E(X)$, or $B \in E_o(X)$, respectively, *an infra-semi-unit*, or *a semi-unit*, or *a unit* of the space X. In the scalar case, $E(X) = E_v(X)$ and both ones are reduced to the set of all multifunctions B_x of type (2), where x is an arbitrary element of the cone of all nonnegative functions in X. In the vector case, $E(X) \neq E_v(X)$. It is obvious that the following embeddings hold:

$$
\begin{array}{lcccccccc}
 & & & & SO(\mathbb{R}^m) & & SO(\mathbb{R}^m) & & SO(\mathbb{R}^m) \\
 & & & & \cup & & \cup & & \cup \\
X & \longrightarrow & \tilde{X} & \subset & \hat{E}_v(X) & \supset & \hat{E}(X) & \supset & \hat{E}_0(X), \\
| & & & & \cup & & \cup & & \cup \\
| & & & & E_v(X) & \supset & E(X) & \supset & E_0(X), \\
| & & & & \cap & & \cap & & \cap \\
| & & & & S\mathcal{B}(\mathbb{R}^m) & & S\mathcal{B}(\mathbb{R}^m) & & S\mathcal{B}(\mathbb{R}^m) \\
| & & & & & & & & \\
\multicolumn{9}{l}{\mathcal{L}_X \in S\mathcal{L}(\mathbb{R}^m) \subset SO(\mathbb{R}^m).}
\end{array}
$$

Theorem 2. *Let X be an ideal space. Then, for every number $\varepsilon>0$, there exists a semi-unit $B_\varepsilon \in E(X)$ such that*

$$\mu_*[\operatorname{supp} X \setminus \operatorname{supp} B_\varepsilon] < \varepsilon.$$

Theorem 3. *Let X be an ideal space. Then, $\hat{E}_v(X)$ $[E_v(X)]$ is an order complete ideal semimodule of countable type; this semimodule is generated by the family $\{B_X: x\in X\}$ in the semimodule $SO(\mathbb{R}^m)$ [respectively, $S\mathcal{B}(\mathbb{R}^m)$].*

3. A normed space $X \subset S(\mathbb{R}^m)$ equipped with a norm $\|\cdot\|_X$ is called *a normed ideal space* (cf. [1]) if the relations $x \in X$ and $\theta \in L_\infty$ imply that both $\theta x \in X$ and $\|\theta x\|_X \leq \|\theta x\|_{L_\infty}\|x\|_X$; *a Banach ideal space* if X is a complete normed linear space. The semimodule $E_v(X)$ equipped with the "norm"

(8) $$\|B\|_X = \sup\{\|x\|_X: x \in S(B)\},$$

is a normed linear semimodule. We define the Δ-support Δ_X belonging to $SO(\mathbb{R}^m)$ by the relation

(9) $$\Delta_X(s) = \sup\{B_x: x \in X, \|x\|_X \leq 1\};$$

this notion plays the significant role in the theory of nonlinear operators in these ideal spaces.

Analogously we can define the notions of seminormed ideal spaces, locally convex ideal spaces, ideal spaces with F-norms and others. We can obtain a sufficiently comprehensive theory for these spaces of vector functions as in the scalar case. In particular, we obtain theorems about separability and regularity of these spaces, compactness and absolute boundedness of sets; the vector analogues of

Luxemburg-Zaanen reflexivity criterion, Mory-Amemyia-Nakano theorems about almost perfectness, Iosida-Hewitt-Luxemburg theorems about Lebesgue decomposition, Lozannovskii-Buchvalov theorems about compact in measure convex sets, Lozannovskii theorem about the representation of dual spaces of ideal spaces of vector-functions, Zabrejko theorems about construction of unit balls of Banach spaces of vector functions.

Now we are in a position to formulate some of these statements.

Theorem 4. *Let X be a normed ideal space. Then X is continuously embedded into $S(\mathbb{R}^m)$. If X is a Banach ideal space, then $E_o(X) \neq \emptyset$.*

Let X be an ideal space. A set $N \subset X$ is called *U-bounded* if there exists $B \in E_v(X)$ such that $N \subset S(B)$. A functional $f:X \to \mathbb{R}$ is called *regular* if f maps every U-bounded subset of X into a bounded subset of $\mathbb{R}$. A regular linear functional f is called *(mo)-continuous* ($f \in X_n^\sim$) if f maps every sequence which is U-bounded in X and convergent to 0 in $S(\mathbb{R}^m)$ into a sequence which converges to 0; *singular* ($f \in X_s^\sim$) if f vanishes on some ideal subspace $Y \subset X$ for which $\mathcal{L}_X = \mathcal{L}_Y$.

We point out that $f \in X_n^\sim$ iff there exists $x' \in S(\mathcal{L}_X)$ such that $f(x) = \langle x,x' \rangle$ $(x \in X)$, where $\langle \cdot,\cdot \rangle$ denotes the usual scalar product of vector functions:

$$\langle x,x' \rangle = \int_\Omega (x(s),x'(s))\, d\mu(s). \tag{10}$$

If X is a normed ideal space then the associate to X ideal space X' of all functions $x' \in S(\mathcal{L}_X)$ for which $|\langle x,x' \rangle| < \infty$ $(x \in X)$ and which is equipped with the norm

$$\|x'\|_{X'} = \sup\{\langle x,x' \rangle: \|x\|_X \le 1\}, \tag{11}$$

is a Banach ideal space isometrically embedded into the usual dual space X^* and $\mathcal{L}_X = \mathcal{L}_{X'}$.

Theorem 5. *Let X be a normed ideal space. Then every element $f \in X$ can be represented (uniquely !) in the form $f = f_c + f_s$, $f_c \in X_n^\sim$, $f_s \in X_s^\sim$.*

Let X be a normed ideal space. Then we denote by X^o the ideal normed subspace of all vector functions $x \in X$ which have so-called absolutely continuous norm:

$$\lim_{\mu_*(D) \to 0} \|P_D x\|_X = 0. \tag{12}$$

We call a normed ideal space X *almost perfect* if $\|x\|_X = \|x\|_{X''}$; *perfect* if X is almost perfect and X = X''; *regular* if X = X°.

Theorem 6. *Let X be a normed ideal space. Then X is reflexive iff X' is regular and X is regular and perfect.*

Theorem 7. *Let X be a Banach ideal space. Then X is separable iff X is regular and the restriction of measure μ on supp X is separable.*

Theorem 8. *Let X be a Banach ideal space. Then a set $\mathcal{N}$ of X° is compact iff $\mathcal{N}$ is compact in $S(\mathbb{R}^m)$ and $\mathcal{N}$ has so-called uniformly absolutely continuous norms:*

(13) $$\lim_{\mu_*(D)\to 0} \sup_{x\in\mathcal{N}} \|P_D x\|_X = 0.$$

Theorem 9. *Let X be a Banach ideal space. Then a set $\mathcal{N} \subset X^\circ$ is absolutely bounded (i.e., bounded and has uniformly absolutely continuous norms with respect to μ_*) iff there exists a nonnegative increasing function $\Phi:[0,\infty] \to [0,\infty]$, $\Phi(0) = 0$, $\lim_{t\to\infty} \Phi(t) = \infty$ such that*

(14) $$\sup_{x\in\mathcal{N}} \|x(\cdot)\Phi(\mathcal{G}_B(\cdot,x(\cdot)))\|_X < \infty$$

for some $B \in E_v(X)$, where

(15) $$\mathcal{G}_B(s,u) = \inf\,\{\lambda>0:\ u \in \lambda B(s)\}.$$

4. We point out that we have obtained a comprehensive theory for linear and nonlinear operators in ideal spaces of vector functions, in particular, the analogues of Banach continuity theorem, theorems of Zabrejko and Gribanov about weak continuity and the existences of associate operators, theorems about regularity, compactness in measure and in norm, the Buchvalov integral representability criterion and some theorems on Urysohn operators. Now we formulate some of these statements.

Let X and Y be two Banach ideal spaces. We consider a linear operator $K:X \to Y$ which has an integral representation

(16) $$Kx(t) = \int_\Omega k(t,s)x(s)d\mu(s),$$

where a kernel k(t,s) is a measurable on $\Omega\times\Omega$ matrix-valued function from $S(\mathbb{R}^m\times\mathbb{R}^m)$.

Theorem 10. *Let X, Y be two Banach ideal spaces and K be an operator (16) acting from X to Y. Then K has an associate operator $K':Y' \to X'$.*

Theorem 11. *Let X be an ideal space and $K:X \to S(\mathbb{R}^m)$ be a linear operator. Then K has an integral representation of the form (16) iff K maps every sequence x_n which is U-bounded in X and convergent in $S(\mathbb{R}^m)$ to 0 into a sequence Kx_n which converges to 0 almost everywhere.*

Theorem 12. *Let X be a Banach ideal space and K be an operator (16) defined on X. Then K is weakly compact in measure (i.e., K maps every U-bounded subset of X into a compact subset of $S(\mathbb{R}^m)$). Moreover, if X' is regular then K is compact in measure (i.e., K maps every bounded subset of X into a compact subset of $S(\mathbb{R}^m)$).*

5. We point out that by some standard modifications we can obtain analogous results for ideal spaces in $S(\mathbb{C}^m)$ and for measure space Ω which have the so-called direct sum property.

REFERENCES

[1] Nguyễn Hồng Thái, *The superposition operators in Orlicz spaces of vector functions*, Doklady Acad. Nauk BSSR, 31 (1987), 191–200.

[2] Nguyễn Hồng Thái, *The Orlicz spaces of vector functions and their applications to nonlinear integral equations*, the Candidate (Ph.D.) dissertation, Minsk, 1987.

[3] Nguyễn Hồng Thái, *Semimodulus of infra-semi-units of vector functions ideal spaces and their applications to the integral representability problem for operators*, Soviet Math. Doklady, (to appear).

[4] Nguyễn Hồng Thái, *Semimodulus of infra-semi-units of vector functions ideal spaces and their applications to the interpolation theory of operators*, Doklady Acad. Nauk BSSR, (to appear).

[5] P.P.Zabrejko, *Ideal spaces of vector function*, Doklady Acad. Nauk BSSR, 31 (1987), 289–301.

[6] P.P.Zabrejko, Nguyễn Hồng Thái, *On the theory of vector functions Orlicz spaces*, Doklady Acad. Nauk BSSR, 31 (1987), 116–119.

[7] P.P.Zabrejko, Nguyễn Hồng Thái, *The linear integral operators in ideal spaces of vector functions*, Doklady Acad.Nauk BSSR, 32 (1988), 587–590.

[8] P.P.Zabrejko, Nguyễn Hồng Thái, *The duality theory of ideal spaces of vector functions*, Soviet Math. Doklady, (to appear).

[9] P.P.Zabrejko, Nguyễn Hồng Thái, *The new existence theorems for operators and integral Hammerstein equations*, Soviet Math. Doklady and Soviet Differential Equations, (to appear).

MARTINGALE INEQUALITIES
IN REARRANGEMENT INVARIANT FUNCTION SPACES

IGOR NOVIKOV

Voronezh, USSR

Classical martingale inequalities of J.L.Doob and B.Davis known in L_p spaces are proved for general rearrangement invariant spaces.

1. Notation.

Let $\mathcal{A}_1, \mathcal{A}_2, \ldots$ be a nondecreasing sequence of σ-fields of Lebesgue measurable subsets of the interval $[0,1]$, $m(A)$ – Lebesgue measure of set A. A sequence of integrable functions $\bar{x} = (x_1, x_2, \ldots)$ is called martingale relative to $\mathcal{A}_1, \mathcal{A}_2, \ldots$ if for every $k = 1,2,\ldots$ x_k is $\mathcal{A}_k$ – measurable and $E_{\mathcal{A}_k} x_{k+1} = x_k$, where $E_{\mathcal{A}} x$ is the conditional expectation of x with respect to $\mathcal{A}$. The sequence $\bar{d} = (d_1, d_2, \ldots)$ defined by equalities $d_1 = x_1$, $d_k = x_k - x_{k-1}$, $k = 2,3,\ldots$ is called *martingale-difference sequence*. The square function of $\bar{x}$ is $P_{\bar{x}} = (\sum_{k=1}^{\infty} d_k^2)^{1/2}$ and the maximal function of $\bar{x}$ is $\bar{x}^{\#} = \sup\limits_{1 \le n < \infty} |x_n|$.

The decreasing rearrangement of function x is defined by equality $x^*(t) = \inf\{\lambda > 0 : m\{s \in (0,1) : |x(s)| > \lambda\} < t\}$ for $t \in [0,1]$. The function x^* is non-increasing, left continuous and has the same distribution function as $|x|$.

A Banach function space X is said to be *a rearrangement invariant (r.i.) space* if the following conditions hold:

1) if $|x| \le |y|$ a.e. and $y \in X$, then $x \in X$ and $\|x\| \le \|y\|$;

2) if $x^* = y^*$ and $y \in X$, then $x \in X$ and $\|x\| = \|y\|$.

The classical examples of r.i. spaces: Orlicz spaces, Lorentz and Marcinkiewicz spaces ([4],[5]).

The operators σ_τ, $\tau > 0$ defined by equality

AMS Subject Classification: 60G42, 46E30

$$(\sigma_\tau x)(t) = \begin{cases} x(t/\tau), & \text{if } 0 \le t < \min(\tau,1); \\ 0, & \text{if } \min(\tau,1) \le t \le 1 \end{cases}$$

are bounded in any r.i. space ([4],p.131).

We say that r.i. space X has Fatou property if $x_n \to x$ a.e. and $\sup_n \|x_n\| < \infty \implies x \in X$ and $\|x\| \le \liminf_{n\to\infty} \|x_n\|$.

For space X and sequence $\bar{x}$ let $\|\bar{x}\|_X = \sup_{1\le n<\infty} \|x_n\|_X$.

2. Inequalities of J.L.Doob and B.Davis.

Doob's inequality [3] $\|\bar{x}\|_{L_p} \le \|\bar{x}^{\#}\|_{L_p} \le \frac{p}{p-1}\cdot\|\bar{x}\|_{L_p}$ holds for any $p \in (1,\infty)$ and any martingale $\bar{x}$.

Davis proved [2] that there are two positive numbers C_1 and C_2 such that for any martingale $\bar{x}$

$$C_1\|P\bar{x}\|_{L_1} \le \|\bar{x}^{\#}\|_{L_1} \le C_2\|P\bar{x}\|_{L_1}.$$

3. Main results.

Theorem 1. *Let X br r.i. space with Fatou property. The following conditions are equivalent:*

1) there is a positive real number C such that for any martingale $\bar{x}$ $\|\bar{x}\|_X \le \|\bar{x}^{\#}\|_X \le C\|\bar{x}\|_X$;

2) $\lim_{\tau\to\infty} \|\sigma_\tau\|_X/\tau = 0$.

Theorem 2. *Let X be r.i. space which is interpolation space between the spaces L_∞, L_1. There are two positive numbers C_1 and C_2 such that for every martingale $\bar{x}$ $C_1\|P\bar{x}\|_X \le \|\bar{x}^{\#}\|_X \le C_2\|P\bar{x}\|_X$ if and only if $\lim_{\tau\to 0} \|\sigma_\tau\|_X = 0$.*

4. Proof of Theorem 1.

Lemma 1. *Let X be r.i. space and $\lim_{\tau\to\infty} \|\sigma_\tau\|_X/\tau = 0$. If x and y are nonnegative functions satisfying*

$$(1) \qquad \lambda\cdot m(t:\ x(t) > \lambda) \le \alpha \int_{\{t:x(t)>\lambda\}} y(s)ds$$

for all $\lambda>0$ and some number $\alpha>0$, then $\|x\|_X \le \alpha\cdot C\cdot\|y\|_X$, where the constant C depends only on X.

Proof. Function $\eta_z(\tau) = m\{t\in(0,1): |z(t)| > \tau\}$, $\tau>0$, is called *the distribution function of z*. The distribution function is non-increasing and right continuous. Moreover the distribution function and the decreasing rearrangement have the following properties:

1) $\eta_z(z^*(t)) \le t$; if $z^*(t)$ is the point of continuity of $\eta_z(t)$ then $\eta_z(z^*(t)) = t$ ([4],p.83);

2) $\int_0^t z^*(s)ds = \sup_{m(e)=t} \int_e |z(s)|ds$ ([4],p.89);

3) function $z^{**}(t) = \frac{1}{t}\int_0^t z^*(s)ds$ is non-increasing ([4],p.170);

4) if $\lim_{\tau\to\infty} \|\sigma_\tau\|_X/\tau = 0$ then ([4],p.187)

$$\|z^{**}\|_X \le \int_0^1 \|\sigma_{1/s}\|_X ds\cdot\|z\|_X.$$

First note that if $\eta_x(x^*(t)) = t$ then inequality (1) gives for $\lambda = x^*(t)$

$$x^*(t)\cdot t \le \alpha\cdot\int_{\{\tau:x(\tau)>x^*(t)\}} y(s)ds.$$

Using property 2) we obtain $x^*(t) \le \alpha\cdot y^{**}(t)$.

Let now $\eta_x(x^*(t)) < t$. By property 1) $\tau_0 = x^*(t)$ is the point of discontinuity of $\eta_x(\tau)$. Let $t_1 = \lim_{\tau\to\tau_0,\tau<\tau_0} \eta_x(\tau)$. Then

$$t \le t_1,\ x^*(t) = x^*(t_1) = \tau. \tag{2}$$

Using inequality (1) and property 2) for $\lambda = \tau_0-\varepsilon$, $\varepsilon>0$, we have

$$(x^*(t_1)-\varepsilon)\cdot\eta_x(\tau_0-\varepsilon) \le \int_0^{\eta_x(\tau_0-\varepsilon)} y^*(s)ds.$$

If $\varepsilon \to 0$, then the last inequality gives

$$x^*(t_1) \le \alpha\cdot y^{**}(t_1). \tag{3}$$

Using property 3), inequalities (2) and (3) we obtain

$$x^*(t) \le \alpha\cdot y^{**}(t).$$

So we have shown that $x^*(t) \le \alpha\cdot y^{**}(t)$ for any $t \in [0,1]$. The last inequality together with property 4) establishes Lemma 1.

Proof of Theorem 1. First we prove 2) $\Longrightarrow$ 1). The left-hand side of the inequality is obvious. To prove the right-hand side let $\bar{x}_n^{\#} = \sup_{1\le k\le n} |x_k|$. It is known ([3]), that

$$\lambda\cdot m\{t: \bar{x}_n^{\#}(t) > \lambda\} \le \int_{\{t:\bar{x}_n^{\#}(t)>\lambda\}} |x_n(s)|ds,\quad \lambda>0.$$

Using Lemma 1 we have $\|\bar{x}_n^{\#}\|_X \le C\|x_n\|_X \le C\|\bar{x}\|_X$. By the Fatou property of X the last inequality implies the right-hand side of the inequality in 2).

The proof of the converse assertion is based on the fact that if $\lim_{\tau\to\infty} \|\sigma_\tau\|_X/\tau > 0$ then for any integer n there are mutually disjoint functions having the same distribution function $\{x_i\}_{i=1}^{2^n}$ which are 2-equivalent to the unit vector basis of $\ell_1^{2^n}$ [5,p.141]. We may assume without loss of generality that each x_i, $1 \le i \le 2^n$ is a finite linear combination of characteristic functions of intervals of the form $[(l-1)2^{-k}, l\cdot 2^{-k})$ for some fixed k independent of i. Hence, by applying a suitable automorphism to [0,1] we may assume that on the first 2^n dyadic intervals of length 2^{-k} each x_i is non-zero on exactly one of these intervals and takes there a value independent of i, that the same is true on the next 2^n dyadic intervals of length 2^{-k} and so on.

Consider now functions $\{y_j\}_{j=1}^{n+1}$ which are defined by

$$y_1 = x_1 + \dots + x_{2^n};$$

$$y_2 = x_1 + \dots + x_{2^{n-1}} - x_{2^{n-1}+1} - \dots - x_{2^n};$$

$$y_3 = (x_1 + \dots + x_{2^{n-2}} - x_{2^{n-2}+1} - \dots - x_{2^{n-1}})\cdot 2;$$

$$y_n = 2^{n-2}(x_1+x_2-x_3-x_4);$$

$$y_{n+1} = 2^{n-1}(x_1-x_2).$$

Note that $\{y_j\}_{j=1}^{n+1}$ is martingale-difference sequence and $\sum_{i=1}^{n+1} y_i = 2^n x_1$;

$$\sup_{1\le k\le n+1} |\textstyle\sum_{i=1}^{k} y_i| = 2^n x_1 + 2^{n-1}x_2 + 2^{n-2}(x_3+x_4) + \dots +$$

$$+ 2(x_{2^{n-2}+1}+\dots+x_{2^{n-1}}) + (x_{2^{n-1}+1}+\dots+x_{2^n}).$$

Since $\{x_i\}_{i=1}^{2^n}$ is 2-equivalent to unit vector basis of $\ell_1^{2^n}$ we have $\|\sum_{i=1}^{n+1} y_i\|_X \le 2^{n+1}$,

$$\|\sup_{1\le k\le n+1} |\textstyle\sum_{i=1}^{k} y_i|\,\|_X \ge n\cdot 2^{n-2} + 2^{n-1}.$$

These inequalities contradict Doob's inequality in X. So implication 1) $\Longrightarrow$ 2) is proved.

5. Proof of Theorem 2.

We need the following results which are of interest in their own rights.

Lemma 2. *Let X be r.i. space. If two nonnegative functions x and y from X satisfy inequality*

$$(4)\qquad m\{t:\ x(t) > a\lambda,\ y(t) \le b\lambda\} \le c\cdot m\{t:\ x(t) > \lambda\}$$

for some positive constants a,b,c and every $\lambda>0$ and there is $\alpha>1$ such that $a\cdot\|\sigma_{\alpha c}\|_X < 1$ then

$$\|x\|_X \le \frac{a\cdot\|\sigma_{\alpha/(\alpha-1)}\|_X}{b(1-a\|\sigma_{\alpha c}\|_X)}\cdot\|y\|_X.$$

Proof. Using inequality (4) we obtain that

$$\eta_x(\lambda) \le c\eta_x(\frac{\lambda}{a}) + \eta_y(\frac{b\lambda}{a}) \quad \text{for any } \lambda>0.$$

These inequality together with definition of the decreasing rearrangement gives

$$x^*(t) \le \inf\left\{\lambda>0:\ c\eta_x(\frac{\lambda}{a}) + \eta_y(\frac{b\lambda}{a}) < t\right\}$$

$$\le \inf\left\{\lambda>0:\ c\eta_x(\frac{\lambda}{a}) < \frac{t}{\alpha},\ \eta_y(\frac{b\lambda}{a}) < \frac{\alpha-1}{\alpha}\cdot t\right\}$$

$$\le \max\left[\inf\left\{\lambda>0:\ c\eta_x(\frac{\lambda}{a}) < \frac{t}{\alpha}\right\},\ \inf\left\{\lambda>0:\ \eta_y(\frac{b\lambda}{a}) < \frac{\alpha-1}{\alpha}\cdot t\right\}\right]$$

$$= \max\left[a\cdot\inf\left\{\lambda>0:\ c\eta_x(\lambda) < \frac{t}{\alpha}\right\},\ \frac{a}{b}\cdot\inf\left\{\tau>0:\ \eta_y(\tau) < \frac{\alpha-1}{\alpha}\cdot t\right\}\right]$$

$$= \max\left[a\cdot x^*(\frac{t}{c\alpha}),\ \frac{a}{b}\cdot y^*(\frac{\alpha-1}{\alpha}\cdot t)\right] \le a\cdot x^*(\frac{t}{c\alpha}) + \frac{a}{b}\cdot y^*(\frac{\alpha-1}{\alpha}\cdot t)$$

for $t < \min(c\alpha,\ \alpha/(\alpha-1))$. That is why

$$\|x\|_X \le a\cdot\|\sigma_{\alpha c}\|_X\cdot\|x\|_X + \frac{a}{b}\cdot\|\sigma_{\alpha/(\alpha-1)}\|_X\cdot\|y\|_X.$$

So Lemma 2 is proved.

Lemma 3. *Let X be r.i. space which is interpolation space between the spaces L_∞, L_1 and $\lim\limits_{\tau\to 0}\|\sigma_\tau\|_X = 0$. If two nonegative functions x and y from X satisfy inequality*

$$(5)\qquad \int_{\{t:x(t)>\lambda\}} x(s)ds \le \lambda\cdot m\{t:\ x(t)>\lambda\} + \int_{\{t:x(t)>\lambda\}} y(s)ds$$

for every $\lambda>0$ then $\|x\|_X \le C\|y\|_X$, where the constant C depends only on X.

Proof. If $\lambda \to 0$ then inequality (5) gives

$$(6)\qquad \|x\|_{L_1} \le \|y\|_{L_1} \le \|y\|_X.$$

By the ideas of the proof of Lemma 1 inequality (5) implies

$$(7)\qquad \int_0^t x^*(s)ds \le tx^*(t) + \int_0^t y^*(s)ds$$

for any $t \in [0,1]$.

Choose $\tau_o \in (0,1)$ so that $\|\sigma_{\tau_o}\|_X < 1$. Note that

$$\int_0^t x^*(s)ds \geq \int_0^{\tau_o t} x^*(s)ds + (1-\tau_o)tx^*(t)$$

$$= \tau_o \cdot \int_0^t x^*(\tau_o s)ds + (1-\tau_o)tx^*(t)$$

$$= \tau_o \cdot \int_0^t \sigma_{1/\tau_o} x^*(s)ds + (1-\tau_o)tx^*(t),$$

$t \in (0,1)$. The last inequality together with (7) establishes

$$\int_0^t \sigma_{1/\tau_o} x^*(s)ds \leq tx^*(t) + \frac{1}{\tau_o} \int_0^t y^*(s)ds \leq \int_0^t \left[x^*(s)ds + \frac{1}{\tau_o} y^*(s)\right]ds$$

for any $t \in (0,1)$. Using description of interpolation spaces between L_∞ and L_1 ([4],p.130) we obtain

(8) $$\|\sigma_{1/\tau_o} x^*\|_X \leq \|x\|_X + \frac{1}{\tau_o} \cdot \|y\|_X.$$

An elementary calculation shows $\sigma_{\tau_o}\sigma_{1/\tau_o}x^* = x^* \cdot \varkappa_{(0,\tau_o)}$, where $\varkappa_e$ is the indicator function of the set e. Consequently

(9) $$\|x\|_X \leq \|\sigma_{\tau_o}\|_X \cdot \|\sigma_{1/\tau_o} x^*\|_X + x^*(\tau_o).$$

The inequalities (8) and (9) imply

$$(1 - \|\sigma_{\tau_o}\|_X)\ \|x\|_X \leq \frac{\|\sigma_{\tau_o}\|_X}{\tau_o} \|y\|_X + x^*(\tau_o).$$

The inequality (6) gives $x^*(\tau_o) \leq \|x\|_{L_1}/\tau_o \leq \|y\|_X/\tau_o$. Finally, we have shown

$$\|x\|_X < \frac{1+\|\sigma_{\tau_o}\|_X}{(1-\|\sigma_{\tau_o}\|_X)\tau_o} \|y\|_X.$$

Lemma 3 is proved.

Proof of Theorem 2. To prove the sufficiency of the condition $\lim_{\tau\to 0} \|\sigma_\tau\|_X = 0$ we use the method of Theorem 15.1 from paper [1].

Let $\bar{x} = (x_1,x_2,\ldots)$ be martingale relative to $\mathcal{A}_1,\mathcal{A}_2,\ldots$; $\bar{d} = (d_1,d_2,\ldots)$ be martingale-difference sequence for $\bar{x}$. Then $\bar{x} = \bar{y} + \bar{z}$, where $\bar{y}$ and $\bar{z}$ are the martingales defined by

$$y_k = \sum_{i=1}^k a_i = f_1 + \sum_{i=2}^k [f_i - E_{\mathcal{A}_{i-1}} f_i];$$

$$z_k = \sum_{i=1}^k b_i = g_1 + \sum_{i=2}^k [g_i - E_{\mathcal{A}_{i-1}} g_i];$$

with $f_k = d_k \cdot \varkappa_{A_k}$, $g_k = d_k - f_k$ and $A_k = \{t: |d_k(t)| \leq 2e_{k-1}\}$,

$e_k = \max_{1\le i\le k} |d_i|$, $e = \sup_{1\le k<\infty} |d_k|$, $e_0 = 0$, $k = 1,2,\dots$. Therefore

(10) $$\bar{x}^\# \le \bar{y}^\# + \bar{z}^\# \le \bar{y}^\# + \sum_{k=1}^{\infty} |b_k|,$$

(11) $$P\bar{y} \le P\bar{x} + P\bar{z} \le P\bar{x} + \sum_{k=1}^{\infty} |b_k|.$$

Note that $|f_k| \le 2e_{k-1}$, $k = 1,2,\dots$ so that $|a_k| \le 4e_{k-1}$. By the result of [1], p.31 we have that if $a>1$ and $0 < b < a-1$ then

$$m\{t:\ \bar{y}^\#(t)>a\lambda,\ \max(P\bar{y}(t),4e(t))\le b\lambda\} \le \frac{2b^2}{(a-b-1)^2}\, m\{t:\ \bar{y}^\#(t)>\lambda\}$$

for $\lambda>0$. Using Lemma 2 for $a=2$, $\alpha=2$ and b satisfying conditions $0 < b < 1$, $\|\sigma_{4b^2(1-b)^{-2}}\|_X < \frac{1}{2}$ we obtain $\|\bar{y}^\#\|_X \le C(\|P\bar{y}\|_X + 4\|e\|_X)$ with the constant C depending only on X. Therefore, by (10), (11) and elementary inequality $e \le P\bar{x}$ we have

(12) $$\|\bar{x}^\#\|_X \le C(\|P\bar{x}\|_X + \|\sum_{k=1}^{\infty} |b_k|\,\|_X).$$

The martingale $\bar{z}$ is controlled by

(13) $$\sum_{k=1}^{\infty} |b_k| \le \sum_{k=1}^{\infty} |g_k| + \sum_{k=2}^{\infty} E_{\mathcal{A}_{k-1}} |g_k|.$$

On the set $\{t:\ |d_k(t)| > 2e_{k-1}(t)\}$, $k = 2,3,\dots$

$$|d_k(t)| + 2e_{k-1}(t) \le 2|d_k(t)| \le 2e_k.$$

Therefore, $|g_k| \le 2(e_k-e_{k-1})$, $k = 2,3,\dots$ and $\sum_{k=1}^{\infty} |g_k| < 2e$.

It is known [6] that the pair $h_1 = \sum_{k=2}^{\infty} E_{\mathcal{A}_{k-1}}(|g_k|)$, $h_2 = \sum_{k=2}^{\infty} |g_k|$ satisfies

$$\int_{\{t:h_1(t)>\lambda\}} h_1(s)ds \le \lambda\cdot m\{t:\ h_1(t)>\lambda\} + \int_{\{t:h_1(t)>\lambda\}} h_2(s)ds,$$

for $\lambda>0$. Using Lemma 3 and inequality (13) we obtain $\|\sum_{k=1}^{\infty} |b_k|\,\|_X \le C\|e\|_X \le C\|P\bar{x}\|_X$. The last inequality together with (12) gives the proof of the right-hand side of the inequality of Theorem 2.

The proof of the left-hand side has exactly the same pattern.

The proof of the necessity of the condition $\lim_{\tau\to 0} \|\sigma_\tau\|_X = 0$ is obtained by the same method as the proof of the implication 1) $\Longrightarrow$ 2) of Theorem 1.

Thus Theorem 2 is proved.

REFERENCES

[1] D.L.Burkholder, *Distribution function inequalities for martingales*, Ann. Prob., 1973, 1, 1, 19–42.

[2] B.Davis, *On the integrability of the martingale square function*, Israel J. Math. 8 (1970), 187–190.

[3] J.L.Doob, *Stochastic Processes*, New York: Wiley, 1953.

[4] S.G.Krein, Ju. I. Petunin, E.M.Semenov, *Interpolation of Linear Operators*, Trans. Math. Monographs Amer. Math. Soc. 54.

[5] J.Lindenstrauss, L. Tzafriri, *Classical Banach Spaces II, Function Spaces*, Berlin, Springer Verlag 1979.

[6] P.L.Meyer, *Martingales and Stochastic Integrals I*, Berlin, Springer Verlag 1972.

THE SPACE OF CONVEX BODIES AND QUASIDIFFERENTIABLE FUNCTIONS

DIETHARD PALLASCHKE, STEFAN SCHOLTES, RYSZARD URBAŃSKI

Karlsruhe, Federal Republic of Germany; Poznań, Poland

SUMMARY. We consider the vector lattice of convex bodies of a Hausdorff topological real vector space and its relation to quasidifferentiable functions.

1. Let (X,τ) be a Hausdorff topological real vector space. By $\mathcal{K}(X)$ we will denote the collection of all convex bodies in X, i.e. of all nonempty convex compact subsets of X. Endowed with the Minikowski-addition $A+B = \{a+b: a\in A,\ b\in B\}$ and the scalar multiplication defined by $\lambda A = \{\lambda a: a\in A\}$, $\mathcal{K}(X)$ is a convex cone with the property

$$\text{(lc)} \qquad \text{if} \quad A+B \subseteq B+C \quad \text{then} \quad A \subseteq C$$

(cf. [2],[8],[9]). This implies in particular the cancellation property, i.e. if $A+B = B+C$ then $A = C$.

On $\mathcal{K}^2(X) = \mathcal{K}(X)\times\mathcal{K}(X)$ an equivalence relation "$\sim$" is defined by $(A,B)\sim(C,D)$ if $A+D = B+C$. Let us denote the elements of the quotient space $\tilde{X} = \mathcal{K}^2(X)/\sim$ by $[A,B]$, where (A,B) is an arbitrary representant of the class $[A,B]$. On $\tilde{X}$ addition is defined by $[A,B]+[C,D] = [A+C,B+D]$ and scalar multiplication by

$$\lambda[A,B] = \begin{cases} [\lambda A, \lambda B], & \text{if } \lambda\geq 0 \\ [-\lambda B, -\lambda A], & \text{if } \lambda < 0. \end{cases}$$

Endowed with these operations, $(\tilde{X},+,\cdot)$ is a real vector space in which $\mathcal{K}(X)$ is isomorphically embedded as a generating cone. The zero of $\tilde{X}$ is given by the class $\tilde{0} = [A,A]$, where A is an arbitrary element of $\mathcal{K}(X)$.

In [6], Pinsker introduced in $\tilde{X}$ the following partial ordering:

$$[A,B] \leq [C,D] \quad \text{if} \quad A+D \subseteq B+C.$$

With respect to this ordering the space $\tilde{X}$ is a vector lattice and the supremum of two elements is given by

$$\sup([A,B],[C,D]) = [(A+D)\vee(C+B),B+D],$$

where $A\vee B$ denotes the convex hull of the union of A and B. Using induction it is not difficult to show that

$$\sup([A_i,B_i]: 1\le i\le n) = \left[\bigvee_{i=1}^{n}(A_i+\sum_{\substack{j=1\\ j\ne i}}^{n} B_j),\ \sum_{l=1}^{n} B_l\right],$$

where $\bigvee_{i=1}^{n} C_i = \mathrm{conv}(\bigcup_{i=1}^{n} C_i)$ (cf. [3]).

Note that Pinsker assumed X to be locally convex. Using the embedding theorem proved in [9], Pinsker's result can immediately be generalized to arbitrary Hausdorff topological real vector spaces.

In [5] we introduced a partial ordering on $\mathcal{K}^2(X)$ by defining $(A,B)\preceq(C,D)$ if $A\subseteq C$ and $B\subseteq D$. It has been shown that in every equivalence class $[A,B]\in\tilde{X}$ there exists a minimal representant with respect to the ordering "$\preceq$". We now quote a result proved by Pinsker (cf. [6]) which implies the existence of the supremum of two equivalent pairs of sets within this equivalence class.

Lemma. *If $A,B \in \mathcal{K}(X)$ then $(A\vee B)+C = (A+C)\vee(B+C)$.*

Observe that if (X,τ) is locally convex and

$$p_A: X^*\to \mathbb{R}$$
$$p_A(f) = \sup_{x\in A} f(x)$$

denotes the support-function of the set $A \in \mathcal{K}(X)$ (cf. [2]), then the above equality is equivalent to the identity

$$\max(p_A,p_B) + p_C = \max(p_A+p_C,p_B+p_C).$$

Using this Lemma we can prove the following theorem:

Theorem. *If $(A,B)\sim(C,D) \in (\mathcal{K}^2(X),\preceq)$ then*

$$\sup((A,B),(C,D)) = (A\vee C,B\vee D) \in [A,B].$$

Proof. It is easy to see that $\sup((A,B),(C,D)) = (A\vee C,B\vee D)$. We have to show that $(A\vee C,B\vee D)\sim(A,B)$. This follows from the Lemma above, namely

$$A+(B\vee D) = (A+B)\vee(A+D) = (A+B)\vee(B+C) = (A\vee C)+B$$

and hence $(A\vee C,B\vee D) \in [A,B]$.

As defined in [5] a pair $(A,B) \in \mathcal{K}^2(X)$ is called *minimal* if it is a minimal element of $[A,B]$, i.e. for every $(C,D)\sim(A,B)$ the relation $(C,D) \preceq (A,B)$ implies $(C,D) = (A,B)$. Some properties of minimal pairs have been investigated in [5].

Proposition 1. *If $(A,B) \in \mathcal{K}^2(X)$ is a minimal pair and $(A,B) = (C,D) + (E,F)$ then (C,D) is a minimal pair.*

Proof. Let $C+D' = D+C'$ and suppose $(C',D') \preceq (C,D)$. Then $A+D+F+D' = B+C+E+D' = B+E+D+C'$. The cancellation law implies that $A+(F+D') = B+(E+C')$ which means that $(A,B) \sim (E+C',F+D')$. But since $A = E+C \supseteq E+C'$ and $B = F+D \supseteq F+D'$ we can deduce from the minimality of (A,B) that $(C',D') = (C,D)$ and hence that the pair (C,D) is minimal.

The investigations of the ordering relation "$\preceq$" on $\mathcal{K}^2(X)$ leads to the problem of characterizing all minimal pairs in a given equivalence class of $\tilde{X}$. It is easy to see that if $X = \mathbb{R}$ two equivalent minimal pairs are translations of each other, i.e. there exists $x \in X$ such that $(A,B) = (C+\{x\},D+\{x\})$. It is still open problem whether this property is true for more general spaces.

2. In [1] Demyanov introduced the notion of quasidifferentiable functions which is strongly related to the Hörmander– Rådström lattice of convex bodies.

Let $\Omega \subseteq \mathbb{R}^n$ be an open set and let $f\colon \Omega \longrightarrow \mathbb{R}$ be a function. Then f is said to be *quasidifferentiable* in $x_o \in \Omega$ if

i) f is directionally differentiable in x_o in every direction $g \in \mathbb{R}^n$, i.e.

$$df|_{x_o}(g) = \lim_{\substack{\alpha \to 0 \\ \alpha > 0}} \frac{1}{\alpha}(f(x_o+\alpha g) - f(x_o))$$

exists for every $g \in \mathbb{R}^n$,

ii) there exist two convex bodies $A,B \in \mathcal{K}(\mathbb{R}^n)$ such that $df|_{x_o}(g) = p_A(g) - p_B(g)$, where p_A and p_B are the support functions of the sets A and B respectively.

The representation of the directional derivative as the difference of two support functions is obviously not unique, but $p_A - p_B = p_C - p_D$ iff $p_A + p_D = p_B + p_C$ iff $A+D = B+C$ iff $(A,B) \sim (C,D)$. Thus we can assign to f at x_o a class $\mathcal{D}f|_{x_o} \in \tilde{\mathbb{R}}^n$ which is called *the quasidifferentaial of f at x_o*. Note that our definition differs slightly from Demyanov's original definition but is more suited to the context of the Hörmander–Rådström lattice.

For quasidifferentiable functions the following calculus rules hold. Let $f,g\colon \Omega \to \mathbb{R}^n$ be quasidifferentiable in $x_o \in \Omega$ and $\alpha,\beta \in \mathbb{R}$, then

i) $\alpha f+\beta g$ is quasidifferentiable in x_o and

$$\mathcal{D}(\alpha f+\beta g)|_{x_o} = \alpha \mathcal{D}f|_{x_o} + \beta \mathcal{D}g|_{x_o},$$

ii) fg is quasidifferentiable in x_o and

$$\mathcal{D}(fg)|_{x_o} = g(x_o)\mathcal{D}f|_{x_o} + f(x_o)\mathcal{D}g|_{x_o}.$$

Moreover if the functions $f_i\colon \Omega \to \mathbb{R}$ $(i = 1,2,\dots,m)$ are quasidifferentiable in x_o then $f = \max\{f_i\colon 1 \le i \le m\}$ is quasidifferentiable in x_o and $\mathcal{D}f|_{x_o} = \sup\{\mathcal{D}f_i|_{x_o}\colon i \in R(x_o)\}$, where $R(x_o) = \{i\colon f_i(x_o) = f(x_o)\}$ and the supremum is taken in the sense of the Pinsker ordering as described in Chapter 1.

Quasidifferentiable functions play an important role in nonsmooth optimization. The following necessary minimality condition is well known and follows immediately from the remarks above.

Proposition 2. *Let $f\colon \Omega \to \mathbb{R}$ be quasidifferentiable in x_o. If x_o is a local minimizer of the function f then $\mathcal{D}f|_{x_o} \ge \tilde{O}$.*

Proof. If x_o is a local minimizer of f then $df|_{x_o}(g) = p_A(g) - p_B(g) \ge 0$ for every $g \in \mathbb{R}^n$ and hence $B \subseteq A$ which is equivalent to $\mathcal{D}f|_{x_o} \ge \tilde{O}$.

The investigation of minimal pairs $(A,B) \in \mathcal{D}f|_{x_o}$ is of considerable interest from the computational point of view. The affirmative answer to the translation problem of Chapter 1 would imply that we can assign to the function f at x_o a pair $(A,B) \in \mathcal{D}f|_{x_o}$ which is unique up to translation.

REFERENCES

[1] V. F. Demyanov, A. M. Rubinov, *Quasidifferential Calculus*, Optimization Software inc., Springer Verlag 1986.

[2] L. Hörmander, *Sur la fonction d'appui des ensembles convexes dans une espace localement convexe*, Arkiv för Math. (1954), 181–186.

[3] H. Hudzik, J. Musielak, R. Urbański, *Lattice properties of the space of convex closed and bounded subsets of a topological vector space*, Bul. Aca. Polon. Sci. Sér. Sci. Math. 27 (1979), 157–162.

[4] D.Pallaschke, P. Recht, R. Urbański, *On locally Lipschitz quasidifferentiable functions in Banach-Space*, Optimization 17,3, (1986), 287-295.

[5] D.Pallaschke, S.Scholtes, R.Urbański, *On minimal pairs of convex compact sets*, Bul. Aca. Polon. Sci. Sér. Sci. Math. (to appear).

[6] A. G. Pinsker, *The space of convex sets of a locally convex space*, Trudy Len. Inzh.-Ekon. Un-to 63 (1966), 13-17.

[7] L.N.Polyakova, *Necessary conditions for an extremum of quasidifferentiable functions*, Vestnik Leningrad Univ. Math. 13 (1981).

[8] H.Rådström, *An embedding theorem for spaces of convex sets*, Proc. Amer. Math. Soc. 3 (1952).

[9] R.Urbański, *A generalization of the Minikowski-Rådström-Hörmander Theorem*, Bul. Aca. Polon. Sci. Sér. Sci. Math. 288 (1976), 709-715.

ADMISSIBLE WEIGHTS AND WEIGHTED BERGMAN FUNCTIONS

ZBIGNIEW PASTERNAK-WINIARSKI

Warszawa, Poland

1. Introduction.

The weighted Bergman functions have been considered in several mathematical papers (see [1,3,4,5,7,8]). Recently they have also appeared in theoretical physics. Namely the functions of this kind play the fundamental role in the model of quantum theory described in [2] and [6]. The weights of integration considered in [6] are assumed to satisfy the following complex Monge-Ampére equation

$$\det \left[\frac{\partial^2 \log \mu(z)}{\partial z_i \partial \bar{z}_j} \right] = \frac{C}{n!} (-1)^{n(n+1)/2} \mu(z) \Big[K(\mu)\Big](z,z), \tag{1.1}$$

where C is a constant, μ is a weight on an open set $\Omega \subset \mathbb{C}^n$ and $K(\mu)$ is a weighted Bergman function defined by μ. It is clear that for investigation of this equation one can know the properties of the transform $\mu \longmapsto K(\mu)$.

The space $AW(\Omega)$ of all admissible weights on the set Ω, which is a domain of the transform K were introduced and investigated in [7] and [8]. In [8] there are given some sufficient and necessary conditions for a weight to be an admissible weight. Here one can also find an example of a weight, which is not an admissible weight.

In [7] it is proved that the transform K is analytic and the explicite formulas for Taylor series are given.

The main purpose of this study is to prove the following two theorems

THEOREM 1.1. *Let Ω be an open bounded nonempty set in $\mathbb{C}^n$. Then the transform K restricted to the space $L^1AW(\Omega)$ of all Lebesgue integrable admissible weights on Ω is an one-to-one map onto its range* (see Sect.3).

Theorem 1.2. *Under the assumptions of Theorem 1.1, if* $\mu \in L^1AW(\Omega)$ *then the total derivative* $DK(\mu)$ *is an one-to-one operator onto its range* (see Sect.4).

We prove the last theorem constructing the inverse of the operator $DK(\mu)$. The suitable differential (analytic) structures on the domain and on the range of the transform K and other preliminary notions are described in the next section.

2. Preliminaries.

Any positive, Lebesgue measurable real function on an open set $\Omega \subset \mathbb{C}^n$ is called *a weight (of integration) on* Ω. The set of all weights on Ω will be denoted by $W(\Omega)$ (we consider two weights as equivalent if they are equal almost everywhere with respect to the lebesgue measure on Ω). If $\mu \in W(\Omega)$ then $L^2(\Omega,\mu)$ denotes the space of all μ-square integrable functions on Ω equipped with the scalar product

$$\langle f|g\rangle_\mu := \int_\Omega \overline{f(z)}g(z)\mu(z)d^{2n}z,$$

where $f,g \in L^2(\Omega,\mu)$. The space of all holomorphic, μ-square integrable functions on Ω will be denoted by $L^2H(\Omega,\mu)$. It is well known that $L^2(\Omega,\mu)$ is a separable Hilbert space. For any $z\in\Omega$ we define the evaluation functional E_z on $L^2H(\Omega,\mu)$ as follows

$$E_zf := f(z), \qquad f \in L^2H(\Omega,\mu). \tag{2.1}$$

Definition 2.1. A weight $\mu \in W(\Omega)$ is called *an admissible weight* if:

(i) $L^2H(\Omega,\mu)$ is a closed subspace of $L^2(\Omega,\mu)$;

(ii) for any $z\in\Omega$ the evaluation functional E_z is continuous on $L^2H(\Omega,\mu)$.

The set of all admissible weights on Ω will be denoted by $AW(\Omega)$ and the set of all Lebesgue integrable admissible weights on Ω will be denoted by $L^1AW(\Omega)$.

In order to endow $W(\Omega)$ with a suitable differential structure let us consider the set

$$U(\Omega) := \{g\in L^\infty_{\mathbb{R}}(\Omega) : \operatorname*{ess\,inf}_{z\in\Omega} g(z) > 0\}.$$

It is an open subset of the Banach space $L^\infty_{\mathbb{R}}(\Omega)$. Let for any $\mu \in W(\Omega)$ the map $\Phi_\mu : U(\Omega) \to W(\Omega)$ be given by the formula

(2.2) $$[\bar{\Phi}_\mu(g)](z) := g(z)\mu(z), \qquad g \in U(\Omega),\ z \in \Omega.$$

It is proved in [8], Proposition 2.3, that for any $\mu \in W(\Omega)$ the map $\bar{\Phi}_\mu$ is a bijection of $U(\Omega)$ onto the set $U(\Omega,\mu) := \Phi_\mu(U(\Omega))$ the family $\{(\Phi_\mu^{-1}U(\Omega,\mu): \mu \in W(\Omega)\}$ forms an analytic atlas of a Banach manifold on $W(\Omega)$. In this construction $W(\Omega)$ is considered as a topological space endowed with the weakest topology for which maps $\bar{\Phi}_\mu^{-1}$ are all continuous, $\mu \in W(\Omega)$. The sets $AW(\Omega)$ and $L^1AW(\Omega)$ are open submanifolds in $W(\Omega)$.

From now on we will consider $W(\Omega)$, $AW(\Omega)$ and $L^1AW(\Omega)$ as analytic manifolds with the structure described above.

Let $\mu \in AW(\Omega)$ and let for any $z\in\Omega$ the function $e_{z,\mu} \in L^2H(\Omega,\mu)$ represents the evaluation functional E_z (in the sense of the Riesz representation theorem).

Definition 2.2. The function $K(\mu)\colon \Omega \times \Omega \longrightarrow \mathbb{C}$ given by the formula

$$[K(\mu)](z,w) := \overline{e_{z,\mu}(w)} \qquad z,w \in \Omega$$

is called the *μ-Bergman function of the set* Ω (see [1], [7] or [8]).

The above definition can be interpreted as a definition of the transform K with the domain $AW(\Omega)$. To characterize the range of this transform we need the following result.

Theorem 2.1. *If $\mu \in AW(\Omega)$ then*

(i) for any $z,w \in \Omega$

(2.3) $$[K(\mu)](z,w) = \overline{[K(\mu)](w,z)}\ ;$$

(ii) the function $[K(\mu)](z,w)$ is analytic in the real sense, holomorphic in z and antiholomorphic in w;

(iii) $K(\mu)$ is the integral kernel of the operator P_μ of $\langle\cdot|\cdot\rangle_\mu$ - orthogonal projection of $L^2(\Omega,\mu)$ onto $L^2H(\Omega,\mu)$, i.e., for any $z\in\Omega$ and any $f \in L^2(\Omega,\mu)$

(2.4) $$[P_\mu f](z) = \int_\Omega [K(\mu)](z,w)f(w)\mu(w)dw^{2n};$$

(iv) the family $\{[K(\mu)](\cdot,w)\colon w \in \Omega\}$ is linearly dense in $L^2H(\Omega,\mu)$ and for any $z,w \in \Omega$

(2.5) $$\langle[K(\mu)](\cdot,z)\,|\,[K(\mu)](\cdot,w)\rangle_\mu = [K(\mu)](z,w).$$

Proof. For the proof of (i), (ii) and (iii) see [7] Th. 2.1 or [8] Th. 2.1. Point (iv) follows immediately from (iii).

Let $HA(\Omega)$ be the real vector space of all complex-valued functions F on $\Omega \times \Omega$, which are analytic in the real sense, holomorphic with respect to the first n variables, antiholomorphic with respect to the last n variables and satisfy the equality

$$F(z,w) = \overline{F(w,z)} \qquad w,z \in \Omega.$$

We endow $HA(\Omega)$ with a Fréchet space topology given by the family of seminorms $\{\|\cdot\|_X: X \subset \Omega, X\text{-compact}\}$, where

$$\|F\|_X := \sup_{(z,w)\in X\times X} |F(z,w)| \qquad F \in HA(\Omega).$$

In [7], Th. 5.1 it is proved that the transform $K: AW(\Omega) \longrightarrow HA(\Omega)$ is analytic. In Section 4 we need the following formula for the total derivative of K

$$(2.7) \qquad [D_g K(g\mu)](z,w) = -\int_\Omega [K(g\mu)](u,w)h(u)[K(g\mu)](z,u)\mu(u)du^{2n},$$

where $g \in U(\Omega)$, $\mu \in AW(\Omega)$, $h \in L^\infty_{\mathbb{R}}(\Omega)$ and $z,w \in \Omega$ (see [7], Th. 5.1).

3. The proof of Theorem 1.1.

Assume that $\mu,\nu \in L^1AW(\Omega)$ and $K(\mu) = K(\nu) =: K$. Since the family $\{K(\cdot,w): w\in\Omega\}$ is linearly dense in $L^2H(\Omega,\mu)$ and in $L^2H(\Omega,\nu)$ and for any $z,w \in \Omega$

$$\langle K(\cdot,z)|K(\cdot,w)\rangle_\mu = K(z,w) = \langle K(\cdot,z)|K(\cdot,w)\rangle_\nu$$

(see Th. 2.1, (iv) we obtain that $L^2H(\Omega,\mu) = L^2H(\Omega,\nu)$).

Let for any multiindex $I = (i_1,i_2,\dots,i_n)$

$$Q_I(z) := z^I := z_1^{i_1}\cdot z_2^{i_2}\cdot\dots z_n^{i_n},$$

where $z = (z_1,z_2,\dots,z_n) \in \Omega$. Notice that any Q_I is an element of $L^2H(\Omega,\mu)$. If $J = (j_1,j_2,\dots j_n)$ is also a multiindex we have

$$\int_\Omega \overline{Q_I}(z)\, Q_J(z)\mu(z)dz^{2n} = \langle Q_I|Q_J\rangle_\mu = \langle Q_I|Q_J\rangle_\nu = \int_\Omega \overline{Q_I}(z)\, Q_J(z)\nu(z)dz^{2n}.$$

On the other hand any real polynomial on $\mathbb{C}^n$ is a linear combination of components of the form $\overline{Q}_I\cdot Q_J$. This implies that for each polynomial W on Ω

$$(3.1) \qquad \int_\Omega W(z)\mu(z)dz^{2n} = \int_\Omega W(z)\nu(z)dz^{2n}$$

The equality $\mu = \nu$ follows from the suitable version of the Lerch theorem (see [10], XII, 6.4; or [9]).

4. The inverse of $DK(\mu)$.

One can prove Theorem 1.2 using the similar arguments as in the previous section. However we prefer some other way.

Let for any $m \in \mathbb{N}$ and $x, w, z, \in \Omega$

$$(4.1) \qquad g_m(z,w,x) := \left(\frac{m^2}{\Pi}\right)^n e^{-m^2\langle z-x|w-x\rangle}$$

where $\langle x|y\rangle = \bar{x}_1y_1 + \bar{x}_2y_2 + \ldots + \bar{x}_ny_n$ for $x = (x_1,x_2,\ldots,x_n)$, $y = (y_1,y_2,\ldots,y_n) \in \mathbb{C}^n$. The function g_m is antiholomorphic in z, holomorphic in w and real-analytic in x. If for any $x \in \Omega$ we denote

$$(4.2) \qquad g_{m,x}(z) := g_m(z,z,x) \qquad z \in \Omega$$

then the sequence $(g_{m,x})$ approximates the Dirac $\delta(z-x)$ distribution on $\mathbb{C}^n$. More precisely we need the following result

Lemma 4.1. *If Ω is bounded and for each $m \in \mathbb{N}$*

$$(4.3) \qquad [A_m f](x) := \int_\Omega g_{m,x}(z) f(z) dz^{2n} \qquad x \in \Omega,\ f \in L^1(\Omega)$$

then all A_m are linear bounded operators on $L^1(\Omega)$ into $L^1(\Omega)$ and for any $f \in L^1(\Omega)$

$$\lim_{m\to\infty} A_m f = f,$$

i.e., the sequence (A_m) strongly converges to the identity on $L^1(\Omega)$.

For the proof see [9].

Let $\mu \in L^1AW(\Omega)$ and $g \in U(\Omega)$. Denote by Y_μ the range of the operator $D_gK(g\mu)$ (it does not depend on g, see (2.7)); $Y_\mu \subset HA(\Omega)$. Define the maps $R_{g\mu,m}$, $R_{g\mu}$: $Y_\mu \longrightarrow L^1(\Omega)$ as follows $(m \in \mathbb{N})$.

$$(4.4) \qquad [R_{g\mu m}F](x) := \int_\Omega g(z)\mu(z)dz^{2n} \int_\Omega g_m(z,w,x)F(z,w)g(w)\mu(w)dw^{2n},$$

$$(4.5) \qquad R_{g\mu}F := \lim_{m\to\infty} R_{g\mu,m}F,$$

where $\lim_{m\to\infty} R_{g\mu,m}F$ is a limit in $L^1(\Omega)$ and the integral in (4.4) is the iterated integral.

Theorem 4.1. *If $\mu \in L^1AW(\Omega)$ and $g \in U(\Omega)$ then the maps $R_{g\mu,m}$, $m \in \mathbb{N}$ and $R_{g\mu}$ are well defined and*

$$(4.6) \qquad (1/\mu)\, R_{g\mu} = [DgK(g\mu)]^{-1}.$$

Proof. If $F \in Y_\mu$ and $h \in L^\infty_{\mathbb{R}}(\Omega)$ is such that

$F = D_g K(g\mu)h$ then by (2.7) and by Th. 2.7

$$F(z,\cdot) = \overline{P_{g\mu}[K(g\mu)](\cdot,z)\cdot(h/g)} \in L^2(\Omega,\mu).$$

Hence the function $F(z,\cdot)g_m(z,\cdot,x)\cdot g\mu$ is integrable on Ω. Using Th. 2.1.(i) and the fact that $P_{g\mu}$ is a self-adjoint operator we obtain

$$\int_\Omega F(z,w)g_m(z,w,x)g(w)\mu(w)dw^{2n}$$
$$= \langle P_{g\mu}[K(g\mu)](\cdot,z)(h/g)\,|\,g_m(z,\cdot,x)\rangle_{g\mu}$$
$$= \langle [K(g\mu)](\cdot,z)(h/g)\,|\,g_m(z,\cdot,x)\rangle_{g\mu}$$

(4.7) $$= \int_\Omega [K(g\mu)](z,u)g_m(z,u,x)h(u)\mu(u)du^{2n}$$

(see [9]). Since for any $u,x,z \in \Omega$

$$\overline{g_m(z,u,x)} = g_m(u,z,x)$$

we see that bounded operator $B_m(x)$ on $L^2(\Omega,\mu)$ given by the formula

$$[B_m(x)f](z) := \int_\Omega [K(g\mu)](u,z)g_m(u,z,x)f(u)g(u)\mu(u)du^{2n}$$

is self-adjoint. Therefore (see (4.4) and (4.7))

$$[R_{g\mu,m}F](x) = \langle B_m(x)\ (h/g)\ |\ \chi_\Omega\ \rangle_{g\mu} = \langle (h/g)\ |\ B_m(x)\ \chi_\Omega\ \rangle_{g\mu},$$

where χ_Ω is the characteristic function of Ω. Using now the reproducing property of $K(g\mu)$ (see Th. 2.1.(iii) for $f \in L^2H(\Omega,\mu)$) we obtain

$$[B_m(x)\chi_\Omega] = \int_\Omega [K(g\mu)](u,z)g_m(u,z,x)g(z)\mu(z)dz^{2n}$$
$$= g_m(u,u,x) = g_{m,x}(u)$$

(see [9]). This implies

$$[R_{g\mu,m}F](x) = \int_\Omega g_{m,x}(u)h(u)\mu(u)du^{2n} = [A_m(h\mu)](x)$$

and the theorem follows from Lemma 4.1.

Acknowledgement. I would like to express my thanks to M.Horowski, A.Jakubiec, A.Karpio, A.Kryszeń and A.Odzijewicz for friendly help and conversations on the topics of the study. I also wish to thank my wife for her help.

REFERENCES

[1] J.Burbea, P. Masani, *Banach und Hilbert Spaces of Vector-valued Functions*, Research Notes in Mathematics, 90, Pitman, Boston (1984).

[2] M.Horovski, A. Kryszeń, A. Odzijewicz, *Classical and quantum mechanics on the unit ball in* $\mathbb{C}^n$, Rep. Math. Phys. 24 (1986), 351-363.

[3] J.P.Jakobsen, M. Vergne, *Wave and Dirac operators and representations of the conformal group*, J. Funct. Anal. 24 (1977), 52-106.

[4] S.Janson, J. Peetre, R. Rochberg, *Hankel forms and the Fock space*, Uppsala University, Dept. of Math., Report 6 (1986).

[5] T.Mazur, *Canonical isometry on weighted Bergman spaces*, Pacific J. Math. 136 (1989), 303-310.

[6] A.Odzijewicz, *On reproducing kernels and quantization of states*, Commun. Math. Phys. 114 (1988), 577-597.

[7] Z.Pasternak-Winiarski, *On the dependence of the reproducing kernel on the weight of integration*, J. Funct. Anal., to appear.

[8] Z.Pasternak-Winiarski, *On weights which admit the reproducing kernel of Bergman type*, Technical University of Warsaw, Inst. of Math., preprint.

[9] Z.Pasternak-Winiarski, *Admissible weights and weighted Bergman functions*, Technical University of Warsaw, Inst. of Math., preprint.

[10] R.Sikorski, *Funkcje Rzeczywiste*, PWN, Warszawa, T.1 (1958), T.2 (1959).

FOURIER TRANSFORM INEQUALITIES WITH MEASURE WEIGHTS II

John J. Benedetto, Hans P. Heinig

College Park, USA, Hamilton, Canada

Abstract. In [2] Fourier transform inequalities of the form $\|\hat{f}\|_{q,\mu} \le C\|f\|_{p,v}$ were proved for measure weights μ on certain moment subspaces dense in $L_v^p(\mathbb{R}^n)$ for $1 < p \le q < \infty$. In this note we establish such estimates in the index range $0 < q < p < \infty$, $p > 1$. The density theorems established in our previous paper carry over in this case, and, as before, the measure-weight conditions for the validity of the inequalities are computable.

1. Introduction and Notation. We shall prove weighted Fourier transform inequalities of the form

$$\|\hat{f}\|_{q,\mu} \le C\|f\|_{p,v}, \quad 0 < q < p < \infty,\ p > 1, \tag{1.1}$$

where μ is a positive measure and $v \ge 0$ is a weight function. In the index range $1 < p \le q < \infty$, such inequalities were recently obtained in [2]. If μ and v are both weights satisfying certain monotonicity conditions, then the weights have been characterized for which (1.1) holds in the range $1 < p \le q < \infty$ (cf., [3] and the literature cited there). Moreover, sufficient conditions on μ and v ensure (1.1) in the range $0 < q < p < \infty$, $p > 1$ ([3],[9, Cor.4.2]). However, these results typically use rearrangement methods which lead to monotone or radially monotone weights. Our goal is to establish (1.1) with minimal or sometimes no monotonicity assumptions on μ and v. This complements our earlier work and yields additional applications.

The notation here is as in [2]; however, to make this paper reasonably self-contained we restate some of the basic notation. X denotes a locally compact subspace of $\mathbb{R}^n$ and $C_c(X)$ is the vector

---

Research supported by NSERC of Canada No. A4837

space of complex valued continuous functions $f: X \to \mathbb{C}$ with compact support, $\operatorname{supp} f \subseteq X$. A measure ν on X is a linear functional on $C_c(X)$ satisfying $\lim_{j\to\infty} \langle \nu, f_j \rangle = 0$ for every sequence $\{f_j\} \subset C_c(X)$ such that $\lim_{j\to\infty} \|f_j\|_\infty = 0$ and $\operatorname{supp} f_j \subseteq K$, where $K \subseteq X$ is a compact set independent of j. M(X) is the space of measures on X and $M_+(X) = \{\nu \in M(X): \langle \nu, f \rangle \geq 0$ for all f, $0 \leq f \in C_c(X)\}$ is the space of positive measures. As usual, $\langle \nu, f \rangle = \int_X f(t)d\nu(t)$ and if $X = \mathbb{R}^n$ we simply write $\langle \nu, f \rangle = \int f(t)d\nu(t)$.

For $p \in (0,\infty)$, $L^p_{loc}(\mathbb{R}^n)$ denotes the set of functions $f: \mathbb{R}^n \to \mathbb{C}$ such that $|f|^p$ is locally integrable with respect to Lebesgue measure. If $\nu \in M_+(\mathbb{R}^n)$ then $L^p_\nu(\mathbb{R}^n)$ is the set of Borel measurable functions f defined ν-a.e. on $\mathbb{R}^n$ such that $\|f\|_{p,\nu} = \left\{ \int |f(t)|^p d\nu(t) \right\}^{1/p} < \infty$. For $v \geq 0$ a Borel measurable function (not necessarily an element of $L^1_{loc}(\mathbb{R}^n)$) we define

$$L^p_v(\mathbb{R}^n) = \left\{ f: \|f\|^p_{p,v} = \int |f(t)|^p v(t)dt < \infty \right\}.$$

Of course if $v \in L^1_{loc}(\mathbb{R}^n)$ is non-negative then $d\nu(t) = v(t)dt$ defines a positive measure.

The conjugate index p' of p is defined by $p' = p/(p-1)$ even if $0 < p < 1$. If $p = 1$ then $p' = \infty$. The Fourier transform $\hat{f}$ of $f \in L^1(\mathbb{R}^n)$ is defined by

$$\hat{f}(\gamma) = \int e^{-2\pi i \gamma \cdot t} f(t)dt \qquad \gamma \in \hat{\mathbb{R}}^n \equiv \mathbb{R}^n$$

where $\gamma \cdot t = \sum_{j=1}^n \gamma_j t_j$ is the inner product on $\mathbb{R}^n$. Constants are denoted by A, B and C and are not necessarily the same at each occurrence; and χ_E is the characteristic function of the set E.

In the next section we give variants of weighted Hardy inequalities while Section 3 contains the Fourier inequalities with measure weights on $\mathbb{R}$. In Section 4 we prove the higher dimensional extensions and apply them to obtain restriction theorems.

2. Hardy's Iequality with Index p,q in the Range $0 < q < p < \infty$, $p > 1$.

The following result is due to Sinnamon [9]. If μ and v are weight functions and $1 < q < p < \infty$ the result is due to Maz'ja and

Rozin [6] and for $0 < q < p$, $p > 1$ to Sawyer [8] with different (but equivalent) weight conditions. The corresponding result for the case $1 < p \le q < \infty$ is Theorem 1 of [2].

Theorem 2.1. *Given $v \in L^1_{loc}(\mathbb{R}^n)$, $v > 0$ a.e. and $\mu \in M_+(\mathbb{R})$. Let $0 < q < p < \infty$, $p > 1$ and $v^{1-p'} \in L^1_{loc}(\mathbb{R})$.*

(a) There is a constant $C > 0$, such that for all $h \in L^1_{loc}(\mathbb{R})$

$$(2.1) \qquad \left\{ \int_{[0,\infty)} \left| \int_0^\gamma h(t)dt \right|^q d\mu(\gamma) \right\}^{1/q} \le C \cdot \left\{ \int_0^\infty |h(t)|^p v(t)dt \right\}^{1/p}$$

if and only if

$$(2.2) \qquad B \equiv \left\{ \int_0^\infty \left[\int_{[y,\infty)} d\mu(\gamma) \right]^{r/q} \left[\int_0^y v(t)^{1-p'}dt \right]^{r/q'} v(y)^{1-p'}dy \right\}^{1/r} < \infty$$

where $1/r = 1/q - 1/p$. Moreover, if C is the best constant in (2.1) and $q > 1$, then

$$\left(\frac{p-q}{p-1}\right)^{1/q'} q^{1/q} \cdot B \le C \le B \cdot q^{1/q}(p')^{1/q'}.$$

(b) There is a constant $C > 0$, such that for all $h \in L^1_{loc}(\mathbb{R})$

$$(2.3) \qquad \left\{ \int_{[0,\infty)} \left| \int_\gamma^\infty h(t)dt \right|^q d\mu(\gamma) \right\}^{1/q} \le C \cdot \left\{ \int_0^\infty |h(t)|^p v(t)dt \right\}^{1/p}$$

if and only if

$$(2.4) \qquad B \equiv \left\{ \int_0^\infty \left[\int_{[0,y]} d\mu(\gamma) \right]^{r/q} \left[\int_y^\infty v(t)^{1-p'}dt \right]^{r/q'} v(y)^{1-p'}dy \right\}^{1/r} < \infty.$$

As before, if $q > 1$ and C is the best constant in (2.3) then the relation between B and C given above holds.

As a consequence of Theorem 2.1 we now can prove the following propositions required in Section 3.

Proposition 2.2. *Suppose $v \in L^1_{loc}(\mathbb{R})$, $v > 0$ a.e. and $\mu \in M_+(\hat{\mathbb{R}})$. Let $1 < q < p < \infty$ and $v^{1-p'} \in L^1_{loc}(\mathbb{R})$. Then for all $h \in L^1_{loc}(\mathbb{R})$, $h \ge 0$ the condition,*

$$(2.5) \quad B \equiv \left\{ \int_{\infty}^{\infty} \left[\int_{|\gamma| < |y|} d\mu(\gamma) \right]^{\frac{r}{q}} \left[\int_{|x| > |y|} v(x)^{1-p'} dx \right]^{\frac{r}{q'}} v(y)^{1-p'}dy \right\}^{\frac{1}{r}} < \infty.$$

where $1/r = 1/q - 1/p$, implies

$$(2.6) \qquad \left\{ \int \left[\int_{|t| > |\gamma|} h(t)dt \right]^q d\mu(\gamma) \right\}^{1/q} \le C \cdot \left\{ \int h(t)^p v(t)dt \right\}^{1/p}.$$

If v is an even weight function then the result also holds in case $0 < q < p$, $p > 1$.

Proof. If $\mu = a\delta$, $a > 0$, then (2.5) implies $v^{1-p'} \in L^1(\mathbb{R})$ and since the left side of (2.6) takes the form $a^{1/q}\int h(t)dt$ the result follows from Hölder's inequality. Next, if $\mu(\{0\}) = 0$, then with $m(\gamma) = \mu(\gamma) + \mu(-\gamma)$

$$\left\{ \int\left[\int_{|t|>|\gamma|} h(t)dt\right]^q d\mu(\gamma) \right\}^{1/q}$$

$$= \left\{ \int_{(0,\infty)} \left[\int_{|t|>\gamma} h(t)dt\right]^q d\mu(\gamma) + \int_{(-\infty,0)} \left[\int_{|t|>-\gamma} h(t)dt\right]^q d\mu(\gamma) \right\}^{1/q}$$

$$= \left\{ \int_{(0,\infty)} \left[\int_{\gamma}^{\infty} h(t)dt + \int_{-\infty}^{-\gamma} h(t)dt\right]^q dm(\gamma) \right\}^{1/q}$$

$$\leq C\cdot\left\{ \int_{(0,\infty)} \left[\int_{\gamma}^{\infty} h(t)dt \right]^q dm(\gamma) \right\}^{1/q} + C\cdot\left\{ \int_{(0,\infty)} \left[\int_{-\infty}^{-\gamma} h(t)dt\right]^q dm(\gamma) \right\}^{1/q}$$

$$\equiv C\cdot[I_1 + I_2].$$

Here we used Minikowski's inequality if $q \geq 1$, so that in this case $C = 1$. If $q < 1$ the estimate is trivial with $C = 2^{(1/q)-1}$. By Theorem 2.1b,

$$I_1 \leq C\left[\int_0^{\infty} h(t)^p v(t)dt\right]^{1/p}$$

if and only if

$$\left\{\int_0^{\infty}\left[\int_{|\gamma|<y} d\mu(\gamma)\right]^{\frac{r}{q}} \left[\int_y^{\infty} v(t)^{1-p'}dt\right]^{\frac{r}{q'}} v(y)^{1-p'}dy\right\}^{\frac{1}{r}} \equiv B_+ < \infty;$$

and also

$$I_2 \leq C\left[\int_0^{\infty} h(-t)^p v(-t)dt\right]^{1/p} = C\left[\int_{-\infty}^{0} h(t)^p v(t)dt\right]^{1/p}$$

if and only if

$$\left\{\int_0^{\infty}\left[\int_{|\gamma|<y} d\mu(\gamma)\right]^{\frac{r}{q}} \left[\int_y^{\infty} v(-t)^{1-p'}dt\right]^{\frac{r}{q'}} v(-y)^{1-p'}dy\right\}^{\frac{1}{r}}$$

$$= \left\{\int_{\infty}^{0}\left[\int_{|\gamma|<-y} d\mu(\gamma)\right]^{\frac{r}{q}} \left[\int_{\infty}^{y} v(t)^{1-p'}dt\right]^{\frac{r}{q'}} v(y)^{1-p'}dy\right\}^{\frac{1}{r}} \equiv B_- < \infty.$$

(2.5) implies that both B_+ and B_- finite provided $q \geq 1$, since then $q' > 0$ (in fact, $q' > 1$). If v is an even function then $B < \infty$ in

(2.5) implies B_+ and B_- finite even in the case $q' < 0$, i.e., $0 < q < 1$. Finally, the result follows since

$$\left[\int_0^\infty h(t)^p v(t)dt\right]^{1/p} + \left[\int_\infty^0 h(t)^p v(t)dt\right]^{1/p} \leq 2\left[\int h(t)^p v(t)dt\right]^{1/p}.$$

Proposition 2.3. *Suppose* $v \in L^1_{loc}(\mathbb{R})$, $v > 0$ *a.e. and* $\mu \in M_+(\hat{\mathbb{R}})$. *Let* $1 < q < p < \infty$, $1/r = 1/q - 1/p$ *and* $v^{1-p'} \in L^1_{loc}(\mathbb{R})$. *Then for all* $h \in L^1_{loc}(\mathbb{R})$, $h \geq 0$ *the condition,*

$$(2.7)\qquad \left\{\int\left[\int_{|\gamma|>|y|} d\mu(\gamma)\right]^{\frac{r}{q}}\left[\int_{|x|<|y|} v(x)^{1-p'}dx\right]^{\frac{r}{q'}} v(y)^{1-p'}dy\right\}^{\frac{1}{r}} \equiv B < \infty.$$

where $1/r = 1/q - 1/p$, *implies*

$$(2.6)\qquad \left\{\int\left[\int_{|t|<|\gamma|} h(t)dt\right]^q d\mu(\gamma)\right\}^{1/q} \leq C\cdot\left\{\int h(t)^p v(t)dt\right\}^{1/p}.$$

If v *is an even weight function then the result holds also for* $0 < q < p$, $p > 1$.

The proof of this result is quite similar to that of Proposition 2.2, only now Theorem 2.1a must be applied. We omit the details.

3. The Fourier Transform Inequality on $\mathbb{R}$.

For this and the next section we need the following definition of the *moment subspaces*:

$$M_n = \{f \in L^1(\mathbb{R}^n): \text{supp } f \text{ is compact and } \hat{f}(0) = 0\}.$$

Theorem 3.1. *Given* $v \in L^1_{loc}(\mathbb{R})$, $v > 0$ *a.e. and* $\mu \in M_+(\hat{\mathbb{R}})$. *Let* $1 < q < p < \infty$, $v^{1-p'} \in L^1_{loc}(\mathbb{R}\setminus[-y,y])$ *for each* $y > 0$, *and*

$$(3.1)\qquad B_1 = \left\{\int\left[\int_{|\gamma|<1/|2\pi s|} |\gamma|^q d\mu(\gamma)\right]^{\frac{r}{q}}\left[\int_{|x|<|s|} |x|^{p'} v(x)^{1-p'}dx\right]^{\frac{r}{q'}} \times |s|^{p'} v(s)^{1-p'}ds\right\}^{\frac{1}{r}} < \infty$$

and

$$(3.2)\qquad B_2 = \left\{\int\left[\int_{|\gamma|>1/|\pi s|} d\mu(\gamma)\right]^{\frac{r}{q}} \left[\int_{|s|<|x|} \upsilon(x)^{1-p'}dx\right]^{\frac{r}{q'}} \times \upsilon(s)^{1-p'}ds\right\}^{\frac{1}{r}} < \infty,$$

where $1/r = 1/q - 1/p$*. Then there is a* $C > 0$*, such that for all* $f \in M_1 \cap L^p_v(\mathbb{R})$, $\|\hat{f}\|_{q,\mu} \le C\|f\|_{p,v}$*. If* υ *is radial, then the result also holds for* $0 < q < p$, $p > 1$.

Proof. Since $f \in M_1$ it follows that

$$|\hat{f}(\gamma)| = |\int (e^{-2\pi i\gamma t}-1)\, f(t)dt| = |-2i\int e^{-\pi i t\gamma}\left[\frac{\sin \pi t\gamma}{\pi t\gamma}\right]\pi t\gamma\, f(t)dt|$$

$$\le 2\pi|\gamma| \int_{\pi|t\gamma|\le 1} |tf(t)|dt + 2\int_{\pi|t\gamma|>1} |f(t)|dt$$

$$= 2|\gamma|/\pi \int_{|\gamma|\le|x|} |x^{-3}f(1/(\pi x))|dx + 2/\pi \int_{|\gamma|>|x|} |x^{-2}f(1/(\pi x))|dx,$$

where $x = 1/(\pi t)$. Now, by Minikowski's inequality if $q \ge 1$ and trivially otherwise we obtain

$$\left\{\int|\hat{f}(\gamma)^q d\mu(\gamma)\right\}^{\frac{1}{q}} \le C\left\{\int|\gamma|^q\left[\int_{|\gamma|\le|x|} |x^{-3}f(1/(\pi x))|dx\right]^q d\mu(\gamma)\right\}^{\frac{1}{q}}$$

$$+ C\left\{\int\left[\int_{|\gamma|>|x|} |x^{-2}f(1/(\pi x))|dx\right]^q d\mu(\gamma)\right\}^{\frac{1}{q}} \equiv C[J_1+J_2].$$

By Proposition 2.2, with $h(x) = |x^{-3}f(1/(\pi x)|$ and $v(x)$ and $d\mu(\gamma)$ replaced by $|x|^{3p-2}v(1/(\pi x))$ and $|\gamma|^q d\mu(\gamma)$, respectively, we obtain

$$J_1 \le C\left\{\int|x^{-3}f(1/(\pi x))|^p|x|^{3p-2}v(1/(\pi x))dx\right\}^{\frac{1}{p}} = C\left\{\int|f(y)|^p v(y)dy\right\}^{\frac{1}{p}}$$

whenever

$$\left\{\int\left[\int_{|\gamma|<|y|}|\gamma|^q d\mu(\gamma)\right]^{\frac{r}{q}} \left[\int_{|t|\ge|y|} (|t|^{3p-2}v(1/(\pi t)))^{1-p'}dt\right]^{\frac{r}{q'}}\times \left[|t|^{3p-2}v(1/(\pi y))\right]^{1-p'}dy\right\}^{\frac{1}{r}}$$

$$= C\left\{\int\left[\int_{|\gamma|<1/|s\pi|}|\gamma|^q d\mu(\gamma)\right]^{\frac{r}{q}} \left[\int_{1/|\pi x|\ge 1/|\pi s|}\left(|x|^{2-3p}v(x)\right)^{1-p'}x^{-2}dx\right]^{\frac{r}{q'}}\times \left[|s|^{2-3p}v(s)\right]^{1-p'}s^{-2}ds\right\}^{\frac{1}{r}}$$

$$= C\left\{\int\left[\int_{|\gamma|<1/|s\pi|}|\gamma|^{q}d\mu(\gamma)\right]^{\frac{r}{q}}\left[\int_{|s|\geq|x|}|x|^{p'}v(x)^{1-p'}dx\right]^{\frac{r}{q'}}\times\right.$$

$$\left.|s|^{p'}v(s)^{1-p'}ds\right\}^{\frac{1}{r}} \equiv CB_1$$

is finite. This is the case by (3.1).

To estimate J_2 we apply Proposition 2.3 with $h(x) = |x^{-2}f(1/(\pi x)|$ and $v(x)$ replaced by $x^{2p-2}v(1/(\pi x))$ to obtain

$$J_2 \le C\left\{\int|x^{-2}f(1/(\pi x))|^{p}|x|^{2p-2}v(1/(\pi x))dx\right\}^{\frac{1}{p}} = C\left\{\int|f(y)|^{p}v(y)dy\right\}^{\frac{1}{p}}$$

whenever

$$\left\{\int\left[\int_{|y|<|\gamma|}d\mu(\gamma)\right]^{\frac{r}{q}}\left[\int_{|t|<|y|}\left(|t|^{2p-2}v(1/(\pi t))\right)^{1-p'}dt\right]^{\frac{r}{q'}}\times\right.$$

$$\left.\left(|y|^{2p-2}v(1/(\pi y))\right)^{1-p'}dy\right\}^{\frac{1}{r}}$$

$$= C\left\{\int\left[\int_{1/|\pi s|<|\gamma|}d\mu(\gamma)\right]^{\frac{r}{q}}\left[\int_{|s|<|x|}v(x)^{1-p'}dx\right]^{\frac{r}{q'}}v(s)^{1-p'}ds\right\}^{\frac{1}{r}} \equiv CB_2$$

is finite. This is the case by (3.2). Hence the result follows.

From this result and a minor modification of the density theorem [2, Theorem 2.2] we obtain the following from a standard density argument.

Theorem 3.2. *Given a radial weight $v \in L^1_{loc}(\mathbb{R})$, $v > 0$ a.e. and $\mu \in M_+(\hat{\mathbb{R}})$. Suppose $0 < q < p < \infty$, $p > 1$, and*

$$v^{1-p'} \in L^1_{loc}(\mathbb{R}\setminus[-y,y])\setminus L^1(\mathbb{R})$$

for each $y > 0$, and let (3.1) and (3.2) be satisfied.

(a) If $f \in L^p_v(\mathbb{R})$, then $\lim_{j\to\infty}\|f_j-f\|_{p,v} = 0$ for a sequence $(f_j) \subseteq M_1 \cap L^p_v$ and $(\hat{f}_j)$ converges in $L^q_\mu(\hat{\mathbb{R}})$ to a function $\hat{f} \in L^q_\mu(\hat{\mathbb{R}})$. $\hat{f}$ is independent of (f_j) and is called the Fourier transform of f.

(b) There is a constant $C > 0$, such that for all $f \in L^p_v(\mathbb{R})$, $\|\hat{f}\|_{q,\mu} \le C\|f\|_{p,v}$.

Remark 3.3. i) If $1 < q < p < \infty$ the condition that v is radial in Theorem 3.2 is not required.

ii) If $d\mu(\gamma) = u(\gamma)d\gamma$, where $u(\gamma) = e^{-|\gamma|}$ and $v(x) = e^{|x|}$ then Theorem 3.1 holds. Similarly the result holds with $u(\gamma) = \chi_{(-1,1)}(\gamma)$ and $v(x)^{1-p'} = |x|^{\alpha-1}$ where $1/2 < q < 1 < p$ and $-1 < \alpha/p' < 1/q - 2$.

4. The Fourier Transform Inequality in $\mathbb{R}^n$.

In this section we require measures μ on $\mathbb{R}^n$. For a discussion we refer to [2]. Here we only note that if $\mu \in M(\hat{\mathbb{R}}^n)$ and $\mu(\{0\}) = 0$ is radial, then there exists a unique measure $\nu \in M(0,\infty)$ such that for all radial $\varphi \in C_c(\hat{\mathbb{R}}^n)$,

$$\langle\mu,\varphi\rangle = \omega_{n-1}\int_{(0,\infty)} \rho^{n-1}\varphi(\rho)d\nu(\rho)$$

where $\omega_{n-1} = 2\pi^{n/2}/\Gamma(n/2)$ is the surface area of the unit sphere Σ_{n-1} of $\hat{\mathbb{R}}^n$ ([2, Proposition 3.4]). Moreover this formula extends to all radial elements of $L^1_\mu(\mathbb{R}^n)$ by Lebesgue's theorem.

The main result of this section is the following:

Theorem 4.1. *Given a radial weight $v \in L^1_{loc}(\mathbb{R}^n)$, $v > 0$ a.e. and a radial measure $\mu \in M_+(\hat{\mathbb{R}}^n)$ satisfying $\mu(\{0\}) = 0$. Let $\nu \in M(0,\infty)$ be the measure obtained from μ as above; and assume $0 < q < p < \infty$, $p > 1$, $1/r = 1/q - 1/p$ and $v^{1-p'} \in L^1_{loc}(\mathbb{R}^n\setminus B(0,y))$ for each $y > 0$. Here $B(0,y)$ is the ball in $\mathbb{R}^n$ centered at zero with radius y. If*

$$B_1 = \left\{\int\left[\int_{(0,1/y)} \gamma^{n-1+q}d\nu(\gamma/\pi)\right]^{\frac{r}{q}} \times \left[\int_0^y t^{n-1+p'}v(t)^{1-p'}dt\right]^{\frac{r}{q'}} y^{n-1+p'}v(y)^{1-p'}dy\right\}^{\frac{1}{r}} < \infty$$

and

$$B_2 = \left\{\int\left[\int_{(1/y,\infty)} \gamma^{n-1}d\nu(\gamma/\pi)\right]^{\frac{r}{q}} \times \left[\int_y^\infty t^{n-1}v(t)^{1-p'}dt\right]^{\frac{r}{q'}} y^{n-1}v(y)^{1-p'}dy\right\}^{\frac{1}{r}} < \infty$$

then for all $f \in M_n \cap L^p_v(\mathbb{R}^n)$, $\|\hat{f}\|_{q,\mu} \le C\|f\|_{p,v}$. Moreover, the *constant C may be taken of the form $C = A[B_1+B_2]$ where $A = A(p,q,n)$ and the condition that v is radial may be omitted if $q > 1$.*

Proof. Since $f \in M_n$ we can write as in the proof of Theorem 3.1 of [2] that

$$|\hat{f}(\gamma)| = |-2i\int e^{-\pi i t\cdot\gamma}\sin(\pi t\cdot\gamma)f(t)dt|$$

$$\leq 2\pi|\gamma|\int_{|t|\leq 1/|\pi\gamma|}|t||f(t)|dt + 2\int_{|t|>1/|\pi\gamma|}|f(t)|dt$$

so that as before

$$\left\{\int|\hat{f}(\gamma)|^q d\mu(\gamma)\right\}^{\frac{1}{q}} \leq C\left\{\int\left[\int_{|t|\leq 1/|\pi\gamma|}|t||f(t)|dt\right]^q|\gamma|^q d\mu(\gamma)\right\}^{\frac{1}{q}}$$

$$+ C\left\{\int\left[\int_{|t|>1/|\pi\gamma|}|f(t)|dt\right]^q d\mu(\gamma)\right\}^{\frac{1}{q}} \equiv C[J_1+J_2].$$

Let $y = \pi|\gamma|$, then the integrand of J_1 takes the form

$$\int_{|t|\leq 1/|\pi\gamma|}|t||f(t)|dt = \int_{\Sigma_{n-1}}\int_0^{1/y}r^n|f(r\theta)|dr\,d\sigma_{n-1}(\theta)$$

$$= \int_{\Sigma_{n-1}}\int_y^{\infty}s^{-n-2}|f(\theta/s)|ds\,d\sigma_{n-1}(\theta)$$

$$= \int_y^{\infty}r^{-n-2}\left[\int_{\Sigma_{n-1}}|f(\theta/r)|d\sigma_{n-1}(\theta)\right]dr,$$

where σ_{n-1} is (n-1)-dimensional area measure on $\hat{\mathbb{R}}^n$. Therefore, writing μ in terms of v and $h(r) = r^{-n-2}\left[\int_{\Sigma_{n-1}}|f(\theta/r)|d\sigma_{n-1}(\theta)\right]$ we get

$$J_1 = \left\{\int\left[\int_{\pi|\gamma|}^{\infty}r^{-n-2}\left[\int_{\Sigma_{n-1}}|f(\theta/r)|d\sigma_{n-1}(\theta)\right]dr\right]^q|\gamma|^q d\mu(\gamma)\right\}^{\frac{1}{q}}$$

$$= \left\{\omega_{n-1}\int_{(0,\infty)}s^{n-1+q}\left[\int_{\pi s}^{\infty}h(r)dr\right]^q d\nu(s)\right\}^{\frac{1}{q}}.$$

By Theorem 2.1b, with $d\mu$ replaced by $s^{n-1+q}d\nu(s)$ and a change of variables, we have

$$J_1 = C_1\omega_{n-1}^{1/q}\,\pi^{-(n-1+q)/q}\left\{\int_0^{\infty}|h(r)|^p V(r)dr\right\}^{\frac{1}{p}}$$

whenever

$$(4.1)\qquad \left\{\int_0^\infty \left[\int_{(0,y)} \gamma^{n-1+q} d\nu(\gamma/\pi)\right]^{\frac{r}{q}} \left[\int_y^\infty V(t)^{1-p'}dt\right]^{\frac{r}{q'}} V(y)^{1-p'}dy\right\}^{\frac{1}{r}} < \infty.$$

Choosing $V(s) = s^{-n-1+p(n+2)}v(1/s)$ it follows from Hölder's inequality that

$$\left\{\int_0^\infty |h(r)|^p V(r)dr\right\}^{\frac{1}{p}} = \left\{\int_0^\infty r^{-(n+2)p}\left[\int_{\Sigma_{n-1}} |f(\theta/r)|d\sigma_{n-1}(\theta)\right]^p V(r)dr\right\}^{\frac{1}{p}}$$

$$= \left\{\int_0^\infty s^{(n+2)p-2}\left[\int_{\Sigma_{n-1}} |f(s\theta)|d\sigma_{n-1}(\theta)\right]^p V(1/s)ds\right\}^{\frac{1}{p}}$$

$$\le \omega_{n-1}^{1/p'}\left\{\int_0^\infty s^{(n+2)p-2}V(1/s)\left[\int_{\Sigma_{n-1}} |f(s\theta)|^p d\sigma_{n-1}(\theta)\right]ds\right\}^{\frac{1}{p}}$$

$$= \omega_{n-1}^{1/p'}\left\{\int_{\Sigma_{n-1}}\int_0^\infty v(s)s^{n-1}|f(s\theta)|^p ds d\sigma_{n-1}(\theta)\right\}^{\frac{1}{p}} = \omega_{n-1}^{1/p'}\|f\|_{p,v}.$$

On substituting $V(s)$ into (4.1) we obtain the constant B_1 of the hypothesis.

In a similar way one estimates $J_2 \le C\omega_{n-1}^{1/p'}B_2\|f\|_{p,v}$, only now Theorem 2.1a is utilized and the finiteness of B_2 is required. We omit the details.

We note that the density theorem [2, Theorem 4.2], properly modified, yields a higher dimensional result corresponding to Theorem 3.2 (cf., Theorem 4.3 of [2] for the case $1 < p \le q < \infty$).

Our final result is a weighted spherical restriction theorem.

THEOREM 4.2. *Given a radial $v \in L^1_{loc}(\mathbb{R}^n)$, $v > 0$ a.e., $0 < q < p < \infty$, $p > 1$, and $v^{1-p'} \in L^1_{loc}(\mathbb{R}^n\setminus B(0,y))$ for all $y > 0$. Then for all $f \in M_n \cap L^p_v(\mathbb{R}^n)$ and $\rho > 0$*

$$\left\{\int_{\Sigma_{n-1}(\rho)} |\hat{f}(\gamma)|^q d\sigma_{n-1}(\gamma)\right\}^{1/q} \le C(p,q,\rho)\|f\|_{p,v}$$

where $\Sigma_{n-1}(\rho) = \{\gamma \in \hat{\mathbb{R}}^n: |\gamma| = \rho\}$ and σ_{n-1} is $(n-1)$-dimensional area measure on $\hat{\mathbb{R}}^n$. The constant $C(p,q,\rho)$ is of the form $C(p,q)(A_1+A_2)$ where

$$A_1 = \rho^{(n-1+q)/q}\left(\int_c^{1/(\rho\pi)} t^{n-1+p'}v(t)^{1-p'}dt\right)^{1/p'}$$

and

$$A_2 = \rho^{(n-1)/q}\left(\int_{1/(\rho\pi)}^{\infty} t^{n-1}v(t)^{1-p'}dt\right)^{1/p'}.$$

Proof. The result is an application of Theorem 4.1. Let μ_ρ be the restriction of σ_{n-1} to $\Sigma_{n-1}(\rho)$; then the measure ν on $(0,\infty)$ corresponding to μ_ρ is δ_ρ (cf., [2, Definition 5.1]). If $\nu = \delta_\rho$, then for all y such that $1/y < \rho\pi$, $\int_{(0,1/y)} \gamma^{n-1+q}d\nu(\gamma/\pi) = 0$, and for all y such that $1/y > \rho\pi$, $\int_{(1/y,\infty)} \gamma^{n-1}d\nu(\gamma/\pi) = 0$. Hence the conditions B_1 and B_2 take the form

$$B_1 = \left\{\int_c^{1/(\rho\pi)}\left[\int_{(0,1/y)} \gamma^{n-1+q}d\nu(\gamma/\pi)\right]^{r/q}\left[\int_0^y t^{n-1+p'}v(t)^{1-p'}dt\right]^{r/q'}\times y^{n-1+p'}v(y)^{1-p'}dy\right\}^{1/r}$$

$$\leq \left(\int_{(0,\infty)} \gamma^{n-1+q}d\nu(\gamma/\pi)\right)^{1/q}\left\{\int_0^{1/(\rho\pi)}\left[\int_0^y t^{n-1+p'}v(t)^{1-p'}dt\right]^{r/q'}\times y^{n-1+p'}v(y)^{1-p'}dy\right\}^{1/r}$$

$$= (\rho\pi)^{(n-1+q)/q}(p'/r)^{1/r}\left(\int_0^{1/(\rho\pi)} t^{n-1+p'}v(t)^{1-p'}dt\right)^{1/p'}$$

$$= (p'/r)^{1/r}\pi^{(n-1+q)/q}A_1$$

and

$$B_2 = \left\{\int_{1/(\rho\pi)}^{\infty}\left[\int_{(1/y,\infty)} \gamma^{n-1}d\nu(\gamma/\pi)\right]^{r/q}\left[\int_y^{\infty} t^{n-1}v(t)^{1-p'}dt\right]^{r/q'}\times y^{n-1}v(y)^{1-p'}dy\right\}^{1/r}$$

$$\leq \left(\int_{(0,\infty)} \gamma^{n-1}d\nu(\gamma/\pi)\right)^{1/q}\left\{\int_{1/(\rho\pi)}^{\infty}\left[\int_y^{\infty} t^{n-1}v(t)^{1-p'}dt\right]^{r/q'}\times y^{n-1}v(y)^{1-p'}dy\right\}^{1/r}$$

$$= (\rho\pi)^{(n-1)/q}(p'/r)^{1/r}\left(\int_{1/(\rho\pi)}^{\infty} t^{n-1}v(t)^{1-p'}dt\right)^{1/p'}$$

$$= (p'/r)^{1/r}\pi^{(n-1)/q}A_2.$$

In each case we integrated and used the fact that $1/r = 1/q - 1/p$. The result now follows from Theorem 4.1.

REFERENCES

[1] JOHN J. BENEDETTO, *Uncertainty principle inequalities and spectrum estimation*, Fourier Analysis and its Applications (J. Byrnes, Ed.) NATO Advanced Study Inst., Il Ciocco (1989) (to appear).

[2] JOHN J. BENEDETTO, HANS P. HEINIG, *Fourier transform inequalities with measure weights*, Advances in Math. (to appear).

[3] JOHN J. BENEDETTO, HANS P. HEINIG, R. JOHNSON, *Weighted Hardy spaces and the Laplace transform II*, Math. Nachr. 132 (1987), 29–55.

[4] JOHN J. BENEDETTO, HANS P. HEINIG, R. JOHNSON, *Fourier inequalities with A_p-weights*, General Inequalities 5 (W.Walter Ed.) ISNM 80 (1987), 217–232.

[5] HANS P. HEINIG, GORDON J. SINNAMON, *Fourier inequalities and integral representations of functions in weighted Bergman spaces over tube domains*, Indiana Univ. Math. J. 38(3) (1989) (to appear).

[6] W. MAZ'JA, *Einbettungssätze für Sobolevsche Räume, Teil 1*, Teubner-Texte zur Math., Teubner, Leipzig, 1979.

[7] C. SADOSKY, R. WHEEDEN, *Some weighted norm inequalities for the Fourier transform of functions with vanishing moments*, Trans. Amer. Math. Soc. 300 (1987), 521–533.

[8] E. T. SAWYER, *Weighted Lebesgue and Lorentz norm inequalities for the Hardy operator*, Trans. Amer. Math. Soc. 281(1) (1984), 329–337.

[9] GORDON J. SINNAMON, *Operators on Lebesgue spaces with general measures*, Ph.D. Thesis, McMaster Univ., Hamilton, Can. 1987.

APPROXIMATION BY BOUNDED PSEUDO-POLYNOMIALS

Claudia Cottin

Duisburg, Federal Republic of Germany

Abstract. In the recent literature, pseudo-polynomials arise in the context of approximation in tensor product spaces and of blending approximation in Computer Aided Geometric Design. Little seems to be known about uniform pseudo-polynomial approximation of discontinuous functions. In this paper, we characterize the closure of the spaces of bounded trigonometric and algebraic pseudo-polynomials as certain spaces of so-called B-continuous functions, and give direct and inverse theorems for the degree of approximation in this setting.

1. Bounded pseudo-polynomials and B-continuous functions.

In the first part of this paper we want to determine the classes of bivariate real valued functions $f(x,y)$ which can be approximated arbitrarily well in the uniform norm by *algebraic or trigonometric pseudo-polynomials*; i.e., by approximants of the form

$$\sum_{i=0}^{m} A_i(y)\cdot x^i + \sum_{j=0}^{n} B_j(x)\cdot y^j ,$$

(1) $$\text{or} \quad \sum_{i=0}^{2m} A_i(y)\cdot \tau_i(x) + \sum_{j=0}^{2n} B_j(x)\cdot \tau_j(y),$$

$$\tau_k(z) = \begin{cases} \sin\frac{k+1}{2}z & \text{if } k \text{ is odd,} \\ \cos\frac{k}{2}z & \text{if } k \text{ is even,} \end{cases}$$

where A_i, B_j are univariate coefficient functions which should be 2π-periodic in the trigonometric case.

This work was supported by "Deutsche Forschungsgemeinschaft", in the form of funds of the "Auswärtiges Amt" for the participation in the conference, and of a postdoc grant during the final preparation of this paper.

Algebraic pseudo-polynomials were already studied in the twenties and thirties of this century by Marchaud [18], [19], and Popoviciu [20], [21]. Of special interest in the framework of this paper is the fact from [20, Chapitre 3] that an algebraic pseudo-polynomial is continuous or bounded if and only if the coefficient functions in (1) are continuous or bounded, respectively. An analogous result holds true for trigonometric pseudo-polynomials. We will denote the space of continuous pseudo-polynomials of degree (m,n) (m,n as in (1)) defined on a compact rectangle R by $P^C_{m,n}(R)$. The corresponding spaces of continuous trigonometric pseudo-polynomials are denoted by $T^C_{m,n}$. Likewise, we define the spaces $P^B_{m,n}(R)$ and $T^B_{m,n}$ of bounded pseudo-polynomials. The notations $P^C(R)$, T^C, $P^B(R)$, and T^B stand for the corresponding spaces of continuous or bounded pseudo—polynomials of arbitrary degree.

Approximation by pseudo-polynomials is connected with approximation in tensor product spaces; see, e.g., [17]. For instance, $T^C_{m,n} = C_{2\pi}\otimes T_m + T_n\otimes C_{2\pi}$, where $C_{2\pi}$ denotes the space of univariate continuous 2π-periodic functions, and T_k the space of trigonometric polynomials of degree k. Moreover, in the recent literature pseudo-polynomials sometimes arise in the context of blending approximation in Computer Aided Geometric Design; see, e.g., [15]. A typical blending method is the approximation of bivariate functions by means of *Boolean sums of parametric extensions of univariate approximation operators* L_m, M_n, i.e., by functions of the form

$$(2) \qquad (L^x_m \oplus M^y_n)f := (L^x_m + M^y_n - L^x_m \circ M^y_n)f.$$

The parametric extension of the operator L_m with respect to the variable x is defined in such a way that it acts on f as if the second variable were fixed; i.e.,

$$L^x_m f(x,y) := L_m f^y(x), \quad f^y(x) := f(x,y).$$

The parametric extension M^y_n is defined similarly. If the underlying operators are polynomial operators, then the approximants in (2) are pseudo-polynomials.

Obviously, the uniform closure of $P^C(R)$ and T^C consists of all continuous bivariate real-valued functions (defined on R, or 2π-periodic with respect to both variables). In order to determine the closure of the spaces $P^B(R)$ and T^B, the notion of B-continuous functions is useful. A function f(x,y) is called *B-continuous* iff for each point (x_o,y_o) in its domain,

$$(3) \qquad \lim_{(x,y)\to(x_o,y_o)} \Delta f[(x,y),(x_o,y_o)] = 0$$

where $\Delta f[(x,y),(x_0,y_0)] := f(x,y) - f(x,y_0) - f(x_0,y) + f(x_0,y_0)$ is a *mixed difference*. Such ("Δ-continuous", in his terminology) functions were introduced by K.Bögel in [7], [8], and therefore, the expression B-continuity (Bögel continuity) was used in a variety of papers of mostly Romanian mathematicians dealing with this subject. We will adopt this terminology (also because of the formal resemblance of (3) to the elementary definition of continuity), although there seems to lack a satisfactory topological description of B-continuous functions in terms of continuity. For some results on a topological explanation of B-continuous functions see [5], [6]. For illustration, we give some examples of B-continuous functions. In the sequel, we will always assume that the domain of a given bivariate function f is a compact rectangle, or the whole plane in the periodic case.

Example 1. *(i) Each continuous function is B-continuous.*

(ii) Each function of the form $f(x,y) = \phi(x) + \psi(y)$ with two arbitrary univariate functions ϕ and ψ is B-continuous, since $\Delta f[(x_0,y_0),(x,y)] = 0$ in this case. Such functions we will call B-constants.

(iii) A function of the form $f(x,y) = g(x)\cdot h(y)$ with bounded univariate functions g and h is B-continuous if g or h is continuous, since

$$|\Delta f[(x_0,y_0),(x,y)]| = |g(x) - g(x_0)|\cdot|h(y) - h(y_0)|.$$

(iv) It can easily be checked that each function of the form

$$f(x,y) = \begin{cases} g(x) & \text{for } x\in\mathbb{R}\setminus\mathbb{Q}, y\in\mathbb{Q} \\ h(y) & \text{for } x\in\mathbb{Q}\quad, y\in\mathbb{R}\setminus\mathbb{Q} \\ g(x)+h(y) & \text{for } x\in\mathbb{Q}\quad, y\in\mathbb{Q}, \\ 0 & \text{for } x\in\mathbb{R}\setminus\mathbb{Q}, y\in\mathbb{R}\setminus\mathbb{Q} \end{cases}$$

is B-continuous if the univariate functions g and h are continuous.

(v) Sums and scalar multiples of B-continuous functions are B-continuous.

By the examples (iii) and (v), a bounded pseudo-polynomial is always B-continuous. The first example shows that there exist unbounded B-continuous functions. However, this is the only example in a certain sense: From [1, Corollary 2.4] and [12, Satz 1.1.8] we know that each B-continuous function f may be written in the form

$$f(x,y) = \tilde{f}(x,y) + b(x,y), \tag{4}$$

where $\tilde{f}$ is bounded and b is a B-constant. Because of this result, we are mainly interested in bounded B-continuous functions. We introduce the function spaces

$$B(R) := \{f:R \to \mathbb{R};\ f \text{ is B-continuous and bounded}\},$$

$$B_{2\pi} := \{f:\mathbb{R}^2 \to \mathbb{R};\ f \text{ is B-continuous, bounded and } 2\pi\text{-periodic with respect to both components}\}.$$

We remark that, on the other hand, B-continuous functions differ "quite a bit" from continuous functions. An example of the fact that $\tilde{f}$ in (4) may not be replaced by a continuous function, was given in [9]. The following example shows that $\tilde{f}$ in (4) even may not be assumed to satisfy partial continuity conditions.

Example 2. *For the function f in Example 1 (iv) with $g(x) = x^2$ and $h(y) = y^2$, the function $\tilde{f}$ in a decomposition of the form (4) may neither be chosen continuous in the variable x, nor in y.*

Proof. Suppose the existence of univariate functions ϕ and ψ with

$$f(x,y) = \tilde{f}(x,y) + \phi(x) + \psi(y)$$

such that $\tilde{f}$ is continuous in y; i.e.,

$$\lim_{y\to y_o} [f(x,y) - \psi(y)] = \lim_{y\to y_o} [\tilde{f}(x,y) + \phi(x)] = \tilde{f}(x,y_o) + \phi(x)$$

exists. Thus, for $x\in\mathbb{R}\setminus\mathbb{Q}$, we obtain

$$\begin{aligned} x^2 &= \limsup_{y\to y_o} f(x,y) - \liminf_{y\to y_o} f(x,y) \\ &= \limsup_{y\to y_o} [\psi(y)+(f(x,y)-\psi(y))] - \liminf_{y\to y_o} [\psi(y)+(f(x,y)-\psi(y))] \\ &= \limsup_{y\to y_o} \psi(y) - \liminf_{y\to y_o} \psi(y). \end{aligned}$$

This is a contradiction, since the right hand side of this equation does not depend on x. The same argumentation can be used with respect to the other variable.

Using the notion of B-continuity, we may formulate the following result:

Theorem 3. *(i) The uniform closure of $P^B(R)$ is $B(R)$.*

(ii) The uniform closure of T^B is $B_{2\pi}$.

Proof. From [3] and [1] we know that functions from $B(R)$ and $B_{2\pi}$ may be approximated arbitrarily well by elements of $P^B(R)$ and T^B, respectively. The above result follows from the observation that the spaces B(R) and $B_{2\pi}$ are closed with respect to the uniform norm: Consider a sequence (f_n) in B(R) or $B_{2\pi}$, uniformly converging to a (bounded) function f. For each $\varepsilon>0$ there is an n_o such that

$$\sup_{(x,y)} |f(x,y)| - f_{n_o}(x,y)| < \varepsilon.$$

The elements of B(R) and $B_{2\pi}$ are even <u>uniformly</u> B-continuous (cf. [9, Satz 7], [1, Lemma 2.5]; i.e., there is a $\delta>0$ such that

$$|f_{n_o}(x,y) - f_{n_o}(x_o,y) - f_{n_o}(x,y_o) + f_{n_o}(x_o,y_o)| < \varepsilon$$

for each (x_o,y_o) in the domain of f_{n_o}, if $\max(|x-x_o|, |y-y_o|) < \delta$. Using the triangle inequality, these facts yield the B-continuity of f.

2. Degree of approximation by bounded pseudo-polynomials.

In this section we give quantitative results on the approximation of $f \in B(R)$ or $B_{2\pi}$ by pseudo-polynomials in terms of the *mixed modulus of smoothness*

$$\omega(f;t_1,t_2) := \sup_{\substack{|h_1|\le t_1\\ |h_2|\le t_2}} \|\Delta f(\cdot,*),(\cdot + h_1,* + h_2)\|_\infty$$

for $f \in B(R)$, the supremum norm in (5) is taken over the largest rectangle on which the mixed difference is sensibly defined. Sometimes it is convenient to denote the above difference by $\Delta_{h_1,h_2}f$. The modulus (5), introduced already in [18], has turned out to be an appropriate measure for the smoothness of a bivariate function in the context of pseudo-polynomial (or, more general, blending) approximation; see, e.g., [2], [4], [11], [12], [13], [15], [16], [22]. Obviously, $\omega(f;t_1,t_2) \to 0$ for $(t_1,t_2) \to (0,0)$ if and only if f is B-continuous. We refer to [4] for some further basic properties of the moduli (5), which we will use in the sequel.

As an alternative to the classical moduli of smoothness, the so-called K-functionals play an important role in quantitative approximation theory. Let us remark that *mixed K-functionals*, as introduced in [14] (for continuous functions) and [12], can also be used in the formulation of quantitative results for pseudo-polynomial (or blending) approximation.

Jackson-type estimates for the approximation of measurable or continuous functions by pseudo-polynomials were proved in [11] and [16]. We state a similar result for the approximation of $f \in B(R)$ and $B_{2\pi}$.

THEOREM 4. *There is a constant c>0 such that for all $m,n \in \mathbb{N}_o$, and $f \in B(R)$ and $B_{2\pi}$,*

$$\inf \|f - \pi_{m,n}\|_{\infty} \le c \cdot \omega(f;\frac{1}{m+1},\frac{1}{n+1}),$$

where the infimum is taken over all $\pi_{m,n}$ in $P^{B}_{m,n}(R)$ or $T^{C}_{m,n}$, respectively.

Proof. For functions in B(R), the theorem was proved in [2]. In the case $f \in B_{2\pi}$, the desired estimate can be obtained by a Boolean sum approximation $\pi_{m,n} = (K^{x}_{m} \oplus K^{y}_{n})f$, where K_k is the univariate polynomial operator considered in the second example of Theorem B in [10]. Since K_k is discretely defined, the Boolean sum may be applied to all $f \in B_{2\pi}$.

Using the estimate

$$|z-u| \le \pi \cdot \sin \frac{|z-u|}{2}, \quad \text{for } |z-u| \le \pi,$$

it is not hard to show that for $(x,y),(s,t) \in \mathbb{R}^2$,

$$|\Delta f[(x,y),(s,t)]|$$
$$\le [1+\pi(m+1)|\sin \frac{x-s}{2}|] \cdot [1+\pi(n+1)|\sin \frac{y-t}{2}|] \cdot \omega(f;\frac{1}{m+1},\frac{1}{n+1}).$$

From [10] we know that K_k is positive and reproduces constants. Therefore,

$$|f(x,y) - (K^{x}_{m} \oplus K^{y}_{n})f(x,y)| = |(K^{x}_{m} \circ K^{y}_{n})(\Delta f[(x,y),(\cdot,*)];x,y)|$$
$$\le [1+\pi(m+1)K_m(|\sin \frac{x-\cdot}{2}|;x)] \cdot [1+\pi(n+1)K_n(|\sin \frac{y-*}{2}|;y)] \cdot \omega(f;\frac{1}{m+1},\frac{1}{n+1}).$$

Thus the assertion follows from the result of [10], that there is a constant $c_1>0$ satisfying

$$|K_k(\gamma_z;z)| = |K_k(\gamma_z;z) - \gamma_z(z)| \le \frac{c_1}{k+1},$$

for $\gamma_z(u) = |\sin \frac{z-u}{2}|$.

In the formulation of inverse theorems we consider approximation by *sequences of pseudo-polynomials of Boolean sum type*

$$(6) \qquad \pi_{m,n}(x,y) = (a_m + b_n + c_{m,n})(x,y)$$

where a_m is a polynomial of degree m in x (with coefficient functions in y independent of n), b_n a polynomial of degree n in y (independent of m), and $c_{m,n}$ a polynomial of degree m in x and n in y. Note that the Jackson order of approximation in Theorem 4 can be obtained by such sequences, as we see from the proof.

An inverse theorem for approximation by sequences (6) should tell us something about the smoothness properties of f, if we assume a certain degree of approximation

(7) $$\|f-\pi_{m,n}\|_{\infty} \le \Phi_{m,n}, \quad m,n \to \infty.$$

Let us first state the following simple result along these lines.

PROPOSITION 5. *If for $f \in B(R)$ or $B_{2\pi}$ there exists a sequence of the form (6) satisfying*

$$\|f-\pi_{m,n}\|_{\infty} \to 0, \text{ for } m \to \infty \text{ (n fixed) and } n \to \infty \text{ (m fixed)},$$

then f is the sum of a continuous function and a B-constant.

Proof. Consider the sequence of polynomials

$$q_{m,n} := -\pi_{m,n} + \pi_{m,o} + \pi_{o,n} - \pi_{o,o} = -c_{m,n} + c_{m,o} + c_{o,n} - c_{o,o}.$$

For $\tilde{f} = f - \pi_{o,o}$,

$$\|\tilde{f} - q_{m,n}\|_{\infty} = \|f + \pi_{m,n} - \pi_{m,o} - \pi_{o,n}\|_{\infty}$$

$$\le \|f - \pi_{m,n}\|_{\infty} + \|f - \pi_{m,o}\|_{\infty} + \|f - \pi_{o,n}\|_{\infty} \to 0, \quad m,n \to \infty;$$

i.e, $(q_{m,n})$ converges uniformly to $\tilde{f}$. thus $\tilde{f}$ is continuous. Since $\pi_{o,o}$ is a B-constant, we have found a suitable decomposition of f.

In [13], it was observed that a general transfer principle allows one to derive inverse theorems for pseudo-polynomial approximation from inverse theorems known from the univariate case, when the degree of approximation in (7) is estimated by certain sequences $\Phi_{m,n} = \phi^{(1)}_m \cdot \phi^{(2)}_n$ of product type. Applications of this principle include inverse theorems of Bernstein-type, as obtained in [11] and [22]. By Proposition 5, we know that such "product type" theorems essentially only handle continuous functions. At the end of this section, we give an inverse theorem of Bernstein-type which characterizes (in combination with Theorem 4) a class of B-continuous functions, containing "non-trivial" discontinuous elements (e.g., functions from Example 1 (iv) with a proper choice of g and h). When dealing with discontinuous functions, it seems not possible to directly exploit the univariate results as done in [11], [13], but we can use some ideas from the univariate case.

THEOREM 6. *If for $f \in B_{2\pi}$ and $0 < \alpha < 1$ there exists a sequence $(\tau_{k,l}) \subset T^B$ of Boolean sum type (6) such that*

(8) $$\max_{k,l \ge m} \|f - \tau_{k,l}\|_{\infty} \le \tilde{c} \cdot (m+1)^{-\alpha}, \quad \text{for } m \in \mathbb{N}_o,$$

then

(9) $$\omega(f;t,t) \le c \cdot t^{\alpha}, \text{ for } t>0.$$

Here, $\tilde{c}$ and c are positive constants.

Proof. For $\lambda > 1$, let $u_\lambda^{(1)} := \tau_{2^\lambda-1,2^\lambda-1} - \tau_{2^{\lambda-1}-1,2^\lambda-1}$, $u_\lambda^{(2)} := \tau_{2^{\lambda-1}-1,2^\lambda-1} - \tau_{2^{\lambda-1}-1,2^{\lambda-1}-1}$, and $u_\lambda := u_\lambda^{(1)} + u_\lambda^{(2)}$. By (8), there is a constant c_1 such that

$$\|u_\lambda\|_\infty \le \|f - \tau_{2^\lambda-1,2^\lambda-1}\|_\infty + \|f - \tau_{2^{\lambda-1}-1,2^{\lambda-1}-1}\|_\infty \le c_1 \cdot 2^{-\lambda\alpha}.$$

Similarly,

$$\max(\|u_\lambda^{(1)}\|_\infty, \|u_\lambda^{(2)}\|_\infty) \le c_1 \cdot 2^{-\lambda\alpha}.$$

Since $u_\lambda^{(1)}$ and $u_\lambda^{(2)}$ are polynomials of degree $2^\lambda-1$ in x and y, respectively, Bernstein's inequality yields,

$$\max(\|D^{1,0}u_\lambda^{(1)}\|_\infty, \|D^{0,1}u_\lambda^{(2)}\|_\infty) \le c_1 \cdot 2^{(1-\alpha)\lambda},$$

where $D^{1,0}$, $D^{0,1}$ denote the first partial derivatives with respect to the first and second variable.

By (8), we know that

$$\|f - \tau_{0,0} - g_M\|_\infty \le \tilde{c} \cdot 2^{-M\alpha}$$

for $g_M := \sum_{\lambda=1}^M u_\lambda = \tau_{2^M-1,2^M-1} - \tau_{0,0}$; i.e., $(g_M)_{M\in\mathbb{N}}$ converges uniformly to $\tilde{f} := f - \tau_{0,0}$. Since $\tau_{0,0}$ is a B-constant,

$$\Delta_{h_1,h_2} f = \Delta_{h_1,h_2}\tilde{f} = \Delta_{h_1,h_2}\left(\sum_{\lambda=1}^\infty u_\lambda \right) = \sum_{\lambda=1}^\infty \Delta_{h_1,h_2} u_\lambda .$$

For $\max(h_1,h_2) \le t < 1/2$ (in the case $t \ge 1/2$, condition (9) is obvious), we select M such that $1/2^{M+1} \le t < 1/2^M$. Then we arrive at the estimates,

$$\left| \sum_{\lambda=1}^M \Delta_{h_1,h_2} u_\lambda(x,y) \right| \le \sum_{\lambda=1}^M \|\Delta_{h_1,h_2} u_\lambda\|_\infty \le \sum_{\lambda=1}^M \omega(u_\lambda;t,t)$$

$$\le \sum_{\lambda=1}^M [\omega(u_\lambda^{(1)};t,t) + \omega(u_\lambda^{(2)};t,t)] \le \sum_{\lambda=1}^M [2t\|D^{1,0}u_\lambda^{(1)}\|_\infty + 2t\|D^{0,1}u_\lambda^{(2)}\|_\infty]$$

$$\le \quad 4c_1 t \cdot \sum_{\lambda=1}^M (2^{1-\alpha})^\lambda \le c_2 t^\alpha,$$

and,

$$\left| \sum_{\lambda=M+1}^\infty \Delta_{h_1,h_2} u_\lambda(x,y) \right| \le \sum_{\lambda=M+1}^\infty \omega(u_\lambda;t,t) \le 4 \sum_{\lambda=M+1}^\infty \|u_\lambda\|_\infty$$

$$\le 4c_1 \sum_{\lambda=M+1}^\infty 2^{-\alpha\lambda} = 4c_1 \sum_{\lambda=0}^\infty 2^{-(\lambda+M+1)\alpha} = 4c_1 2^{-(M+1)\alpha} \cdot \sum_{\lambda=0}^\infty 2^{-\lambda\alpha} \le c_3 t^\alpha,$$

with certain constants c_2 and c_3.

Since

$$\omega(f;t,t) = \sup_{|h_1|,\,|h_2|\le t} \|\Delta_{h_1,h_2} f\|_\infty$$

$$\leq \sup_{|h_1|,|h_2|\leq t} \left\{ \| \sum_{\lambda=1}^{M} \Delta_{h_1,h_2} u_\lambda \|_\infty + \| \sum_{\lambda=M+1}^{\infty} \Delta_{h_1,h_2} u_\lambda \|_\infty \right\} \leq c \cdot t^\alpha,$$

the assertion is proved.

Let us finally remark, that (8) could be replaced by the equivalent condition

$$\|f-\tau_{k,l}\|_\infty \leq \tilde{c} \cdot \frac{1}{(\min(k,l)+1)^\alpha}, \quad \text{for } k,l \in \mathbb{N}_0,$$

and (9) by

$$\omega(f;t_1,t_2) \leq c^* \cdot (\max(t_1,t_2))^\alpha,, \quad \text{for } t_1,t_2>0$$

with a suitable constant c^*.

REFERENCES

[1] C. Badea, I. Badea, C. Cottin, *A Korovkin-type theorem for generalizations of Boolean sum operators and approximation by trigonometric pseudopolynomials*. Anal. Numér. Théor. Approx. 17 (1988), 7–17.

[2] C. Badea, I. Badea, C. Cottin, H.H. Gonska, *Notes on the degree of approximation of B-continuous and B-differentiable functions*, J. Approx. Theory Appl. 4 (1988), 95–108.

[3] C. Badea, I. Badea, H.H. Gonska, *A test function theorem and approximation by pseudopolynomials*, Bull. Austral. Math. Soc. 34 (1986), 53–64.

[4] I. Badea, *Modulul de continuitate în sens Bögel şi unele aplicaţii în approximarea printr-un operator Bernstein*, Studia Univ. Babeş-Bolyai, Ser. Math.-Mech. 18 (1973), 69–78.

[5] T. Balan, *Topological structures in bidimensional analysis*, Rev. Roumaine Math. Pures Appl. 19 (1974), 849–856.

[6] T. Balan, *A topological study in bidimensional continuity*, Rev. Roumaine Math. Pures Appl. 20 (1975), 181–187.

[7] K. Bögel, *Mehrdimensionale Differentiation von Funktionen mehrerer Veränderlicher*, J. Reine Angew. Math. 170 (1934), 197–217.

[8] K. Bögel, *Uber die mehrdimensionale Differetiation, Integration und beschränkte Variation*, J. Reine Angew. Math. 173 (1935), 5–29.

[9] K.Bögel, *Über die mehrdimensionale Differetiation*, Jber. DMV 65 (1962), 45–71.

[10] R.Bojanic, O. Shisha, *Approximation of continuous, periodic functions by discrete positive linear operators*, J. Approx. Theory, 11 (1974), 231–235.

[11] Yu. A. Brudnyĭ, *Approximation of functions of n variables by quasipolynomials*, Math. USSR-Izv. 4 (1970), 568–586.

[12] C.Cottin, *Quantitative Aussagen zur Blending-Typ-Approximation*, Dissertation, Universität Duisburg, 1988.

[13] C.Cottin, *Inverse theorems for blending-type approximation*, to appear in Approximation Theory VI, Proc. Conf. College Station, Texas, 1989.

[14] C.Cottin, *Mixed K-functionals: A measure of smoothness for blending-type approximation*, submitted for publication.

[15] H.H.Gonska, *Quantitative Approximation in C(X)*, Habilitationsschrift, Universität Duisburg, 1985.

[16] H.H.Gonska, K.Jetter, *Jackson-type theorems on approximation by trigonometric and algebraic pseudopolynomials*, J. Approx. Theory 48 (1986), 396–406.

[17] W.A.Light, E. W. Cheney, *Approximation Theory in Tensor Product Spaces*, Lecture Notes in Math. 1169, New York, Springer- Verlag, 1985.

[18] A.Marchaud, *Différences et dérivées d'une fonction de deux variables*, C. R. Acad. Sci. 178 (1924), 1467–1470.

[19] A.Marchaud, *Sur les dérivées et sur les différences des fonctions de variables réelles*, J. Math. Pures Appl. 6 (1927), 337–425.

[20] T.Popoviciu, *Sur quelques propriétés des fonctions d'une ou deux variables réelles*, Mathematica (Cluj) 8 (1934), 1–85.

[21] T.Popoviciu, *Sur les solutions bornées et les solutions mesurables de certaines equations fonctionelles*, Mathematica (Cluj) 14 (1938), 47–106.

[22] M.K.Potapov, *Investigation of certain classes of functions by "angular" approximation*, in: Proc. Steklov Inst. Math. 117 (1972), 301–342; providence, R.I.: Amer. Math. Soc. 1974.

IMBEDDINGS OF WEIGHTED SOBOLEV SPACES AND THE MULTIPLICATIVE INEQUALITY

PETR GURKA, BOHUMÍR OPIC

Praha, Czechoslovakia

0. Introduction.

In this paper we present a short survey of some results achieved in our former articles [7,8] and we make use of them to obtain necessary and sufficient conditions for the validity of the weighted multiplicative inequality (3.3).

Throughout the paper we suppose that Ω is a bounded domain in $\mathbb{R}^n$ with a Lipschitz boundary $\partial\Omega$ (notation $\Omega \in C^{0,1}$, cf [8]). If $x\in\Omega$ then we set $d(x) = \mathrm{dist}(x,\partial\Omega)$. For $\alpha\in\mathbb{R}$ and $1 \le q < \infty$ the weighted Lebesgue space $L^q(\Omega;\alpha)$ is the set of all measurable functions u defined on Ω with a finite norm

$$\|u\|_{q,\Omega,\alpha} = \left(\int_\Omega |u(x)|^q d^\alpha(x)\, dx\right)^{1/q}. \tag{0.1}$$

Let $\beta,\gamma \in R$ and $1 \le p < \infty$. We define the weighted Sobolev space $W^{1,p}(\Omega;\gamma,\beta)$ as the set of all functions $u \in L^p(\Omega;\gamma)$ which have distributional derivatives $\frac{\partial u}{\partial x_i} \in L^p(\Omega;\beta)$, $i = 1,\dots,N$. The space $W^{1,p}(\Omega;\gamma,\beta)$ with the norm

$$\|u\|_{1,p,\Omega,\gamma,\beta} = \left(\|u\|^p_{p,\Omega,\gamma} + \sum_{i=1}^N \left\|\frac{\partial u}{\partial x_i}\right\|^p_{p,\Omega,\beta}\right)^{1/p} \tag{0.2}$$

is a Banach space, see [10]. Finally, we define the space $W_0^{1,p}(\Omega;\gamma,\beta)$ as the closure of the set $C_0^\infty(\Omega)$ with respect to the norm (0.2).

For two Banach spaces X, Y, we write $X \hookrightarrow Y$ or $X \hookrightarrow\hookrightarrow Y$ if $X \subset Y$ and the natural injection of X into Y is continuous or compact, respectively.

1. Imbedding theorems.

Our main results concerning imbeddings of weighted Sobolev spaces into weighted Lebesgue spaces (in the case of power-type weights) read as follows:[*)]

Theorem 1.1. *Suppose that* $1 \le p \le q < \infty$. *Then*

$$W^{1,p}(\Omega;\beta-p,\beta) \hookrightarrow L^q(\Omega;\alpha) \tag{1.1}$$

(or

$$W^{1,p}(\Omega;\beta-p,\beta) \hookrightarrow\hookrightarrow L^q(\Omega;\alpha) \;) \tag{1.2}$$

if and only if

$$\frac{N}{q} - \frac{N}{p} + 1 \ge 0, \qquad \frac{\alpha}{q} - \frac{\beta}{p} + \frac{N}{q} - \frac{N}{p} + 1 \ge 0 \tag{1.3}$$

(or

$$\frac{N}{q} - \frac{N}{p} + 1 > 0, \qquad \frac{\alpha}{q} - \frac{\beta}{p} + \frac{N}{q} - \frac{N}{p} + 1 > 0 \;). \tag{1.4}$$

Remarks 1.2. (i) Theorem 1.1 holds even for an arbitrary bounded domain Ω in $\mathbb{R}^n$ (see [7], Example 5.1). The analogy of this theorem for an unbounded domain can be found in [15].

(ii) It is possible to prove (even for an arbitrary bounded domain Ω in $\mathbb{R}^n$) that $W^{1,p}(\Omega;\beta-p,\beta) = W_o^{1,p}(\Omega;\beta-p,\beta)$, $\beta\in\mathbb{R}$. Consequently Theorem 1.1 holds with the space $W_o^{1,p}(\Omega;\beta-p,\beta)$ instead of $W^{1,p}(\Omega;\beta-p,\beta)$.

Theorem 1.3. *Suppose* $1 \le p \le q < \infty$, $\beta > p-1$ *(or* $\beta \ne p-1$*). Then*

$$W^{1,p}(\Omega;\beta,\beta) \hookrightarrow L^q(\Omega;\alpha) \tag{1.5}$$

(or

$$W_o^{1,p}(\Omega;\beta,\beta) \hookrightarrow L^q(\Omega;\alpha) \;) \tag{1.6}$$

if and only if condition (1.3) is fulfilled.

Theorem 1.4. *Suppose* $1 \le p \le q < \infty$, $\beta > p-1$ *(or* $\beta \ne p-1$*). Then*

$$W^{1,p}(\Omega;\beta,\beta) \hookrightarrow\hookrightarrow L^q(\Omega;\alpha) \tag{1.7}$$

---

*) The proofs can be found in [7,8]. As far as imbedding theorems with general weight functions are concerned, the reader is referred to [7].

(or

(1.8) $$W_o^{1,p}(\Omega;\beta,\beta) \hookrightarrow\hookrightarrow L^q(\Omega;\alpha) \;)$$

if and only if condition (1.4) is fulfilled.

The following two theorems imply that in the case $1 \le q < p < \infty$ the situation is quite different.

Theorem 1.5. *Suppose $1 \le q < p < \infty$. Then the following three conditions are equivalent:*

(1.9) $$W_o^{1,p}(\Omega;\beta,\beta) \hookrightarrow L^q(\Omega;\alpha),$$

(1.10) $$W_o^{1,p}(\Omega;\beta,\beta) \hookrightarrow\hookrightarrow L^q(\Omega;\alpha),$$

(1.11) $$\frac{\alpha}{q} - \frac{\beta}{p} + \frac{1}{q} - \frac{1}{p} + 1 > 0.$$

Theorem 1.6. *Suppose $1 \le q < p < \infty$. Then the following three conditions are equivalent:*

(1.12) $$W^{1,p}(\Omega;\beta,\beta) \hookrightarrow L^q(\Omega;\alpha),$$

(1.13) $$W^{1,p}(\Omega;\beta,\beta) \hookrightarrow\hookrightarrow L^q(\Omega;\alpha),$$

(1.14) $$\begin{cases} \frac{\alpha}{q} - \frac{\beta}{p} + \frac{1}{q} - \frac{1}{p} + 1 > 0 & \text{for } \beta \in \mathbb{R}\setminus(-1,p-1], \\ \alpha > -1 & \text{for } \beta \in (-1,p-1]. \end{cases}$$

2. N-dimensional Hardy inequality.

As a consequence of the theorem on equivalent norms on the space $W_o^{1,p}(\Omega;\beta,\beta)$ (see [9], Proposition 9.2) and imbedding theorems from Section 1 we obtain the following theorem.

Theorem 2.1. *Let $1 \le p,q < \infty$, $\beta < p-1$. Then there exists a positive constant C such that the inequality*

(2.1) $$\left(\int_\Omega |u(x)|^q d^\alpha(x)dx\right)^{1/q} \le C \left(\int_\Omega |\nabla u(x)|^p d^\beta(x)dx\right)^{1/p} \;{}^{*)}$$

holds for all $u \in W_o^{1,p}(\Omega;\beta,\beta)$ if and only if either

---

*) We set $|\nabla u(x)|^p = \sum_{i=1}^N |\frac{\partial u}{\partial x_i}(x)|^p$.

$$(2.2)\qquad 1 \le p \le q < \infty, \quad \frac{N}{q} - \frac{N}{p} + 1 \ge 0, \quad \frac{\alpha}{q} - \frac{\beta}{p} + \frac{N}{q} - \frac{N}{p} + 1 \ge 0$$

or

$$(2.3)\qquad 1 \le q \le p < \infty, \quad \frac{\alpha}{q} - \frac{\beta}{p} + \frac{1}{q} - \frac{1}{p} + 1 > 0.$$

Proof can be found in [8].

Remark 2.2. The inequality (2.1) is sometimes called "N-dimensional Hardy Inequality" (cf. [14]).

3. Multiplicative inequality.

Theorem 2.1 enable us to establish the following assertion.

Theorem 3.1. *Suppose* $\beta, \gamma, \eta \in \mathbb{R}$. *Let* $r \ge 1$, $q \ge 1$, $0 < a < 1$ *be such numbers that*

$$(3.1)\qquad \bar{p} := \frac{a\, r\, q}{q+(a-1)r} \in [1,\infty).$$

(i) *Let*

$$(3.2)\qquad \max\left\{1, \frac{N\bar{p}}{N+\bar{p}}\right\} \le p \le \bar{p}, \qquad \beta < p-1.$$

Then there exists a constant $C > 0$ *such that the inequality*

$$(3.3)\qquad \|u\|_{r,\Omega,\gamma} \le C\, \|\,|\nabla u|\,\|^{a}_{p,\Omega,\beta}\, \|u\|^{1-a}_{q,\Omega,\eta}$$

holds for all functions $u \in C_0^\infty(\Omega)$ *if and only if*

$$(3.4)\qquad \frac{N+\gamma}{r} - a\,\frac{N+\beta-p}{p} - (1-a)\,\frac{N+\eta}{q} \ge 0.$$

(ii) *Let*

$$(3.5)\qquad \bar{p} < p < \infty, \qquad \beta < p-1.$$

Then the inequality (3.3) holds for every $u \in C_0^\infty(\Omega)$ *(with* $C>0$ *independent of* u*) if*

$$(3.6)\qquad \frac{1+\gamma}{r} - a\,\frac{1+\beta-p}{p} - (1-a)\,\frac{1+\eta}{q} > 0.$$

Conversely, if the inequality (3.3) holds for every function $u \in C_0^\infty(\Omega)$, *then*

$$(3.6^*)\qquad \frac{1+\gamma}{r} - a\,\frac{1+\beta-p}{p} - (1-a)\,\frac{1+\eta}{q} \ge 0.$$

Remark. 3.2. The inequality (3.3) is called *the multiplicative inequality*. For $\beta = \gamma = \eta = 0$ (unweighted case) it has been investigated by many authors, see e.g. [1,11] etc. The weighted multiplicative inequality has been studied by Caffarelli, Kohn and Nirenberg [6] and Lin [12] (with $\Omega = \mathbb{R}^n$ and weights of the type $|x|^\varepsilon$, $x\in\Omega$, $\varepsilon\in\mathbb{R}^n$) and Brown and Hinton [2–5]. Theorem 3.1 is an analogue of the above mentioned results for bounded domains Ω.

Proof of Theorem 3.1. (I) Sufficiency. By (3.1) we have

$$a\,\frac{r}{p} + (1-a)\,\frac{r}{q} = 1.$$

Using Hölder's inequality we obtain for $u \in C_0^\infty(\Omega)$ the estimate

$$\text{(3.7)}\qquad \left(\int_\Omega |u(x)|^r d^\gamma(x)\,dx\right)^{1/r} = \left(\int_\Omega |u(x)|^{ra}\left[d(x)\right]^{\frac{\gamma q-\eta r(1-a)}{q}}\cdot |u(x)|^{r(1-a)}\left[d(x)\right]^{\frac{\eta r(1-a)}{q}}dx\right)^{1/r}$$

$$\le \left(\int_\Omega |u(x)|^{\bar p}\left[d(x)\right]^{\bar p\,\frac{\gamma q-\eta r(1-a)}{a\,r\,q}}dx\right)^{a/\bar p}\cdot\left(\int_\Omega |u(x)|^q\, d^\eta(x)\,dx\right)^{(1-a)/q}$$

Now we make use of Theorem 2.1 (with $\bar p$ instead of q and $\bar p\cdot\frac{\gamma q-\eta r(1-a)}{a\,r\,q}$ instead of α) to estimate the first term at the right hand side of (3.7). (Let us remark that the condition (2.2) follows from (3.2) and (3.4) while the condition (2.3) is satisfied due to (3.5) and (3.6).) Consequently we have

$$\left(\int_\Omega |u(x)|^r d^\gamma(x)dx\right)^{1/r} \le$$

$$\le C\left(\int_\Omega |\nabla u(x)|^p d^\beta(x)dx\right)^{a/p}\cdot\left(\int_\Omega |u(x)|^p d^\eta(x)dx\right)^{(1-a)/q}$$

which is the inequality (3.3).

(II) Necessity. We prove that the condition (3.4) follows from the validity of (3.3) for every $u \in C_0^\infty(\Omega)$. Concerning the validity of (3.6*) we give only a sketch of the proof.

(II-1) Let the inequality (3.3) hold for every $u \in C_0^\infty(\Omega)$. Take the sequence $\{x_k\} \subset \Omega$ such that $d(x_k)\searrow 0$ for $k \to \infty$ and define the functions

$$u_k = R_{d(x_k)/16}\cdot\chi_{(3/4)B_k},\qquad k \in \mathbb{N}.$$

(Here, B_k is the ball with the center x_k and the radius $d(x_k)/2$; for ε positive εB_k means the ball with the same center and the radius $\varepsilon\cdot d(x_k)/2$, while R_ε is a mollifier with the radius ε and $\chi_{(3/4)B_k}$ is the characteristic function of the set $\frac{3}{4}B_k$.) The functions u_k possess the following properties:

(a) $u_k \in C_o^\infty(\Omega)$, $\operatorname{supp} u_k \subset B_k$, $0 \le u_k \le 1$;

(b) $u_k = 1$ on $\frac{1}{2} B_k$;

(c) $\exists\, c_o > 0 \implies \left|\frac{\partial u_k}{\partial x_i}(x)\right| \le c_o d^{-1}(x_k)$, $x \in \Omega$, $i = 1,\dots,N$.

After an easy calculation we obtain for $k \in \mathbb{N}$

$$\|u_k\|_{r,\Omega,\gamma} \ge c_1\,[d(x_k)]^{(N+\gamma)/r},$$

$$\|\nabla u_k\|^a_{p,\Omega,\beta} \le c_2\,[d(x_k)]^{a(N+\beta-p)/p},$$

$$\|u_k\|^{1-a}_{q,\Omega,\eta} \le c_3\,[d(x_k)]^{(1-a)(N+\eta)/q},$$

(with $c_1, c_2, c_3 > 0$ independent of k). Hence, from (3.3) and the condition $d(x_k) \searrow 0$ for $k \to \infty$ the inequality (3.4) follows immediately.

(II–ii) Sketch of the proof of (3.6*). Using local coordinates (cf. [8], the proof of Th. 9.5) we consider functions $u \in C_o^\infty(\Omega)$ in the form $u(x) = u(z_1,\dots z_{N-1},z_N) = \varphi(z_1,\dots,z_{N-1})\cdot v(z_N)$, where $\varphi \in C_o^\infty(\mathbb{R}^{N-1})$ and $v \in C_o^\infty((0,\lambda))$ (with suitable $\lambda>0$); local variables $z_1,\dots,z_{N-1}$ are independent of $d(x)$, while z_N is equivalent to $d(x)$. From the validity of the inequality (3.4) for such functions u we derive that

$$\|v\|_{r,Q,\gamma} \le C\,\|v'\|^a_{p,Q,\beta} \le \|v\|^{1-a}_{q,Q,\eta} \tag{3.8}$$

for every $v \in C_o^\infty(Q)$ with $Q = (0,\lambda)$. But (3.8) is the inequality of the type (3.3) with $N = 1$. Consequently, the validity of (3.8) for every $v \in C_o^\infty(Q)$ implies the condition (3.6*).

Remark 3.3. The necessity of (3.4) and (3.6*) was derived from (3.3) without any additional assumptions. Therefore we have practically two necessary conditions – (3.4) and (3.6*). Moreover, we easily obtain

$$\frac{N+\gamma}{r} - a\,\frac{N+\beta-p}{p} - (1-a)\,\frac{N+\eta}{q} = \frac{1+\gamma}{r} - a\,\frac{1+\beta-p}{p} - (1-a)\,\frac{1+\eta}{q} + (N-1)\,s,$$

where

$$s = \frac{1}{r} - \frac{a}{p} - \frac{1+a}{q}\,.$$

As $s \le 0$ for $p \le \bar{p}$ (and $s > 0$ for $p > \bar{p}$), the condition (3.4) is stronger (weaker) than (3.6*) provided $p \le \bar{p}$ (provided $p > \bar{p}$).

Acknowledgement. We thank Professor R.C.Brown for bringing the papers [6,12] to our attention.

REFERENCES

[1] O.V.Besov, V. P. Il'in, S. M. Nikol'skij, *Integral Representations of Functions and Embedding Theorems*, Nauka Moskva 1975 (in Russian).

[2] R. C. Brown, D. B. Hinton, *Sufficient conditions for weighted Gabushin inequalities*, Časopis Pěst. Mat. 111 (1986), 113–122.

[3] R. C. Brown, D. B. Hinton, *Necessary and sufficient conditions for one variable weighted interpolation inequalities*, J. London Math. Soc. (2) 35 (1987), 439–453.

[4] R. C. Brown, D. B. Hinton, *Weighted interpolation inequalities of sum and product form in* $\mathbb{R}^n$, Proc. London Math. Soc. (3) 56 (1988), 261–280.

[5] R. C. Brown, D. B. Hinton, *Weighted interpolation inequalities and embeddings in* $\mathbb{R}^n$, preprint.

[6] L.Caffarelli, R.Kohn, L. Nirenberg, *First order interpolation inequalities with weights*, Composito Math. 53 (1984), 259–275.

[7] P.Gurka, B.Opic, *Continuous and compact imbeddings of weighted Sobolev spaces I*, Czechoslovak Math. J. 38 (113) (1988), 730–744.

[8] P.Gurka, B.Opic, *Continuous and compact imbeddings of weighted Sobolev spaces II*, Czechoslovak Math. J. 39 (114) (1989), 78–94.

[9] A.Kufner, *Weighted Sobolev Spaces*, John Wiley & Sons, Chichester–New York–Brisbane–Toronto–Singapore 1985.

[10] A.Kufner, B. Opic, *How to define reasonably weighted Sobolev spaces*, Comment. Math. Univ. Carolinae 25 (3) (1984), 537–554.

[11] O. A. Ladyženskaja, V. A. Solonnikov, N. N. Ural'ceva, *Linear and Quasilinear Equations of the Parabolic Type*, Nauka Moskva 1967 (in Russian).

[12] Ch. S. Lin, *Interpolation inequalities with weights*, preprint.

[13] B.Opic, P. Gurka, *Imbeddings of weighted Sobolev spaces on unbounded domains and the N-dimensional Hardy inequality*, Proc. Int. Conf. "Summer school on function spaces, differential operators and non-linear analysis", Sodankylä (Finland) 1988, to appear.

[14] B. Opic, A. Kufner, *Weighted Sobolev spaces and the N-dimensional Hardy inequality*, Trudy Sem. S.L.Soboleva, No. 1 (1983), 108–117 (in Russian).

[15] B.Opic, J. Rákosník, *Estimates for mixed derivatives from anisotropic Sobolev-Slobodeckij spaces with weights*, to appear.

ON THE SPACE OF HENSTOCK INTEGRABLE FUNCTIONS

LEE PENG YEE

Singapore

1. Introduction.

The Henstock integral is known to be equivalent to the Denjoy-Perron integral [7]. It is also known as the Kurzweil integral or the generalized Riemann integral. In this paper, we characterize the continuous orthogonally additive functionals on the space of all Henstock integrable functions and on some of its subspaces. Theorem 2 and 4 below are due to Chew.

A real-valued function f defined on [a,b] is said to be *Henstock integrable on [a,b]*, if there is a real number A such that for every $\varepsilon>0$ there is $\delta(\xi) > 0$ such that for every division of [a,b] given by

$$a = x_0 < x_1 < \dots < x_n = b \quad \text{and} \quad (\xi_1, \xi_2, \dots, \xi_n)$$

satisfying $\xi_i \in [x_{i-1}, x_i] \subset (\xi_i - \delta(\xi_i), \xi_i + \delta(\xi_i))$ for $i = 1,2,\dots,n$ we have

$$\left| \sum_{i=1}^{n} f(\xi_i)(x_i - x_{i-1}) - A \right| < \varepsilon.$$

Let D denote the space of all Henstock integrable functions on [a,b]. A norm can de defined in D as follows:

$$\|f\|_D = \sup \left\{ \left| \int_a^x f \right| : a \le x \le b \right\}.$$

Alexiewicz [1] showed that a continuous linear functional T defined on D can be represented by

$$T(f) = \int_b^a f(x)g(x)dx$$

for all $f \in D$ and for some g of bounded variation on [a,b]. However the space D is not complete under the above norm, though Sargent [9] was still able to prove the Banach-Steinhaus theorem for the space D. As we shall see later, the key to the characterization of continuous orthogonally additive functionals on the space D is to provide a suitable convergence in D in place of the above norm convergence.

2. Representation theorems.

Let X be a subspace of the space D. A functional T defined on X is said to be orthogonally additive if

$$T(f+g) = T(f) + T(g)$$

whenever $f, g \in X$ and have disjoint supports. A Caratheodory function $k(t,x)$ is one such that $k(\cdot,x)$ is continuous for almost all x and $k(t,\cdot)$ is measurable for all t. The following two theorems are well-known [3,5].

Theorem 1. *Let X be an Orlicz function space L_φ with φ satisfying the Δ_2 condition. Then a functional T is continuous and orthogonally additive on X if and only if there is a Caratheodory function $k(t,x)$ with $k(0,x) = 0$ for almost all x such that $k(f(\cdot),\cdot)$ is Henstock integrable on [a,b] whenever $f \in X$ and*

$$T(f) = \int_b^a k(f(x),x)dx \qquad \text{for } f \in X.$$

In fact, in the above theorem $k(f(\cdot),\cdot)$ is Henstock integrable for $f \in L_\varphi$ if and only if

$$|k(t,x)| \le g(x) + \alpha\varphi(t)$$

for almost all x, for all t, for some constant α, and for some Lebesgue integrable function g. In the case when φ does not satisfy the Δ_2 condition, Chew [3] gave two versions of the above inequality, one for the Orlicz space using modular convergence and another for the largest linear space in the Orlicz class using norm convergence. When $\varphi(t) = t^p$ for $1 \le p \le \infty$, the theorem reduces to the case for the L_p space.

Let X be the L_∞ space with the convergence in L_∞ defined as follows. A sequence (f_n) in L_∞ is said to be *boundedly convergent* to f if $f_n(x) \to f(x)$ almost everywhere in [a,b] as $n \to \infty$ and (f_n) is uniformly bounded almost everywhere by a constant. A functional T defined on L_∞ is said to be *continuous* if $T(f_n) \to T(f)$ as $n \to \infty$ whenever (f_n) is boundedly convergent to f.

THEOREM 2. *Let X be the L_∞ space provided with the bounded convergence. Then Theorem 1 holds.*

The sufficiency is easy. The necessity follows from a theorem of Drewnowski and Orlicz [5,7].

3. Controlled convergence.

Let $X \subset [a,b]$. A function F is said to be $AC^*(X)$ if for every $\varepsilon>0$ there is $\eta>0$ such that every finite or infinite sequence of non-overlapping intervals $([a_i,b_i])$ with $a_i,b_i \in X$

$$\sum_i |b_i-a_i| < \eta \quad \text{implies} \quad \sum_i \omega(F_i[a_i,b_i]) < \varepsilon$$

where ω denotes the oscillation of F over $[a_i,b_i]$. A function F is said to be ACG^* if $[a,b]$ is the union of a sequence of sets (X_i) such that on each X_i the function F is $AC^*(X_i)$. Lee and Chew proved the following controlled convergence theorem [7].

THEOREM 3. *If the following conditions are satisfied:*

(i) $f_n(x) \to f(x)$ almost everywhere in $[a,b]$ as $n \to \infty$ where f_n is Henstock integrable on $[a,b]$;

(ii) the primitives F_n of f_n, $n = 1,2,....$ are ACG^* uniformly in n;

(iii) F_n converges uniformly to F on $[a,b]$,

then f is Henstock integrable on $[a,b]$ with the primitive F.

A sequence (f_n) in D is said to be *convergent in the controlled sense* if the conditions of Theorem 3 are satisfied. For convenience, we call it *the controlled convergence*. It is well-known [7, p.59] that if f is Henstock integrable on [a,b] then there is a sequence (f_n) in L_∞ which is convergent to f in the controlled sense. Applying Theorems 2 and 3, we can prove (see [4] or [7, p.103]).

THEOREM 4. *Let X be the space D provided with the controlled convergence. Then Theorem 1 holds.*

Note that if T is linear, then Theorem 4 reduces to the classical result of Alexiewicz. It is known [7, p.103] that for linear functionals T on D the norm continuity and the controlled continuity of T coincide. Further, we remark that the above theorem is not included in a similar result of Pluciennik [8] on vector valued functions. The proof of Płuciennik bases on that of Hiai [6] which did not make use of the theorem of Drewnowski and Orlicz.

4. An Orlicz scale of function spaces.

Let φ be an Orlicz function satisfying the Δ condition.Let D denote a subspace of D satisfying the property that if $f \in D_\varphi$ then the primitive F of f is of bounded φ-variation on [a,b], i.e.

$$\sup \sum_{i=1}^{n} \varphi(|F(x_i) - F(x_{i-1})|) < +\infty$$

where the supremum is over all the divisions $a = x_0 < \ldots < x_n = b$. The D_φ is a linear space. A sequence (f_n) in D_φ is said to be *φ-convergent in the controlled sense* if the conditions of Theorem 3 are satisfied and, in addition, the primitives F_n of f_n, $n = 1,2,\ldots$, are of bounded φ-variation on [a,b] uniformly in n. For convenience, we call it *the controlled α-convergence*. Then it is easy to see that the following controlled convergence theorem for the space D_φ holds. Furthermore, if $f \in D_\varphi$ then there is a sequence (f_n) in L_∞ which is φ-convergent to f in the controlled sense in D_φ.

Theorem 5. *If a sequence (f_n) in D_φ is φ-convergent to f in the controlled sense, then f is Henstock integrable on [a,b] with $f \in D_\varphi$ and*

$$\int_a^b f_n(x)dx \longrightarrow \int_a^b f(x)dx \qquad as \quad n \to \infty.$$

We remark that the space D_φ when $\varphi(t) = t^{1/\alpha}$ for $0 < \alpha < 1$ has been considered by Burkill and Gehering [2]. The corresponding results of [2] for D_φ have been discussed by Soeparna [10].

Theorem 6. *Let X be the space D_φ provided with the controlled φ-convergence in D_φ. Then Theorem 1 holds.*

The proof is similar to that of Theorem 4 and therefore omitted. Roughly, X includes L_∞ and the bounded convergence in L_∞ implies that in X. Therefore a continuous functional T on X is also continuous on L_∞ and we may apply Theorem 2. To extend the domain of the functional T from L_∞ to X, we apply the corresponding convergence theorem, namely, Theorem 5 when $X = D_\varphi$. Note that this is the general pattern of proof for the previous representation theorems as well.

Consequently, we have characterized the continuous orthogonally additive functionals on X when X is the L_p space for $1 \le p \le \infty$, the Orlicz space, the space D, or D_φ, and when a suitable convergence is defined in X for each case.

REFERENCES

[1] A.Alexiewicz *Linear functionals on Denjoy integrable functions*, Coll. Math. 1 (1948), 289–293.

[2] J.C.Burkill, F. W. Gehring, *A scale of integrals from Lebesgue's to Denjoy's*, Quart. J. Math. Oxford (2) 4 (1953), 210–220.

[3] Chew Tuan Seng, *Orthogonally additive functionals*, Ph.D. thesis 1985.

[4] Chew Tuan Seng, *Nonlinear Henstock-Kurzweil integrals and representations theorems*, SEA Bull. Math. 12 (1988), 97–108.

[5] L.Drewnowski, W.Orlicz, *Continuity and representation of orthogonally additive functionals*, Bull. Acad. Polon. Sci., Ser. Sci. Math. Astr. et Phys. 17 (1969), 647–653.

[6] F.Hiai, *Representations of additive functionals on vector-valued normed Köthe spaces*, Kodai Math. J. 2 (1979), 300–313.

[7] Lee Peng Yee, *Lanzhou Lectures on Henstock integration*, World Scientific 1989.

[8] R.Płuciennik, *Representation of additive functionals on Musielak-Orlicz space of vector valued functions*, Kodai Math. J. 10 (1987), 49–54.

[9] W.L.C.Sargent, *On some theorems of Hahn, Banach and Steinhaus*, J. London Math. Soc. 28 (1953), 438–451.

[10] D.Soeparna, *An Orlicz scale of integrals*, SEA Bull. Math. 12 (1988), 123–133.

SOME PROPERTIES OF X^p-SPACES

LJUDMILA I. NIKOLOVA, LARS-ERIK PERSSON

Sofia, Bulgaria; Luleå, Sweden

ABSTRACT. Let $X = (X, \|\cdot\|_X)$ be a Banach function space on a σ-finite measure (Ω,Σ,μ). The space X^p, $0<p<\infty$, consists of all measurable functions $x = x(s)$, $s\in\Omega$, satisfying $\| |x|^p \|_X < \infty$. In this paper we present some new properties of the X^p-spaces, the more general $X^p(A,\omega)$-spaces or families of spaces of this kind. In particular we prove an equality, which may be considered as a strong generalization of some classical inequalities including those by Hölder, Minikowski and Beckenbach-Dresher. We prove some results concerning τ-uniformly PL-convexity of the spaces $X^p(A,\omega)$ and also concerning compact or quasi-compact operators between spaces of this kind. We discuss the close connection between some of our results and the theory of generalized Gini means and present some applications.

0. Introduction.

The X^p-spaces and more general spaces of this type have been treated and applied in some recent papers (see e.g. [10], [12], [13] and [15]). In this paper we present some additional results in connection to (generalized) X^p-spaces. The paper is organized in the following way: In Section 1 we present some preliminaries including the necessary definitions and basic facts concerning (generalized) X^p-spaces. In Section 2 we discuss and prove some results in connection to the τ-uniformly PL-convexity of the spaces $X^p(A,\omega)$. Section 3 is used to present and prove some results concerning compact or quasi-compact operators between X^p-spaces. In Section 4 we prove a general inequality (see Theorem 4.1) which may be regarded as a strong generalization of some classical inequalities including those by Hölder, Minikowski and Backenbach-Dresher. In Section 5 we discuss some relations to generalized Gini means of the type treated in Peetre–Persson [14]. We point out some applications and further possibilities to develop this theory.

1. Preliminaries.

1.1 Conventions. (Ω,Σ,μ) denotes a σ-finite complete measure space and $L^{\circ}(\Omega)$ denotes the space of all complex-valued μ-measurable functions on Ω. For given quasi-Banach spaces A and B the notation $A = B$ $(A \equiv B)$ means that A and B coincides as sets and the corresponding quasi-norms are equivalent (equal). $\mathbb{R}$ and $\mathbb{C}$ denotes the set of real and complex numbers, respectively. Moreover, X denotes a Banach function space in the following sense:

1.2 Banach function spaces. A Banach function space $X = (X,\|\cdot\|_X)$ (on (Ω,Σ,μ)) is a Banach subspace of $L^{\circ}(\Omega)$ satisfying that, for every $x,y \in L^{\circ}(\Omega)$, the following lattice property holds:

$$y \in X,\ |x| \le |y|\ \mu\text{-a.e.} \implies x \in X \text{ and } \|x\|_X \le \|y\|_X.$$

1.3 Generalized X^p-spaces. Let X be a Banach function space, let A denote a Banach function space on Ω and let $\omega = \omega(s)$, $s\in\Omega$, be a positive weight function on Ω. The $X^p(A,\omega)$, $0<p<\infty$, consists of all strongly measurable functions (cross-sections) $x = x(s)$ satisfying

$$\|x\|_{X^p(A,\omega)} = (\|[\|x(s)\|_A\omega(s)]^p\|_X)^{1/p} < \infty,$$

see [13]. For the cases $A = \mathbb{R}$ and $\omega = 1$ we use the abbreviated notations $X^p(\omega)$ and $X^p(A)$, respectively. In particular the spaces $X^p(\mathbb{R},1)$ coincides with the usual X^p-spaces. It is easy to prove that the X^p-spaces are Banach function spaces if $p \ge 1$ and at least quasi-Banach spaces for all $p>0$ (see [15]).

1.4 Generalized Gini means. Let Γ be a linear class of real-valued functions $x = x(t)$ defined on a non-empty set E. If $L\colon \Gamma \to \mathbb{R}$, is an isotonic linear functional (i.e., if for every $x,y \in \Gamma$ such that $x(t) \ge y(t)$ on E it holds that $L(x) \ge L(y)$), then the corresponding generalized Gini mean $G(\alpha,\beta;x)$, $-\infty < \alpha,\beta < \infty$, is defined by

$$G(\alpha,\beta;x) = G_L(\alpha,\beta;x) = \begin{cases} (L(x^{\alpha})/L(x^{\beta}))^{1/(\alpha-\beta)} & ,\ \alpha \ne \beta, \\ \exp\left(\left\{\dfrac{d}{da}(\ln L(x^{a}))\right\}_{a=\alpha}\right) & ,\ \alpha = \beta, \end{cases}$$

whenever $x^\alpha, x^\beta \in \Gamma$ and $0 < L(x^\alpha), L(x^\beta) < \infty$. Some basic facts concerning these means can be found in [14].

1.5. Further definitions. We say that a Banach space X is *α-concave* or satisfies a *lower α-estimate*, $0 < \alpha < \infty$, if there exists a positive constant $M = M_\alpha(X)$ such that, for every finite set $x_1, x_2, \ldots, x_N$ of elements in X,

$$\left[\sum_1^N (\|x_i\|_X)^\alpha\right]^{1/\alpha} \le M\left\|\left[\sum_1^N |x_i|^\alpha\right]^{1/\alpha}\right\|_X \quad \text{or} \quad \left[\sum_1^N (\|x_i\|_X)^\alpha\right]^{1/\alpha} \le M\left\|\sum_1^N |x_i|\right\|_X ,$$

respectively. Let $E = (E, \|\cdot\|_E)$ denote a *continuously quasi-normed space*, i.e., let E be a quasi-normed space for which the norm $\|\cdot\|_E$ is uniformly bounded on bounded sets of E. For such a space the following family of moduli was introduced in [3]:
If $0 < p \le \infty$ and $\varepsilon > 0$, then

$$H_p^E(\varepsilon) = \inf\left\{\left[\frac{1}{2\pi}\int_0^{2\pi}(\|x+e^{i\theta}y\|_E)^p d\theta\right]^{1/p} - 1: \|x\|_E = 1, \ \|y\|_E = \varepsilon\right\}, \quad p<\infty,$$

$$H_\infty^E(\varepsilon) = \inf\left\{\sup(\|x+e^{i\theta}y\|_E: 0 \le \theta \le 2\pi) - 1: \|x\|_E = 1, \ \|y\|_E = \varepsilon\right\}, \quad p=\infty.$$

A function inverse to $H_\infty^E(\varepsilon)$ was introduced by Globevnik [5].

In accordance to the terminology used in [3] we say that the space E is *uniformly PL-convex* if $H_1^E(\varepsilon) > 0$ and *τ-uniformly PL-convex* if $H_1^E(\varepsilon) > \varepsilon^\tau$, $2 \le \tau < \infty$, for all $\varepsilon>0$ (τ-uniformly H_∞ PL-convexity is defined in a similar way).

2. Uniform PL-convexity of the $X^p(A,\omega)$-spaces.

It is well-known (see [3,p.117]) that if $0<p<\infty$ and $2 \le \tau < \infty$, then the continuously quasi-normed space E is τ-uniformly PL-convex if and only if there exists $\lambda>0$ such that

$$\left[\frac{1}{2\pi}\int_0^{2\pi}(\|x+e^{i\theta}y\|_E)^p d\theta\right]^{1/p} \ge \left[\|x\|_E^\tau + \lambda\|y\|_E^\tau\right]^{1/\tau}$$

for all $x,y \in E$; the largest possible value of λ is denoted by $I_{\tau,p}(E)$.

The following close relation between the τ-uniformly PL-convexities of the spaces A and $X^p(A,\omega)$ holds:

Theorem 2.1. *Let $0 < p \le \tau$, $\alpha = \tau/p$ and let ω denote a weight function on Ω. If A is a τ-uniformly PL-convex quasi-normed space and if X is an α-concave Banach function space with $M_\alpha(X) = 1$, then the space $X^p(A,\omega)$ is τ-uniformly PL-convex and $I_{\tau,p}(X^p(A,\omega)) = I_{\tau,p}(A)$.*

Proof. Using similar arguments as those used in the proof of Proposition 2.1 in [3] we easily see that $X^p(A,\omega)$ is at least a continuously quasi-normed space. Moreover, by assumption,

$$((\|a\|_X)^\alpha + (\|b\|_X)^\alpha)^{1/\alpha} \le \|(|a|^\alpha + |b|^\alpha)^{1/\alpha}\|_X.$$

Using this inequality, a generalized version of the triangle inequality (see [8,p.110]) and the properties of X and A we obtain that

$$\frac{1}{2\pi}\int_0^{2\pi}(\|x+e^{i\theta}y\|_{X^p(A,\omega)})^p d\theta = \frac{1}{2\pi}\int_0^{2\pi}\|(\|x(s)\omega(s)+e^{i\theta}y(s)\omega(s)\|_A)^p\|_X d\theta$$

$$\ge \frac{1}{2\pi}\Big\|\int_0^{2\pi}(\|x(s)\omega(s)+e^{i\theta}y(s)\omega(s)\|_A)^p d\theta\Big\|_X$$

$$\ge \|((\|x(s)\omega(s)\|_A)^\tau + I_{\tau,p}(A)(\|y(s)\omega(s)\|_A)^\tau)^{p/\tau}\|_X$$

$$\ge ((\|(\|x(s)\omega(s)\|_A)^p\|_X)^\alpha + (\|(\|(I_{\tau,p}(A))^{1/\tau}y(s)\omega(s)\|_A)^p\|_X)^\alpha)^{1/\alpha}$$

$$= ((\|x\|_{X^p(A,\omega)})^\tau + I_{\tau,p}(A)(\|y\|_{X^p(A,\omega)})^\tau)^{p/\tau}.$$

This proves that the space $X^p(A,\omega)$ is τ-uniformly PL-convex and that $I_{\tau,p}(X^p(A,\omega)) \ge I_{\tau,p}(A)$. Moreover, by considering $x(s) = a(s)x$, $y(s) = a(s)y$, where $x,y \in A$, $a(s) \in X^p(\omega)$, we find that the converse inequality is obvious and the proof is complete.

REMARK. For the case $X = L$ and $\omega \equiv 1$ Theorem 2.1 coincides with Theorem 4.1 in [3].

By applying Theorem 2.1 with $A = \mathbb{R}$ and $\omega \equiv 1$ we obtain the following generalization of some results obtained in [3]:

COROLLARY. *(a) if $0 < p \le 2$ and if X is 2/p-concave with $M_{2/p}(X) = 1$, then X^p is 2-uniformly PL-convex and $I_{2,p}(X^p) = I_{2,p}(\mathbb{C})$.*

(b) If $2 \le p \le \tau < \infty$ and if X is τ/p-concave with $M_{\tau/p}(X) = 1$, then the space X^p is τ-uniformly PL-convex and $I_{\tau,p}(X^p) = I_{\tau,p}(\mathbb{C})$.

REMARK. It is well-known that $I_{2,1}(\mathbb{C}) = 0.5$ (see [3,Proposition 3.1]). Therefore Corollary 2.2(a) contains the information that if X is 2-concave with $M_2(X) = 1$, then X is 2-uniformly PL-convex and $I_{2,1}(X) = 0.5$. This fact is also stated in [3,Theorem 7.1].

We say that X is an *AL-space* if $\||x|+|y|\|_X = \|x\|_X+\|y\|_X$ for all $x,y \in X$. It is well-known that every AL-space X is α-concave with $M_\alpha(X) = 1$ for every $\alpha \ge 1$ (see [15,Theorem 4.1]). Hence Corollary 2.2

implies our next example.

Example 2.1. Let X be an AL-space. If $0 < p \leq 2$, then X^p is 2-uniformly PL-convex with $I_{2,p}(X^p) = I_{2,p}(\mathbb{C})$. If $p \geq 2$ then X^p is p-uniformly PL-convex with $I_{p,p}(X^p) = I_{p,p}(\mathbb{C})$.

Remark. For the case $X = L$ Example 2.1 coincides with Corollary 4.2 in [3].

Proposition 2.3. *Let $1 \leq p < \infty$. If $q \geq 1$ and if X satisfies a lower q-estimate, then $X^p(\omega)$ satisfies a lower qp-estimate. In particular, if, for some $q \geq 2$, X is q-uniformly PL-convex, then there exists an equivalent norm of X under which the space $X^p(\omega)$ is qp-uniformly PL-convex.*

Proof. By using the assumption that X satisfies a lower q-estimate and the definition of $X^p(\omega)$-spaces we obtain

$$\left(\sum_1^N(\|x_i\|_{X^p(\omega)})^{pq}\right)^{1/pq} = \left(\sum_1^N(\|\,|x_i|^p\omega^p\|_X)^q\right)^{1/pq} \leq M\left(\|\sum_1^N|x_i|^p\omega^p\|_X\right)^{1/p}$$

$$\leq M\left(\|(\sum_1^N|x_i|)^p\omega^p\|_X\right)^{1/p} = M\;\|\sum_1^N|x_i|\|_{X^p(\omega)}$$

and the first statement is proved. The second statement follows from the first statement and Corollaries 7.1 and 7.4 in [3] (which ensure this connection between q-uniformly PL-convexity and that a lower q-estimate is satisfied).

Remark. In view of the proof above we see that if A is an AL-space, then Proposition 2.3 holds with $X^p(\omega)$ replaced by $X^p(A,\omega)$.

Remark. It is well-known that the space X^p, $p\geq1$, can be identified with the Calderon-Lozanovskyi space $X^{1-\theta}(L^\infty)^\theta$, $\theta = 1-1/p$ (see e.g. [12]), and thus, under some additional assumptions, with the complex interpolation space $[X,L^\infty]_\theta$ (see [1,sec.13.6]). Hence Proposition 2.3 implies that if X is q-uniformly PL-convex (and some additional assumptions are satisfied), then the space $[X,L^\infty]_\theta$ is qp-uniformly PL-convex ($\theta = 1-1/p$). On the other hand, according to a counter example of G.Pisier (see [4,p.503]), there exists a complex interpolation pair for which one of the spaces is 2-uniformly PL-convex but all intermediate complex interpolation spaces do not admit an equivalent uniformly PL-convex norm.

The last remark ought to be compared with the well-known fact that if one of the spaces X_o, X_1 is **uniformly convex** (in the usual sense), then all the complex interpolation spaces $[X_o,X_1]_\theta$, $0<\theta<1$, are also uniformly convex (see Cwikel-Reisner [2]).

3. On (quasi-) compact operators between X^p-spaces.

We state the following generalization of Theorem 3.10 in [7]:

Proposition 3.1. *Let $0 < \theta < 1$, $1 \le p_o,p_1,q_o,q_1 < \infty$, $1/p_\theta = (1-\theta)/p_o + \theta/p_1$ and $1/q_\theta = (1-\theta)/q_o + \theta/q_1$. If X is separable symmetric space and if $\Gamma:X^{p_o} \to X^{p_1}$ is a compact operator and $\Gamma:X^{q_o} \to X^{q_1}$ is a bounded operator, then $\Gamma:X^{p_\theta} \to X^{q_\theta}$ is a compact operator.*

Proof. First we note that $(X^{p_o},X^{p_1},X^{p_\theta})$ is a "good" interpolation triple relative to the triple $(X^{q_o},X^{q_1},X^{q_\theta})$ in the sense as defined in [6,p.21]. Let P_n, $n\in\mathbb{Z}_+$, denote the projection of the linear hull of the first n elements of the Haar system. In particular, the operators P_n are finite-dimensional operators acting in X^{q_o} and X^{q_1}. Since the Haar system is a basis in every separably symmetric space, we find that $P_n x \to x$ for every $x \in X^{q_o}$. Moreover, by using the estimates and interpolation arguments used in [6,p.179] we find that $\{P_n\}_{n\in\mathbb{Z}_+}$ is uniformly bounded (with the constant 1) in X^{q_i}. Therefore all assumptions in Theorem 4.1 in [6] are satisfied and the proof follows.

We will also present a similar result for quasi-compact operators. We recall that an operator $\Gamma:E \to E$ is *quasi-compact* if there exists a compact operator $K:E \to E$ and a positive integer m such that $\|\Gamma^m-K\|_{E\to E} < 1$. We need the following lemma:

Lemma 3.1. *Let X be a Banach function space over Ω such that $\|1_\Omega\|_X<\infty$. If $p_o < p_1$, then $X^{p_o} \supset X^{p_1}$.*

Proof. It is well-known that $\|xy\|_{X^{p_o}} \le \|x\|_{X^q}\|y\|_{X^{p_1}}$, $1/p_1+1/q = 1/p_o$ (see e.g. [15,Theorem 3.1]). We insert $y = 1_\Omega(s)$ into this estimate and the proof follows.

Proposition 3.2. *Let $p_o < p_1$ and let X be a separable symmetric space such that $\|1_\Omega\| < \infty$. If Γ is a quasi-compact operator in X^{p_o} and a bounded operator in X^{p_1}, then Γ is a quasi-compact operator in X^{p_θ}, $1/p_\theta = (1-\theta)/p_o + \theta/p_1$.*

Proof. The triple $(X^{p_o},X^{p_1},X^{p_\theta})$ is an interpolation triple of the type θ. Let P_n, $n\in\mathbb{Z}_+$, denote the operators defined in the proof of Proposition 3.1. Then, according to the arguments used in that proof, we find that $\{P_n\}_{n\in\mathbb{Z}_+}$ satisfies the appropriate approximation condition (H) in [11]. Therefore the proof follows by using Lemma 3.1 and Theorem 4 in [11].

In particular, the statement above implies that if X satisfies the conditions in Proposition 3.2 and if Γ is a Riesz operator in X^{p_o} which is bounded in X^{p_1}, then Γ is a Riesz operator in X^{p_θ} too.

4. A general inequality.

Let $p(t)$ and $W(t)$ denote positive functions on $(0,b)$, $b>0$, and let p be defined by

$$\frac{1}{p} = \frac{1}{I_W}\int_0^b \frac{1}{p(t)}\, W(t)dt, \qquad \text{where } I_W = \int_0^b W(t)dt < \infty. \tag{4.1}$$

Moreover, let $\{\omega_t(s)\}$, $t\in[0,b)$, $s\in\Omega$, denote a family of weight functions such that $\omega_t(s)$ is measurable on $[0,b)\times\Omega$ and consider the generalized geometric mean

$$\omega = \omega(s) = \exp\left[\frac{1}{I_W}\int_0^b \log\omega_t(s)W(t)dt\right].$$

The following generalization of Theorem 1 in [13] may be regarded as an extension of Hölder's inequality and also of some other classical inequalities.

Theorem 4.1. *Let X be a Banach function space such that $X'' \equiv X$ and let $Z_t = X^{p(t)}(A,\omega_t)$. Assume that $x_t(s)\in Z_t$ and $W(t)\log\|x_t(s)\|_{Z_t}$ and $W(t)\log\|x_t(s)\|_A$, $s\in\Omega$, are integrable over $[0,b)$. Then*

$$y = y(s) = \exp\left[\frac{1}{I_W}\int_0^b \log\|x_t(s)\|_A W(t)dt\right]$$

belongs to $X^p(\omega)$ and

$$\|y\|_{X^p(\omega)} \le \exp\left[\frac{1}{I_W}\int_0^b \log\|x_t(s)\|_{Z_t} W(t)dt\right].$$

Proof. First we note that it is sufficient to prove that

$$\left\| \exp \left(\frac{1}{I_W} \int_0^b \log \alpha_t(s) W(t) dt \right) \right\|_{X^p(\omega)} \le 1, \tag{4.2}$$

where, by definition,

$$\alpha_t = \alpha_t(s) = \|x_t(s)\|_A / \|x_t(s)\|_{Z_t}.$$

Since

$$\left[\exp \left(\frac{1}{I_W} \int_0^b \log \alpha_t(s) W(t) dt \right) \omega(s) \right]^p$$

$$= \left[\exp \left(\frac{1}{I_W} \int_0^b \log (\alpha_t(s) \omega_t(s)) W(t) dt \right) \right]^p$$

$$= \exp \left(\frac{1}{I_W} \int_0^b \log (\alpha_t(s) \omega_t(s))^{p(t)} \frac{p}{p(t)} W(t) dt \right),$$

the function exp is convex and

$$\frac{1}{I_W} \int_0^b \frac{p}{p(t)} W(t) dt = 1 \tag{4.3}$$

(see (4.1)) we can use Jensen's inequality (see e.g. [8,p.133]) to obtain that

$$\left[\exp \left(\frac{1}{I_W} \int_0^b \log \alpha_t(s) W(t) dt \right) \omega(s) \right]^p \tag{4.4}$$

$$\le \frac{1}{I_W} \int_0^b (\alpha_t(s) \omega_t(s))^{p(t)} \frac{p}{p(t)} W(t) dt.$$

We consider the function

$$F(t,s) = \frac{1}{I_W} (\alpha_t(s) \omega_t(s))^{p(t)} \frac{p}{p(t)} W(t)$$

and note that

$$\|F(t,s)\|_X = \frac{1}{I_W} \frac{p}{p(t)} W(t) \left\| (\alpha_t(s) \omega_t(s))^{p(t)} \right\|_X \tag{4.5}$$

$$= \frac{1}{I_W} \frac{p}{p(t)} W(t) \frac{\| (\|x_t(s)\|_A \omega_t(s))^{p(t)} \|_X}{(\|x_t(s)\|_{Z_t})^{p(t)}} = \frac{1}{I_W} \frac{p}{p(t)} W(t).$$

Now using (4.3)-(4.5), the lattice property of X and a generalization of Minikowski's integral inequality (see e.g. [6,p.45]) we find that

$$\left(\left\| \exp \left(\frac{1}{I_W} \int_0^b \log \alpha_t(s) W(t) dt \right) \right\|_{X^p(\omega)} \right)^p \le \left\| \int_0^b F(t,s) dt \right\|_{X''}$$

$$\le \int_0^b \|F(t,s)\|_X dt = \int_0^b \frac{1}{I_W} \frac{p}{p(t)} W(t) dt = 1.$$

Thus (4.2) yields and the proof is complete.

Remark. In view of the proof above and a well-known generalization of the triangle inequality (see e.g. [8,p.110]) we find that if $(\alpha_t(s)\omega_t(s))^{p(t)}$, $t \in [0,2\pi)$, are strongly measurable functions (taking values in X), then Theorem 4.1 holds also without the assumption $X'' \equiv X$.

Corollary 4.2. *Let $z_i \in X^{p_i}(A)$, $0 < p_i < \infty$, $0 < \alpha_i < 1$, $i = 1,2,\dots,N$, $\sum_1^N \alpha_i = 1$ and $\frac{1}{p} = \sum_1^N \frac{\alpha_i}{p_i}$. Then*

$$\left\|\prod_1^N(\|z_i(s)\|_A)^{\alpha_i}\right\|_{X^p} \le \prod_1^N(\|z_i(s)\|_{X^{p_i}(A)})^{\alpha_i}. \tag{4.6}$$

Proof. Let $0 = a_1 < a_2 < \dots < a_{N+1} = b$ where $\alpha_i = (a_{i+1}-a_i)/b$, $i = 1,2,\dots,N$. Apply Theorem 4.1 and the last remark with $\omega(s) \equiv 1$, $W(t) \equiv 1$, $p(t) = p_i$ and $x_t(s,a) = z_i(s,a)$ on $[a_i,a_{i+1})$, $i = 1,2,\dots,N$ and the proof follows.

Example 4.1. In particular, if $A = \mathbb{R}$, $p_i/\alpha_i = q_i$, $|z_i|^{\alpha_i} = |x_i|$, $i = 1,2,\dots,N$, then (4.6) can be written as

$$\left\|\prod_1^N x_i\right\|_{X^p} \le \prod_1^N \|x_i\|_{X^{q_i}}, \tag{4.7}$$

where $\frac{1}{p} = \sum_1^N \frac{1}{q_i}$, $0 < q_i < \infty$.

Remark. Another proof of the inequality (4.7) (and similar symmetric forms of Hölder's inequality) can be found in [15].

We close this section by giving a surprisingly simple proof of the following generalization of Beckenbach-Dresher's inequality:

Corollary 4.3. *Let X and Y be Banach function spaces on (Ω,Σ,μ). If $s \ge 1$, $p \ge 1 \ge r > 0$, $p \ne r$, and x and y are functions on Ω satisfying*

$$\|x+y\|_{Y^r} \ge \|x\|_{Y^r}+\|y\|_{Y^r}, \tag{4.8}$$

then

$$\frac{(\|x+y\|_{X^p})^s}{(\|x+y\|_{Y^r})^{s-1}} \le \frac{(\|x\|_{X^p})^s}{(\|x\|_{Y^r})^{s-1}} + \frac{(\|y\|_{X^p})^s}{(\|y\|_{Y^r})^{s-1}}. \tag{4.9}$$

Proof. If $\alpha_1,\alpha_2,\beta_1$ and β_2 are positive numbers, then, by Hölder's inequality (a special case of (4.7)), we find that

(4.10) $$\frac{(\alpha_1+\alpha_2)^s}{(\beta_1+\beta_2)^{s-1}} \le \frac{\alpha_1^s}{\beta_1^{s-1}} + \frac{\alpha_2^s}{\beta_2^{s-1}}$$

Moreover, (4.7) (with N=2, r=1) implies also Minikowski's inequality

(4.11) $$\|x+y\|_{X^p} \le \|x\|_{X^p}+\|y\|_{X^p}.$$

The proof of (4.9) follows by combining (4.10) with (4.8) and (4.11).

REMARK It is well-known that if (4.8) holds for $r=r_o$, then (4.8) holds also for every r, $0 < r \le r_o$ (see [15,Theorem 4.1]). In particular this means that if X is a AL-space of non-negative functions, i.e., if, for all $x,y \in X$, $\|x+y\|_X = \|x\|_X+\|y\|_X$, then Corollary 4.3 holds also without the assumption (4.8).

REMARK. For the case X = Y and s = p/(p-r) (4.9) can be written

(4.12) $$\left[\frac{(\|x+y\|_{X^p})^p}{(\|x+y\|_{X^r})^r}\right]^{\frac{1}{p-r}} \le \left[\frac{(\|x\|_{X^p})^p}{(\|x\|_{X^r})^r}\right]^{\frac{1}{p-r}} + \left[\frac{(\|y\|_{X^p})^p}{(\|y\|_{X^r})^r}\right]^{\frac{1}{p-r}},$$

where $0 < r \le 1 \le p$, $p \ne r$. Another proof of this generalization of Beckenbach-Dresher's inequality has been presented in [15]. A generalization in a quite another (many-valued) direction can be found in [14]. For other generalizations and historical remarks we refer to the lists of references in [14] and [15].

5. Some connections to generalized Gini means and concluding remarks.

Let $-\infty < \alpha,\beta < \infty$ and consider the generalized Gini mean $G_L(\alpha,\beta;x)$ corresponding to the functional $L(x) = \|x\|_X$, i.e.,

$$G_L(\alpha,\beta;x) = \begin{cases} \left[\dfrac{(\|x\|_{X^\alpha})^\alpha}{(\|x\|_{X^\beta})^\beta}\right]^{\frac{1}{\alpha-\beta}} & ,\alpha\ne\beta,\ \ \alpha,\beta\ne0, \\[2ex] \exp\left(\left\{\dfrac{d}{da}(\ln\| \, |x^a| \, \|_X)\right\}_{a=\alpha}\right) & ,\ \alpha = \beta, \end{cases}$$

Here we assume that $0 < \|x\|_{X^\alpha}, \|x\|_{X^\beta} < \infty$. For $\beta=0$ and $\alpha\ne0$ we assume that $\|1_\Omega\|_X < \infty$ and put

$$G_L(\alpha,0;x) = \left[\frac{(\|x\|_{X^\alpha})^\alpha}{\|1_\Omega\|_X}\right]^{1/\alpha}.$$

We introduce the spaces X_β^α as

$$X_\beta^\alpha = \{x \in L^\circ(\Omega) : \|x\|_{X_\beta^\alpha} = G_L(\alpha,\beta;x) < \infty\}$$

and collect some of its properties in the following proposition:

Proposition 5.1. *Let* $-\infty < \alpha,\beta < \infty$. *Then*

(a) $X_o^\alpha = X^\alpha$, $\alpha \neq 0$.

(b) $X_\beta^\alpha = X_\alpha^\beta$.

(c) $\|x+y\|_{X_\beta^\alpha} \le \|x\|_{X_\beta^\alpha} + \|y\|_{X_\beta^\alpha}$, *if* $0 \le \beta \le 1 \le \alpha$.

(d) $\|x+y\|_{X_\beta^\alpha} \ge \|x\|_{X_\beta^\alpha} + \|y\|_{X_\beta^\alpha}$, *if* $\beta < 0 < \alpha \le 1$.

(e) $\lim\limits_{\alpha \to \beta} \|x\|_{X_\beta^\alpha} = \|x\|_{X_\beta^\beta}$.

(f) $\|x\|_{X_\alpha^\beta} = \exp\left(\frac{1}{\beta-\alpha}\int_\alpha^\beta \ln\|x\|_{X_a^a}\,da\right)$, $\alpha \neq \beta$.

Proof. The statements in (a) and (b) are trivially true. The estimate in (c) is only a reformulation of our estimate (4.12) and (d) can be proved in a similar way (see also [14,Corollary 3.2]). The continuity property in (e) is an immediate consequence of the definitions above and the representation formula (f) is a special case of Proposition 4.1 in [14].

Remark. The representation formula in (f) means that the norms in the off-diagonal cases can be written as a generalized geometric mean of the norms in the natural intermediate diagonal cases.

Next we note that y(s) and the function on the right hand side of the estimate in Theorem 4.1 can be written as the generalized Gini means $G_{L_o}(0,0;\|x_t(s,a)\|_A)$, $s\in\Omega$, and $G_{L_o}(0,0;\|x_t(s,a)\|_{Z_t})$, respectively, where $L_o(g(t)) = \int_0^b g(t)W(t)dt$.

According to the well-known properties of generalized Gini means (see [14]), we note that also the functionals $G_{L_o}(\alpha,\beta;g(t))$ has properties corresponding to (a) - (f) in Proposition 5.1.

Finally we remark that well-known results concerning (generalized) Gini means already have implied several applications in data compression, the theory of generalized entropies and Information Theory (see [9],[14] and the references given there). We think that some of the results obtained in this paper will imply new applications of this kind.

REFERENCES

[1] A.P.CALDERON, *Intermediate spaces and interpolation, the complex method*, Studia Math. 24 (1964), 133–190.

[2] M.CWIKEL, S. REISNER, *Interpolation of uniformly convex Banach spaces*, Proc Amer. Math. Soc. 84 (1982), 55–59.

[3] W.J.DAVIS, D. J. H. GARLING, N. TOMCZAK-JAEGERMAN, *The complex convexity of quasi normed linear spaces*, J. Functional Anal. 55 (1984), 110–150.

[4] S.J.DILWORTH, *Complex convexity and the geometry of Banach spaces*, Math. Proc. Camb. Phil. Soc. 99 (1986), 495–506.

[5] J.GLOBEVNIK, *On the complex strict and uniform convexity*, Proc. Amer. Math. Soc. 47 (1975), 175–178.

[6] S.G.KREIN, YU. I. PETUNIN, E. M. SEMENOV, *Interpolation of Linear Operators*, Nauka, Moscow 1978 (in Russian); English translation, AMS, Providence 1982.

[7] M.A.KRASNOSELSKII, P. P. ZABREJKO, E. I. PUSTYLNIK, P. E. SOBOLEVSKII, *Integral Operators in Spaces of Summable Functions*, Moscow 1966.

[8] A.KUFNER, O. JOHN, S. FUCIK, *Function Spaces*, Noordhoff International Publishing, Leiden 1977.

[9] T.KOSKI, L–E. PERSSON, *Some properties of generalized entropies with applications to data compression*, Research report 2, Dept. of Math., Luleå University, 1989 (submitted).

[10] L.MALIGRANDA, L–E. PERSSON, *Generalized duality of some Banach functions spaces*, Indagationes Math. (to appear).

[11] L.I.NIKOLOVA, *On the interpolation of compactness property in families of Banach spaces*, (submitted).

[12] P.NILSSON, *Interpolation of Banach lattices*, Studia Math. 82 (1985), 135–154.

[13] L.I.NIKOLOVA, L–E. PERSSON, *On interpolation between X^p-spaces*, Pitman's Lecture Notes, (to appear).

[14] J.PEETRE, L–E. PERSSON, *A general Beckenbach's inequality with appllications*, Pitman's Lecture Notes, (to appear).

[15] L–E. PERSSON, *Some elementary inequalities in connection with X^p-spaces*, Publishing House of the Bulgarian Academy of Sciences, 1988, 367–376.

APPROXIMATION OF ALMOST PERIODIC FUNCTIONS INTEGRABLE IN THE DENJOY-PERRON SENSE

PAULINA PYCH-TABERSKA

Poznań, Poland

1. Preliminaries.

Let D^*_{loc} be the class of all complex-valued functions defined on the real line $\mathbb{R} = (-\infty,\infty)$ and integrable in the Denjoy-Perron sense on each finite interval. Introduce, for $f \in D^*_{loc}$, the quantity

$$\text{(1)} \qquad \|f\| \equiv \|f(\cdot)\| := \sup_{-\infty<x<\infty} \left\{ \sup_{0\le\eta\le1} \left| \int_x^{x+\eta} f(t)dt \right| \right\}.$$

Clearly, this quantity may be finite or infinite.

A function f of class D^*_{loc} is said to be *S^*-almost periodic* if it possesses the following property: to each $\varepsilon>0$ there corresponds a positive number $l=l(\varepsilon)$ such that in each interval of length l there exists at least one number τ for which $\|f(\cdot+\tau)-f(\cdot)\| < \varepsilon$.

Denote by S^* the set of all functions almost periodic in the above sense. The elementary properties of functions belonging to S^* are very similar to those of Stepanov's almost periodic functions ([4]) and can be found in [6]. In particular, S^* is a normed linear space and the functional $\|f\|_{S^*} := \|f\|$ defined for $f\in S^*$ by (1) is a norm in this space.

Use the symbol L for the space of all complex-valued functions Lebesgue-integrable on $\mathbb{R}$, with the usual norm

$$\|f\|_L := \int_{-\infty}^{\infty} |f(t)|dt.$$

Let B_σ $(\sigma>0)$ be the class of all entire functions of exponential type σ, bounded on $\mathbb{R}$, and let $\tilde{B}_\sigma$ be the subclass of these functions belonging to B_σ which are uniformly almost periodic on $\mathbb{R}$.

In this paper we present some estimates for the best approximation of $f\in S^*$ by the entire functions of class $\tilde{B}_\sigma$, i.e., for the quantity

$$E_\sigma(f)_{S^*} := \inf_{G\in \tilde{B}_\sigma} \|f - G\|_{S^*}.$$

Moreover, the uniqueness theorem for S^*-almost periodic functions is deduced.

2. Auxiliary results.

As is known ([3],p.43 or [7]), if a function f is integrable in the Denjoy-Perron sense on a finite interval $[\alpha,\beta]$ and if φ denotes an arbitrary function of bounded variation on this interval, then the product $f\varphi$ is Denjoy-integrable on $[\alpha,\beta]$. Furthermore, for every interval $[\gamma,\delta] \subset [\alpha,\beta]$,

$$(2) \qquad \left|\int_\gamma^\delta f(t)\varphi(t)dt\right| \le \left\{\sup_{\gamma\le t\le\delta} |\varphi(t)| + \operatorname*{var}_{\gamma\le t\le\delta} \varphi(t)\right\} \times \max_{\gamma\le\xi\le\delta} \left|\int_\alpha^\xi f(t)dt - \int_\alpha^\gamma f(t)dt\right|.$$

Given any function $f\in D^*_{loc}$ and any function g of complex variable $z=x+iy$, with $\operatorname*{var}_{-\infty<u<\infty} g(u+iy) < \infty$ for some fixed $y\in\mathbb{R}$, let us consider their convolution $f*g$ defined, formally, by the improper Denjoy-Perron integral

$$(3) \qquad (f*g)(z) := \int_{\to-\infty}^{\to\infty} f(t)g(z-t)dt \equiv \lim_{\substack{a\to-\infty\\ b\to\infty}} \left(\int_a^0+\int_0^b\right) f(t)g(z-t)dt.$$

Considering a (complex-valued) function φ bounded on finite intervals we adopt the convenient notation

$$((\varphi)) := \sum_{k=-\infty}^{\infty} \sup_{k\le u\le k+1} |\varphi(u)|.$$

Lemma 1. *Let f be of class D^*_{loc}, with $\|f\|<\infty$. Suppose that $g\in B_\sigma$ and that series*

$$(4) \qquad \sum_{k=-\infty}^{\infty}\left\{ \sup_{k\le u\le k+1} |g(u+iv)| + \sup_{k\le u\le k+1} |g'(u+iv)|\right\}$$

is uniformly convergent in v on every finite interval of $\mathbb{R}$. Then $f*g \in B_\sigma$.

Proof. Given any $\psi\in B_\sigma$, let

$$((\psi(\cdot+iv)))^\circ := \sup_{0\le h<1} \sum_{k=-\infty}^{\infty} \sup_{k+h\le u\le k+h+1} |\psi(u+iv)| \qquad (v\in\mathbb{R});$$

write

$$((\psi))^{\circ} \equiv ((\psi(\cdot)))^{\circ} := ((\psi(\cdot+i0)))^{\circ}.$$

The assumption concerning series (4) ensures that $((g))<\infty$. Further,

$$((g)) \le ((g))^{\circ} \le 2((g))$$

and

$$((g(\cdot+s)))^{\circ} = ((g(\cdot)))^{\circ} \qquad \text{whenever} \quad s\in\mathbb{R}.$$

Therefore

$$((g'))^{\circ} \le \sigma((g))^{\circ}, \qquad ((g''))^{\circ} \le \sigma((g'))^{\circ} \le \sigma^{2}((g))^{\circ}, \quad \text{etc.}$$

Applying the well known Taylor's expansion, we get

$$((g(\cdot+iv)))^{\circ} \le ((g))^{\circ}e^{\sigma|v|}, \qquad ((g'(\cdot+iv)))^{\circ} \le \sigma((g))^{\circ}e^{\sigma|v|}$$

for each real v (see [5],pp. 114–115). Consequently,

$$(5) \quad \sum_{k=-\infty}^{\infty}\Big\{ \sup_{k\le u\le k+1} |g(u+iv)| + \sup_{k\le u\le k+1} |g'(u+iv)|\Big\} \le 2(1+\sigma)((g))e^{\sigma|v|}.$$

Let $l_1<l_2$ be two positive numbers and let $n_1=[l_1]$, $n_2=[l_2]$ be their integral parts. Then, by (2), for complex z=x+iy,

$$\Big|\int_{l_1}^{l_2} f(t)g(z-t)dt\Big| = \Big|\Big(\int_{l_1}^{n_1+1} + \sum_{k=n_1+1}^{n_2-1}\int_{k}^{k+1} + \int_{n_2}^{l_2}\Big) f(t)g(z-t)dt\Big|$$

$$\le 2 \sum_{k=n_1}^{n_2}\Big\{ \sup_{k\le t\le k+1} |g(x-t+iy)| + \operatorname*{var}_{k\le t\le k+1} g(x-t+iy)\Big\} \max_{k\le \xi\le k+1} \Big|\int_{k}^{\xi} f(t)dt\Big|$$

$$\le 4 \sum_{\nu=m-n_1-1}^{m-n_2}\Big\{ \sup_{\nu\le u\le \nu+1} |g(u+iy)| + \sup_{\nu\le u\le \nu+1} |g'(u+iy)|\Big\}\|f\|,$$

where $m:=[x]$. This estimate together with the assumption on series (4) ensures that

$$\lim_{b\to\infty} \int_{0}^{b} f(t)g(z-t)dt$$

is finite for fixed z=x+iy. Obviously, the same is true for the limit $\int_a^0 \ldots$ occurring in the definition (3). Furthermore, the improper integral in (3) is convergent uniformly with respect to z on every bounded set of the complex plane.

Let us write

$$(6) \qquad (f*g)(z) = \lim_{n\to\infty} \int_{-n}^{n} f(t)g(z-t)dt = \sum_{k=-\infty}^{\infty} I_k(z),$$

where

$$I_k(z) := \int_{k}^{k+1} f(t)g(z-t)dt.$$

Every integral I_k (k=0,±1,±2,...) is an entire function. Indeed, given complex numbers z, w $(0<|w|\le 1)$ we have, by (2),

$$\left|\frac{I_k(z+w)-I_k(z)}{w} - \int_k^{k+1} f(t)g'(z-t)dt\right|$$

$$= \left|\int_k^{k+1} f(t) \left\{ \frac{1}{w} \int_0^w \left[\int_0^v g''(z+u-t)du\right] dv\right\} dt\right|$$

$$\leq \left\{ \sup_{\substack{k\leq t\leq k+1 \\ |v|\leq|w|}} \left|\int_0^v g''(z+u-t)du\right| + \int_k^{k+1} \left|\frac{1}{w} \int_0^w \left[\int_0^v g'''(z+u-t)du\right] dv\right| dt\right\} \|f\|$$

$$\leq \{ \sup_{\zeta\in\Omega} |g''(\zeta)| + \sup_{\zeta\in\Omega} |g'''(\zeta)| \} |w| \|f\|,$$

Ω being a bounded domain independent of w. Consequently, as $w \to 0$, we obtain

$$I_k'(z) = \int_k^{k+1} f(t)g'(z-t)dt.$$

The series occurring in (6) converges uniformly on each bounded set of complex numbers z. Therefore $f*g$ is an analytic function on the whole complex plane. Moreover, in view of (2) and (5),

$$|(f*g)(z)| \leq \sum_{k=-\infty}^{\infty} \left\{ \sup_{k\leq t\leq k+1} |g(z-t)| + \operatorname*{var}_{k\leq t\leq k+1} g(z-t)\right\} \max_{k\leq\xi\leq k+1} \left|\int_k^{\xi} f(t)dt\right|$$

$$\leq 4(1+\sigma) ((g)) \|f\|e^{\sigma|y|}$$

for all $z=x+iy$; i.e., $f*g \in B_\sigma$.

Denote by V the class of all (complex-valued) functions of bounded variation on $\mathbb{R}$.

Lemma 2. *Suppose that $f\in S^*$, $g\in V$ and $((g)) < \infty$. Then, the convolution $f*g$ is a uniformly almost periodic function and*

(7) $$\sup_{-\infty<x<\infty} |(f*g)(x)| \leq 2 \|f\|_{S^*} \|g\|_V ,$$

where $\|g\|_V := ((g)) + \operatorname*{var}_{-\infty<u<\infty} g(u)$. Moreover,

(8) $$\|f*g\|_{S^*} \leq \|f\|_{S^*} \|g\|_L.$$

Proof. It is easy to verify that, under the assumptions $\|f\|<\infty$ and $((g))<\infty$, the improper integral

$$\int_{\to-\infty}^{\to\infty} f(x-u)g(u)du \equiv \lim_{\substack{a\to-\infty \\ b\to\infty}} \left(\int_a^0 + \int_0^b \right) f(x-u)g(u)du$$

exists uniformly in $x\in\mathbb{R}$ and is equal to $(f*g)(x)$. Hence, for each real number τ,

$$(f*g)(x+\tau) - (f*g)(x) = \lim_{n\to\infty} \int_{-n}^{n} \{f(x-u+\tau) - f(x-u)\} g(u)du$$

uniformly in x. In view of (2),

$$\left|\int_{-n}^{n}\{f(x-u+\tau)-f(x-u)\}\, g(u)du\right| \le \sum_{k=-n}^{n-1}\left|\int_{k}^{k+1}\{f(x-u+\tau)-f(x-u)\}\, g(u)du\right|$$

$$\le \sum_{k=-n}^{n-1}\left\{\sup_{k\le u\le k+1}|g(u)| + \operatorname*{var}_{k\le u\le k+1} g(u)\right\} \max_{k\le\xi\le k+1}\left|\int_{x-\xi}^{x-k}\{f(s+\tau)-f(s)\}ds\right|.$$

Thus we obtain the estimate

$$|(f*g)(x+\tau) - (f*g)(x)| \le 2\, \|f(\cdot+\tau) - f(\cdot)\|_{S^*}\, \|g\|_V ,$$

which implies the uniform almost periodicity of the convolution $f*g$. Analogous calculation leads to inequality (7).

Next, given any real x and any $\eta \in [0,1]$, we have

$$\left|\int_{x}^{x+\eta}(f*g)(t)dt\right| = \left|\int_{x}^{x+\eta}\left\{\lim_{n\to\infty}\int_{-n}^{n} f(t-u)g(u)\right\} dt\right|$$

$$\le \lim_{n\to\infty}\left|\int_{x}^{x+\eta}\left\{\int_{-n}^{n} f(t-u)g(u)du\right\} dt\right|.$$

In view of the known property of the Denjoy-Perron integral ([3], Chap.I, Theorem 58 or [7], Lemma 1), the right-hand side of the last inequality can be rewritten in the form

$$\lim_{n\to\infty}\left|\int_{-n}^{n}\left\{\int_{x}^{x+\eta} f(t-u)dt\right\} g(u)du\right|$$

and can be estimated from above by

$$\lim_{n\to\infty}\int_{-n}^{n}|g(u)|du\cdot\|f\|_{S^*} ,$$

which implies (8).

From Lemma 1 and 2 it follows

Corollary 1. *If f is of class S^* and g is an entire function of exponential type σ, for which series (4) converges uniformly on bounded sets, then the convolution $f*g$ belongs to the class $\tilde{B}_\sigma$.*

Given a positive number c and a positive integer r, let p be an even real-valued function continuous with derivatives p', p'' on $\mathbb{R}$, such that the quotients

$$p(u)/u^{r+2}, \quad p'(u)/u^{r+1}, \quad p''(u)/u^{r}$$

are bounded on the interval (0,c), and p(u)=1 for all u≥c. Introduce the Krejn's kernels Φ_m defined by the improper Lebesgue integrals:

$$\Phi_m(t) := \frac{1}{2\pi}\lim_{\alpha\to\infty}\int_{-\alpha}^{\alpha}\frac{p(u)}{(iu)^m}\, e^{itu}du \qquad (t\in\mathbb{R};\ m=1,\dots,r)$$

([1], p.238). The function Φ_1 is odd, continuous and bounded with its derivative Φ'_1 on $(0,\infty)$, $\Phi_1(0) = 0$, $\Phi_1(0+) = 1/2$, $\Phi_1(0-) = -1/2$. Further, Φ_2 is continuous on $\mathbb{R}$ and $\Phi'_2(t) = \Phi_1(t)$ for all real $t\neq 0$. In the case $r>2$, Φ_r is continuous on $\mathbb{R}$ and $\Phi'_r(t) = \Phi_{r-1}(t)$ for all real t. Moreover, the partial integrations show that

$$\Phi_r(t) = O(t^{-2}), \qquad \Phi'_r = O(t^{-2}) \qquad \text{as } t \to \pm\infty \quad (r\geq 1).$$

Lemma 3. *Let a function f have the dervative $f^{(r-1)}$ $(r\in\mathbb{N})$ absolutely continuous in the generalized sense on each finite interval, i.e., $f^{(r-1)} \in ACG^*_{loc}$. Suppose, further, that*

$$\int_{-\infty}^{\infty}\left(\frac{1}{1+|t|}\ |f^{(r-1)}(t)|\ \right)^2 dt < \infty$$

and $\|f^{(r)}\| < \infty$. Then, for every real x,

(9) $$f(x) = (f^{(r)}*\Phi_r)(x) + F_c(x),$$

F_c being an entire function of exponential type c.

Proof. In view of the known representation formula ([1], Sect. 101), there is an entire function H_c of exponential type c, such that for every real x,

$$\int_0^x f(t)dt = \int_{-\infty}^{\infty}\left\{f^{(r-1)}(x-t) - f^{(r-1)}(-t)\right\} \Phi_r(t)dt + H_c(x)$$

$$= \lim_{n\to\infty} \int_{-n}^{n}\left\{\int_0^x f^{(r)}(s-t)ds\right\} \Phi_r(t)dt + H_c(x).$$

Clearly, the function Φ_r $(r\geq 1)$ belongs to V and $((\Phi_r)) < \infty$. Hence (see the proof of Lemma 2),

$$(f^{(r)}*\Phi_r)(s) = \lim_{n\to\infty} \int_{-n}^{n} f^{(r)}(s-t)\Phi_r(t)dt$$

uniformly in $s\in\mathbb{R}$. Consequently,

$$\int_0^x f(t)dt = \int_0^x (f^{(r)}*\Phi_r)(s)ds + H_c(x) \qquad (x\in\mathbb{R}).$$

This formula is equivalent to the one to be proved.

3. Jackson type theorems.

Let us start with the following

Theorem 1. *Suppose that a function f is bounded on $\mathbb{R}$ with its derivatives f', f'',..., $f^{(r-1)}$ $(r\in\mathbb{N})$, and that $f^{(r-1)}\in ACG^*_{loc}$.*

Moreover, let $f^{(r-1)}$ be uniformly continuous on $\mathbb{R}$ and $f^{(r)} \in S^*$. Then $f \in S^*$ and

$$E_\sigma(f)_{S^*} \leq K_r \sigma^{-r} \|f^{(r)}\|_{S^*} \qquad \text{for each } \sigma > 0,$$

where K_r denote the well-known Favard's constants.

Proof. Let $0<c<\sigma$ and let Φ_r be the Krejn's kernel. The function f satisfies all conditions of Lemma 3; whence it can be represented by formula (9). Clearly, $f^{(r)} * \Phi_r$ is a uniformly almost periodic function (see Lemma 2) and so is the derivative $f^{(r-1)}$ as a bounded and uniformly continuous indefinite Denjoy-Perron integral of $f^{(r)}$. Also, the uniform almost periodicity of f is evident (see [4], Theorem 1.2.1). Consequently, the function F_c occurring in (9) is of class $\tilde{B}_\sigma$.

Consider the operator U_σ defined for all complex z by

$$U_\sigma(z) := (f^{(r)} * G_\sigma)(z) + F_c(z),$$

where G_σ is an entire function of exponential type σ, such that

$$\|\Phi_r - G_\sigma\|_L = K_r \sigma^{-r}$$

(see [1], Sect. 101). It can be shown that, for all real v,

$$|G_\sigma(u+iv)| \leq M_1(\sigma) e^{\sigma|v|}, \qquad |G_\sigma'(u+iv)| \leq M_2(\sigma) e^{\sigma|v|} \qquad \text{if } |u| \leq 1,$$

$$|G_\sigma(u+iv)| \leq M_3(\sigma) u^{-2} e^{\sigma|v|}, \qquad |G_\sigma'(u+iv)| \leq M_4(\sigma) u^{-2} e^{\sigma|v|}, \qquad \text{if } |u| \geq 1,$$

$M_j(\sigma)$ ($j=1,2,3,4$) being some positive constants depending only on σ. These estimates ensure that G_σ satisfies all conditions demanded of g in Lemmas 1 and 2. Hence, $f^{(r)} * G_\sigma$ belongs to the class $\tilde{B}_\sigma$, and so does the function U_σ.

In view of (9),

$$f(x) - U_\sigma(x) = (f^{(r)} * (\Phi_r - G_\sigma))(x)$$

for every real x. Using inequality (8) we get

$$\|f - U_\sigma\|_{S^*} \leq \|f^{(r)}\|_{S^*} \|\Phi_r - G_\sigma\|_L$$

and the result follows.

Given any function $f \in S^*$, let us introduce its *modulus of continuity*

$$\omega(\delta; f)_{S^*} = \sup_{|h| \leq \delta} \|f(\cdot + h) - f(\cdot)\|_{S^*} \qquad (\delta > 0).$$

It is easy to verify that it possesses all standard properties of the ordinary modulus of continuity. In particular, by S^*-almost periodicity of f, $\omega(\delta; f)_{S^*} \to 0$ as $\delta \to 0+$.

Theorem 2. *Under the assumptions of Theorem 1,*

$$E_\sigma(f)_{S^*} \le 3\ \sigma^{-r}\omega(\tfrac{1}{\sigma};f^{(r)})_{S^*} \qquad \textit{for each } \sigma>0.$$

This estimate is also true for $r=0$, *whenever* $f\in S^*$.

Proof. Take $h = 2/\sigma$ and consider the Steklov's function

$$f_h(x) := \frac{1}{h}\int_{-h/2}^{h/2} f(x+t)dt \qquad (x\in\mathbb{R}).$$

Clearly, this function possesses bounded derivatives f_h', f_h'',..., $f_h^{(r)}\in ACG^*_{loc}$. Since

$$f_h^{(r+1)}(x) = \frac{1}{h}\left\{f^{(r)}(x+\tfrac{h}{2}) - f^{(r)}(x-\tfrac{h}{2})\right\}$$

almost everywhere, we have

$$\|f_h^{(r+1)}\|_{S^*} \le \frac{1}{h}\ \omega(h;f^{(r)})_{S^*}.$$

Moreover, it is easy to see that

$$\|f^{(r)}-f_h^{(r)}\|_{S^*} = \|f^{(r)}-(f^{(r)})_h\|_{S^*} \le \omega(\tfrac{h}{2};f^{(r)})_{S^*}.$$

Consequently, in view of Theorem 1,

$$\begin{aligned} E_\sigma(f)_{S^*} &\le E_\sigma(f-f_h)_{S^*} + E_\sigma(f_h)_{S^*} \\ &\le K_r\sigma^{-r}\|f^{(r)}-f_h^{(r)}\|_{S^*} + K_{r+1}\sigma^{-r-1}\|f_h^{(r+1)}\|_{S^*} \\ &\le (K_r+K_{r+1})\ \sigma^{-r}\ \omega(\tfrac{1}{\sigma};f^{(r)})_{S^*}. \end{aligned}$$

The result follows because $K_r+K_{r+1} \le 4/\pi + \pi/2 < 3$ (see [1], Sect. 101).

Now, an estimate of $E_\sigma(f)_{S^*}$ in terms of the modulus of smoothness of f will be given. If m is a positive integer, then the *m-th modulus* of f is defined by

$$\omega_m(\delta;f)_{S^*} := \sup_{|h|\le\delta} \|\Delta_h^m f\|_{S^*} \qquad (\delta>0),$$

where

$$\Delta_h^m f(x) := \sum_{\nu=0}^{m} (-1)^{m-\nu}\binom{m}{\nu} f(x+h\nu).$$

Theorem 3. *Suppose that* $f\in S^*$. *Then, for every positive integer m and for every* $\sigma>0$,

$$E_\sigma(f)_{S^*} \le c(m)\ \omega_m(\tfrac{1}{\sigma};f)_{S^*},$$

$c(m)$ *being a positive constant depending only on m.*

Proof. Let ρ be positive integer such that $2\rho > m+1$. Introduce the entire functions of class B_σ:

$$P(\zeta) := \left[\frac{1}{\zeta} \sin \frac{\sigma\zeta}{2\rho} \right]^{2\rho}, \qquad Q(z) := \sum_{\nu=1}^{m} (-1)^{m-\nu} \frac{1}{\nu} \binom{m}{\nu} P\left(\frac{z}{\nu}\right).$$

From the obvious inequalities

$$|P(u+iv)| \le \begin{cases} e^{\sigma|v|} & \text{if } |u|\le 1,\ |v|\ge 1, \\ |u|^{-2\rho} e^{\sigma|v|} & \text{if } |u|\ge 1,\ v\in\mathbb{R}, \end{cases}$$

$$|P'(u+iv)| \le \begin{cases} (\sigma+2\rho)e^{\sigma|v|} & \text{if } |u|\le 1,\ |v|\ge 1, \\ (\sigma+2\rho)|u|^{-2\rho} e^{\sigma|v|} & \text{if } |u|\ge 1,\ v\in\mathbb{R}, \end{cases}$$

it follows that the condition of Lemma 1 concerning series (4) is satisfied for $g=Q$. Therefore the function

$$J_\sigma(z) := (-1)^{m+1}(\gamma_{\sigma,\rho})^{-1}(f*Q)(z),$$

where

$$\gamma_{\sigma,\rho} := \int_{-\infty}^{\infty} P(t)dt,$$

belongs to the class $\tilde{B}_\sigma$.

Clearly, for real t,

$$J_\sigma(t) = (-1)^{m+1}(\gamma_{\sigma,\rho})^{-1} \lim_{n\to\infty} \int_{-n}^{n} f(t-u) \sum_{\nu=1}^{m} (-1)^{m-\nu} \frac{1}{\nu} \binom{m}{\nu} P\left(-\frac{u}{\nu}\right)du$$

$$= (-1)^{m+1}(\gamma_{\sigma,\rho})^{-1} \lim_{n\to\infty} \int_{-n}^{n} P(s) \sum_{\nu=1}^{m} (-1)^{m-\nu} \binom{m}{\nu} f(t+\nu s)ds.$$

Hence

$$f(t) - J_\sigma(t) = (-1)^{m}(\gamma_{\sigma,\rho})^{-1} \lim_{n\to\infty} \int_{-n}^{n} P(s)\, \Delta_s^m f(t)ds,$$

and this relation holds uniformly in $t\in\mathbb{R}$.

Given any number x and any $\eta \in [0,1]$, we have

$$\int_{x}^{x+\eta} (f(t)-J_\sigma(t))dt = (-1)^{m}(\gamma_{\sigma,\rho})^{-1} \lim_{n\to\infty} \int_{x}^{x+\eta} \left[\int_{-n}^{n} P(s)\, \Delta_s^m f(t)ds \right] dt.$$

Changing the successiveness of integration in the above repeated Denjoy-Perron integral ([7], Lemma 1), we obtain

$$\|f-J_\sigma\|_{s*} \le (\gamma_{\sigma,\rho})^{-1} \lim_{n\to\infty} \int_{-n}^{n} \omega_m(|s|;f)_{s*} P(s)ds$$

$$\le \sigma^m(\gamma_{\sigma,\rho})^{-1}\, \omega_m\left(\frac{1}{\sigma};f\right)_{s*} \int_{-\infty}^{\infty} \left(\frac{1}{\sigma} + |s|\right)^m P(s)ds$$

$$\le \omega_m\left(\frac{1}{\sigma};f\right)_{s*} \left\{ 2^m + \sigma^m(\gamma_{\sigma,\rho})^{-1} \int_{|s|\ge 1/\sigma} |s|^m P(s)ds \right\}.$$

The result follows because the expression in the curly bracket is O(1) uniformly in σ (see [8], p.274).

4. Appendix.

Easy calculation shows that, for every S^*-almost periodic function f, the mean values

$$a_f(\Lambda) := \lim_{T\to\infty} \frac{1}{T}\int_0^T f(t)e^{-i\Lambda t}dt \qquad (\Lambda\in\mathbb{R})$$

are finite, and they are different from zero for some enumerable set of values Λ, at most. Therefore, to each function of class S^* there corresponds a Fourier series

$$f(x) \sim \sum_k a_f(\lambda_k)e^{i\lambda_k x},$$

with real exponents λ_k for which the coefficients $a_f(\lambda_k) \neq 0$.

Let $0 \le \lambda < \mu$ and let (see [2])

$$\Psi_{\lambda,\mu}(z) := \frac{16}{\pi(\mu-\lambda)^2}\,\frac{\sin((\mu+\lambda)z/2)}{z}\left[\frac{\sin((\mu-\lambda)z/4)}{z}\right]^2.$$

The function $\Psi_{\lambda,\mu}$ belongs to B_μ and, by Lemmas 1, 2, the convolution $f*\Psi_{\lambda,\mu}$ is of class $\tilde{B}_\mu$ whenever $f\in S^*$. Moreover, it is easy to verify (as in [4], Lemma 1.10.2) that

$$\lim_{T\to\infty}\frac{1}{2T}\int_{-T}^{T}(f*\Psi_{\lambda,\mu}(t)e^{-i\Lambda t}dt = a_f(\Lambda)\varphi_{\lambda,\mu}(\Lambda) \qquad (\Lambda\in\mathbb{R}),$$

where $\varphi_{\lambda,\mu}$ is the inverse Fourier transform of $\Psi_{\lambda,\mu}$. Consequently,

$$(f*\Psi_{\lambda,\mu})(x) \sim \sum_k a_f(\lambda_k)\varphi_{\lambda,\mu}(\lambda_k)e^{i\lambda_k x}. \tag{10}$$

Since, for an arbitrary function H of class B_λ,

$$H*\Psi_{\lambda,\mu} = H$$

(see [2], Theorem 3), we have

$$f*\Psi_{\lambda,\mu} - f = (f-H)*\Psi_{\lambda,\mu} + H - f$$

and, by (8),

$$\|f*\Psi_{\lambda,\mu} - f\|_{S^*} \le \|f-H\|_{S^*}(\|\Psi_{\lambda,\mu}\|_L + 1).$$

This inequality and the known estimate for $\|\Psi_{\lambda,\mu}\|_L$ ([2]) yield

Theorem 4. *If $f\in S^*$, then*

$$\|f*\Psi_{\lambda,\mu} - f\|_{S^*} \le \left[\frac{3}{\pi} + 1 + \frac{2}{\pi}\cdot\log\frac{2(\mu+\lambda)}{\mu-\lambda}\right]E_\lambda(f)_{S^*} \qquad (0\le\lambda<\mu).$$

In particular

$$\lim_{\mu\to\infty}\|f*\Psi_{\mu/2,\mu} - f\|_{S^*} = 0.$$

Taking into account (10) and the uniqueness theorem for

uniformly almost periodic functions ([4], p.45) we obtain

COROLLARY 2. *If all Fourier coefficients of a function* $f \in S^*$ *are equal to zero, then* $f(x) = 0$ *almost everywhere.*

REFERENCES

[1] N.I.ACHIEZER, *Lectures on the theory of approximation*, Moscow 1965 (in Russian).

[2] E.A.BREDICHINA, *On approximation of Stepanov's almost periodic functions*, Dokl. Akad. Nauk SSSR 164 (1965), 255-258 (in Russian).

[3] V.G.ČELIDZE, A.G.DŽVARŠEJŠVILI, *Theory of Denjoy integral and some its applications*, Tbilisi 1978 (in Russian).

[4] B.M.LEVITAN, *Almost periodic functions*, *Moscow 1953* (in Russian).

[5] S.M.NIKOL'SKIJ, *Approximation of functions of several variables and embedding theorems*, Moscow 1977 (in Russian).

[6] B.K.PAL, S.N.MUKHOPADHYAY, *Denjoy-Bochner almost periodic functions*, J.Aust. Math. Soc., Ser.A. 37 (1984), 205-222.

[7] P.PYCH, *The Denjoy integral in some approximation problems, I*, Functiones et Approximatio 1 (1974), 91-105.

[8] A.F.TIMAN, *Theory of approximation of functions of real variable*, Moscow 1960 (in Russian).

MODIFIED WEYL SPACES AND EXPONENTIAL APPROXIMATION

ROMAN TABERSKI

Poznań, Poland

1. Preliminaries.

Given any positive number p, let $L^p(a,b)$ be the class of all measurable (complex-valued) functions Lebesgue-integrable with p-th power on the interval (a,b). Write L^p instead of $L^p(-\infty,\infty)$. Let L^p_{loc} be the class of all functions belonging to every class $L^p(a,b)$ with finite a, b (a < b). Introduce also the collection AC^m_{loc} of (complex-valued) functions having absolutely continuous derivatives of non-negative integer order m on each finite (closed) interval [a,b] (by convention, $AC_{loc} = AC^0_{loc}$). Denote by $\mathbb{N}$ the set of all positive integers; write $\mathbb{R} := (-\infty,\infty)$.

Under the assumptions $f \in L^p_{loc}$ $(0 < p < \infty)$ and $0 < l < \infty$, the quantity

$$\|f\|_{p,l} \equiv \|f(\cdot)\|_{p,l} := \sup_{u\in\mathbb{R}} \left\{ \frac{1}{l} \int_u^{u+l} |f(x)|^p dx \right\}^{1/p}$$

coincides with a positive number or $+\infty$. Furthermore, there exists a finite or infinite limit of $\|f\|_{p,l}$ as $l \to \infty$ (see e.g. [2], p. 220–222). Putting

$$\|f\|_p \equiv \|f(\cdot)\|_p := \sup_{l\geq 1} \|f\|_{p,l},$$

we have

$$\|f\|_{p,1} \leq \|f\|_p \leq 2^{1/p}\|f\|_{p,1} \tag{1}$$

(see [2], p. 198).

All these functions f of class L^p_{loc} for which $\|f\|_{p,1} < \infty$ form so-called *modified Weyl space* W^p with the norm $\|\cdot\|_p$ [paranorm $\|\cdot\|_p^p$] if $1 \leq p < \infty$ [$0 < p < 1$]. By inequalities (1) and Theorem 5.2.1 of [2], p. 199, W^p is a complete space for each finite $p > 0$.

Obviously, an arbitrary function f of class L^p belongs to the space W^p. Further, $W^p \subset W^q$ when $0 < q < p < \infty$. This inclusion follows at once from the trivial inequality

$$\|f\|_{p,l} \geq \|f\|_{q,l} \quad (l > 0).$$

Let E_σ be the class of all entire functions of exponential type, of order σ at most. Write B_σ for the collection of these functions $F \in E_\sigma$ which are bounded on $\mathbb{R}$. Denote by $C_{\sigma,p}(f)$ the set of all functions $G \in E_\sigma$ such that $G - f \in W^p$. Introduce the quantity

$$A_\sigma(f)_p := \begin{cases} \inf\limits_{G \in C_{\sigma,p}(f)} \|f - G\|_p & \text{if } C_{\sigma,p}(f) \text{ is not empty,} \\ \infty & \text{otherwise,} \end{cases}$$

called *the best exponential approximation of f* by entire functions belonging to $C_{\sigma,p}(f)$ in W^p-metric.

With an arbitrary function $f \in L^p_{loc}$ $(0 < p < \infty)$ is associated the characteristic

$$\omega(\delta;f)_p := \sup_{0 \leq \eta \leq \delta} \|f(\cdot+\eta) - f(\cdot)\|_p.$$

It may be finite or infinite for positive numbers δ. In the first case, the non-negative function $\omega(\cdot;f)_p$ is called *W^p-modulus of continuity of f*. Clearly, this modulus of continuity non-decreases on the interval $[0,\infty)$. Moreover,

(2) $$\lim_{\delta \to 0+} \omega(\delta;f)_p = 0$$

if and only if

(3) $$\lim_{\eta \to 0+} \|f(\cdot+\eta) - f(\cdot)\|_{p,1} = 0.$$

In case $1 \leq p < \infty$, $0 \leq \delta < \infty$ and $n \in \mathbb{N}$,

(4) $$\omega(n\delta;f)_p \leq n\omega(\delta;f)_p;$$

if, in addition, $f \in AC_{loc}$ and $f' \in W^p$, then

$$\omega(\delta;f)_p \leq \|f'\|_p \delta.$$

We notice that relation (3) holds for any function f uniformly continuous on $\mathbb{R}$ or S^p-almost periodic (for Stepanov's definition and details see [2], p. 200–202). If $f \in L^p$ $(0 < p < \infty)$, then $\omega(\delta;f)_p$ does not exceed its L^p-analogue

$$\omega(\delta;f)_{L^p} := \sup_{0 \leq \eta \leq \delta} \left\{\int_{-\infty}^{\infty} |f(x+\eta) - f(x)|^p dx\right\}^{1/p} \quad (0 \leq \delta < \infty).$$

Hence, in these cases, condition (2) is fulfilled.

Given any $f \in L^p_{loc}$ $(1 \leq p < \infty)$, let us introduce the Steklov functions defined on $\mathbb{R}$ by

(5) $$f_\lambda(x) := \frac{1}{\lambda} \int_{x-\lambda/2}^{x+\lambda/2} f(t)dt \quad (0 < \lambda < \infty);$$

they belong to the class AC_{loc}. As in Sect. 95 of [1], it may be shown that

(6) $$\|f_\lambda - f\|_p \le \omega(\lambda/2;f)_p$$

and

(7) $$\|f'_\lambda\|_p < \lambda^{-1}\omega(\lambda;f)_p.$$

Below, the W^p-analogues of other well-known theorems for L^p-metrics will be presented.

2. An analogue of the Nikolskii inequality.

Consider a (complex-valued) function g defined on the real line. Write

$$\|g\| := \sup_{x\in\mathbb{R}} |g(x)|.$$

It is evident that, for each positive number p,

$$\|g\| \ge \|g\|_p \ge \|g\|_{p,1}.$$

Also we have the useful

Theorem 1. *Let g be of class E_σ $(0 < \sigma < \infty)$, and let $g \in W^p$ for some finite $p \ge 1$. Then*

(8) $$\|g\| \le (1+\sigma)\|g\|_{p,1}.$$

Proof. By the mean-value theorem, there is a number ξ in [0,1], such that

$$\int_0^1 |g(x)|^p dx = |g(\xi)|^p.$$

The identity

$$g(\xi) - g(0) = \int_0^\xi g'(t)dt$$

implies

$$|g(0)| - |g(\xi)| \le \int_0^\xi |g'(t)|dt;$$

whence

(9) $$|g(0)| \le \left\{\int_0^1 |g(x)|^p dx\right\}^{1/p} + \int_0^1 |g'(t)|dt.$$

Clearly, estimate (9) also holds for functions $g(\cdot+v)$ $(v \in \mathbb{R})$ instead of $g(\cdot)$. In view of the Hölder inequality,

$$\int_0^1 |g'(t+v)|dt \le \left\{\int_0^1 |g'(t+v)|^p dt\right\}^{1/p}.$$

Consequently, for every real v,

$$|g(v)\| \le \|g\|_{p,1} + \|g'\|_{p,1} \le (1+\sigma)\|g\|_{p,1},$$

by the Bernstein type inequality (see Sect. 84 of [1]).

Thus the desired result is obtained.

Remark 1. If g is of class E_0 and $g \in W^p$ $(p \ge 1)$, then the classical Bernstein's inequality and (8) give

$$\|g'\| \le \sigma\|g\| \le \sigma(1+\sigma)\|g\|_{p,1} \quad \text{for each } \sigma > 0.$$

Consequently, $g'(x) = 0$ for all $x \in \mathbb{R}$, i.e., $g(x) = \text{const}$. Hence, for $\sigma = 0$, Theorem 1 remains valid (cf. Theorems 3.3.1, 3.3.5 of [3], p. 122–126).

3. Existence of an element of best exponential approximation.

There holds the following

Theorem 2. *Given any* $f \in L^p_{loc}$ $(1 \le p < \infty)$ *with* $A_\sigma(f)_p < \infty$, $(0 < \sigma < \infty)$, *there is an entire function* $F \in E_\sigma$ *such that*

(10) $$\|f - F\|_p = A_\sigma(f)_p.$$

Proof. Clearly, there exist entire functions $G_k \in E_\sigma$ satisfying the inequalities

$$\|f - G_k\|_p \le A_\sigma(f)_p + (2^k\sigma)^{-1} \qquad (k = 0,1,\dots).$$

Putting

$$\varphi(x) := f(x) - G_0(x), \quad g_\nu(z) := G_\nu(z) - G_{\nu-1}(z) \qquad (z=x+iy)$$

and

$$S_k(z) := \sum_{\nu=1}^k g_\nu(z),$$

we can write

(11) $$\|\varphi - S_k\|_p = \|f-G_k\|_p \le A_\sigma(f)_p + (2^k\sigma)^{-1} \qquad (k = 1,2,\dots).$$

Further, $g_\nu \in E_\sigma$ and

$$\|g_\nu\|_p \le \|G_\nu - f\|_p + \|f-G_{\nu-1}\|_p \le 2\cdot A_\sigma(f)_p + \frac{1}{2^\nu\sigma} + \frac{1}{2^{\nu-1}\sigma}$$

$(\nu = 1,2,\dots)$. By inequality (8),

$$\sup_{t\in\mathbb{R}} |g_\nu(t)| \le (1+\sigma)\|g_\nu\|_p.$$

Consequently,

$$g_\nu \in B_\sigma \quad (\nu\ge1) \quad \text{and} \quad S_k \in B_\sigma \quad (k\ge1).$$

Also this inequality together with (11) gives

(12) $$\sup_{t\in\mathbb{R}} |S_k(t)| \le (1+\sigma)\{\|\varphi\|_p + A_\sigma(f)_p + (2^k\sigma)^{-1}\} \quad (k = 1,2,\ldots).$$

Hence, all functions S_k are uniformly bounded on $\mathbb{R}$.

From the Taylor formula

$$S_k(z) = \sum_{\mu=0}^{\infty} \frac{S_k^{(\mu)}(x)}{\mu!}(iy)^\mu \qquad (z=x+iy)$$

and the classical Bernstein inequality it follows that

(13) $$|S_k(z)| \le \sup_{t\in\mathbb{R}} |S_k(t)| e^{\sigma|y|} \qquad (k = 1,2,\ldots).$$

Therefore, the entire functions S_k are uniformly bounded on every strip $Q_a := \{\xi=u+iv: -\infty < u < \infty, -a < v < a\}$ $(0 < a < \infty)$.

For arbitrary complex numbers $\zeta_1, \zeta_2 \in Q_a$,

$$|S_k(\zeta_1) - S_k(\zeta_2)| \le |\zeta_1 - \zeta_2| \sup_{\zeta\in Q_a} |S_k'(\zeta)|.$$

Applying the expansion

$$S_k'(z) = \sum_{\mu=0}^{\infty} \frac{S_k^{(\mu+1)}(x)}{\mu!}(iy)^\mu$$

and Bernstein's inequality, we obtain

$$|S_k(\zeta_1) - S_k(\zeta_2)| \le |\zeta_1 - \zeta_2| \sup_{t\in\mathbb{R}} |S_k(t)|\sigma e^{\sigma a} \qquad (k \in \mathbb{N}).$$

This together with (12) ensures that all S_k are equicontinuous on Q_a.

Applying the diagonal method we can construct a subsequence $\{S_{n,n}(z)\}$ of $\{S_k(z)\}$ having the finite limit

$$\lim_{n\to\infty} S_{n,n}(z) =: S(z)$$

for each $z = x+iy$ with rational $x,y \in \mathbb{R}$. Further, by Arzela's theorem, for every positive number a, there exists a subsequence $\{S_{k_j(a)}(z)\}$ of $\{S_{n,n}(z)\}$ tending uniformly on any bounded closed domain $D \subset Q_a$ to some (complex) $T_a(z)$. The function T_a is continuous and $T_a(z) = S(z)$ for rational $z \in D$. Therefore, if $z \in D$, $F(z) := T_a(z)$ is independent of a. Consequently, F coincides with a certain entire function.

In view of (13) and (11), $F \in E_\sigma$ and

$$\left\{\frac{1}{l}\int_u^{u+l} |\varphi(x) - F(x)|^p dx\right\}^{1/p} \le A_\sigma(f)_p$$

for all finite $l \ge 1$, $u \in \mathbb{R}$. This immediately implies (10).

4. Estimates of the Jackson type.

Given a positive number c and a positive integer r, let ρ be an even, real-valued function continuous with its derivatives ρ', ρ'' on $\mathbb{R}$, such that the quotients

$$\rho(t)/t^{r+2}, \quad \rho'(t)/t^{r+1}, \quad \rho''(t)/t^{r}$$

are bounded on the interval $(0,c)$ and

$$\rho(t) = 1 \qquad \text{for all } t \geq c.$$

Write

$$\Phi_r(x) := \frac{1}{2\pi} \lim_{\alpha\to\infty} \int_{-\alpha}^{\alpha} \frac{\rho(t)}{(it)^r} e^{itx}\, dt \qquad (x \in \mathbb{R}).$$

As is known ([1], Sects. 88, 99, 101), the function Φ_r called *Krein's kerenel* is real valued, bounded and Lebesgue-integrable on $\mathbb{R}$. Moreover, for every finite $\sigma > 0$, there exists an entire function $G_\sigma \in E_\sigma$ for which

$$\int_{-\infty}^{\infty} |\Phi_r(x) - G_\sigma(x)|\, dx = K_r/\sigma^r, \tag{14}$$

where K_r means the Favard constant defined by

$$K_r := \frac{4}{\pi} \sum_{k=0}^{\infty} \frac{(-1)^{k(r+1)}}{(2k+1)^{r+1}}.$$

If a function f is of class AC_{loc}^{r-1} $(r \in \mathbb{N})$, and if the condition

$$\int_{-\infty}^{\infty} \left|\frac{f^{(r)}(t)}{1+|t|}\right|^2 dt < \infty \tag{15}$$

is fulfilled, then, for each $x \in \mathbb{R}$,

$$f(x) = \int_{-\infty}^{\infty} f^{(r)}(t)\, \Phi_r(x-t)dt + F(x), \tag{16}$$

where F denotes some entire function of class E_c. The last formula also holds for every $x \in \mathbb{R}$ when, instead of (15), $f^{(r)}$ belongs to W^1 and

$$\int_{-\infty}^{\infty} \left|\frac{1}{1+|t|} \int_0^t f^{(r)}(s)\, ds\right|^2 dt < \infty \tag{17}$$

see [1], Sects. 101, 102).

Under the assumptions $f \in AC_{loc}^{r-1}$ $(r \in \mathbb{N})$, $f^{(r)} \in W^p$ $(1 \leq p < \infty)$ and (15) or (17), from (16) and (14) it follows that

$$A_\sigma(f)_p \leq \frac{K_r}{\sigma^r} \|f^{(r)}\|_p \qquad \text{if } 0 < \sigma < \infty \tag{18}$$

(see Sects. 99 and 102 of [1]).

We will now prove three other Jackson's type theorems. They correspond to the suitable estimates of Sect. 105 in [1].

Theorem 3. *Suppose that $f \in AC_{loc}^{r-1}$ $(r \in \mathbb{N})$, $f^{(r)} \in L_{loc}^{p} \cap W^{1}$ $(1 \le p < \infty)$ and $\omega(\delta; f^{(r)})_p < \infty$ for every positive number δ. Suppose, further, that condition (17) is fulfilled. Then*

$$A_\sigma(f)_p \le \frac{K_r + K_{r+1}}{\sigma^r} \cdot \omega(\tfrac{1}{\sigma}; f^{(r)})_p \le \frac{3}{\sigma^r} \cdot \omega(\tfrac{1}{\sigma}; f^{(r)})_p \quad \text{if } \sigma > 0.$$

Proof. Considering the Steklov functions f_λ $(0<\lambda<\infty)$ defined by (5), we have

$$A_\sigma(f)_p \le A_\sigma(f - f_\lambda)_p + A_\sigma(f_\lambda)_p \quad (0<\sigma<\infty). \tag{19}$$

The assumptions ensure that

$$\int_{-\infty}^{\infty} \left| \frac{f_\lambda^{(r)}(t)}{1+|t|} \right|^2 dt < \infty \quad \text{and} \quad \int_{-\infty}^{\infty} \left| \frac{1}{1+|t|} \int_0^t f_\lambda^{(r+1)}(s)ds \right|^2 dt < \infty.$$

Hence, in view of (16), for all real x,

$$f(x) - f_\lambda(x) = \int_{-\infty}^{\infty} \left\{ f^{(r)}(t) - f_\lambda^{(r)}(t) \right\} \Phi_r(x-t)dt + T(x)$$

and

$$f_\lambda(x) = \int_{-\infty}^{\infty} f_\lambda^{(r+1)}(t) \cdot \Phi_{r+1}(x-t)dt + U(x),$$

where $T, U \in E_c$. These formulae lead to

$$A_\sigma(f - f_\lambda)_p \le \frac{K_r}{\sigma^r} \, \| f^{(r)} - f_\lambda^{(r)} \|_p \le \frac{K_r}{\sigma^r} \, \omega(\tfrac{\lambda}{2}; f^{(r)})_p, \tag{20}$$

$$A_\sigma(f_\lambda)_p \le \frac{K_{r+1}}{\sigma^{r+1}} \, \| f_\lambda^{(r+1)} \|_p \le \frac{2K_{r+1}}{\sigma^{r+1}\lambda} \, \omega(\tfrac{\lambda}{2}; f^{(r)})_p \tag{21}$$

(see (18), (6), (7) and (4)).

From inequalities (19) – (21), in which $\lambda = 2/\sigma$, the desired assertion follows.

Theorem 4. *Let f be a function of class L_{loc}^{p} $(2 \le p < \infty)$, with $\omega(\delta; f)_p < \infty$ for all finite $\delta > 0$. Then*

$$A_\sigma(f)_p \le (1 + \tfrac{\pi}{2}) \, \omega(\tfrac{1}{\sigma}; f)_p \quad \text{if } 0<\sigma<\infty.$$

Proof. In view of (6), for any Steklov function f_λ $(\lambda>0)$,

$$A_\sigma(f - f_\lambda)_p \le \| f - f_\lambda \|_p \le \omega(\lambda/2; f)_p. \tag{22}$$

Further,

$$\left(\int_1^{\infty} + \int_{-\infty}^{-1} \right) \frac{1}{t^2} \left| f(t+\tfrac{\lambda}{2}) - f(t-\tfrac{\lambda}{2}) \right|^2 dt$$

$$\le \sum_{\nu=0}^{\infty} \frac{1}{2^{2\nu}} \left(\int_{2^\nu}^{2^{\nu+1}} + \int_{-2^{\nu+1}}^{-2^\nu} \right) \left| f(t+\tfrac{\lambda}{2}) - f(t-\tfrac{\lambda}{2}) \right|^2 dt$$

$$\leq 2 \sum_{\nu=0}^{\infty} \frac{1}{2^{\nu}} \left\{\omega(\lambda;f)_p\right\}^2 = 4 \left\{\omega(\lambda;f)_p\right\}^2 < \infty.$$

Hence

$$\int_{-\infty}^{\infty} \left\{\frac{|f'_{\lambda}(t)|}{1+|t|}\right\}^2 dt = \frac{1}{\lambda^2} \int_{-\infty}^{\infty} \left\{\frac{|f(t+\lambda/2) - f(t-\lambda/2)|}{1+|t|}\right\}^2 dt < \infty.$$

Applying (18) i (7), we obtain

$$A_{\sigma}(f_{\lambda})_p \leq \frac{K_1}{\sigma} \|f'_{\lambda}\|_p \leq \frac{\pi}{2\sigma\lambda} \omega(\lambda;f)_p. \tag{23}$$

Inequalities (19), (22), (23) and (4) lead to

$$A_{\sigma}(f)_p \leq (1 + \frac{\pi}{\sigma\lambda}) \omega(\lambda/2;f)_p \quad (0 < \lambda,\sigma < \infty).$$

Putting $\lambda = 2/\sigma$, we get at once our thesis.

Theorem 5. *Suppose that* $f \in L^p_{loc}$ $(1 \leq p < 2)$ *and that* $\omega(\delta;f)_p < \infty$ *whenever* $0<\delta<\infty$. *If, moreover,*

$$\int_{-\infty}^{\infty} \left|\frac{f(u+\eta) - f(u)}{1+|u|}\right|^2 du < \infty \tag{24}$$

or

$$\int_{-\infty}^{\infty} \left|\int_{-\eta}^{\eta} \frac{f(u+t) - f(u)}{1+|t|} du\right|^2 dt < \infty \tag{25}$$

for every positive number η, *then*

$$A_{\sigma}(f)_p \leq (1 + \frac{\pi}{2}) \omega(\frac{1}{\sigma};f)_p$$

for each finite $\sigma>0$.

Proof. Let, as before, f_{λ} be the Steklov function with $\lambda = 2/\sigma$ $(0<\sigma<\infty)$.

Obviously, in case (24),

$$\int_{-\infty}^{\infty} \left\{\frac{|f'_{\lambda}(t)|}{1+|t|}\right\}^2 dt = \frac{1}{\lambda^2} \int_{-\infty}^{\infty} \left\{\frac{|f(u+\lambda) - f(u)|}{1+|u+\frac{\lambda}{2}|}\right\}^2 du < \infty;$$

in case (25),

$$\int_{-\infty}^{\infty} \left| \frac{1}{1+|t|} \int_{0}^{t} f'_{\lambda}(s)ds \right|^2 dt$$

$$= \int_{-\infty}^{\infty} \left| \frac{1}{(1+|t|)\lambda} \int_{-\lambda/2}^{\lambda/2} \left\{f(u+t)-f(u)\right\}du \right|^2 dt < \infty.$$

Each of these inequalities implies (23).

Since inequalities (19) and (23) remain valid, the desired result is established.

REMARK 2. Condition (25) can be replaced by a more strong restriction:

$$\int_1^\infty \left\{\frac{1}{t}\ \omega(t;f)_p\right\}^2 dt < \infty.$$

REFERENCES

[1] N.I.AKHIEZER, *Lectures on the theory of approximation*, Moscow 1965 (in Russian).

[2] B.M.LEVITAN, *Almost-periodic functions*, Moscow 1953, (in Russian).

[3] S.M.NIKOLSKIĬ, *Approximation of functions of several variables and embedding theorems*, Moscow 1977 (in Russian).

THE TRACES OF DIFFERENTIABLE FUNCTIONS TO CLOSED SUBSETS OF $\mathbb{R}^n$

Ju. A. Brudnyi, P. A. Shvartsman

Jaroslavl, USSR

1. In this paper we consider the problem of the description of the trace spaces of differentiable functions to arbitrary closed subset of multidimensional space.

Let $C^k(\mathbb{R}^n)$ be the usual space of k-times continuously differentiable functions defined on $\mathbb{R}^n$ and F be a closed subset of $\mathbb{R}^n$. In the problem of the description of the trace space*)

$$C^k(\mathbb{R}^n)|_F := \{f|_F : f \in C^k(\mathbb{R}^n)\}$$

we see two important questions.

Problem 1. Characterize the trace space $C^k(\mathbb{R}^n)|_F$, namely find necessary and sufficient conditions for a function f to be a trace of a function $\tilde{f}$ belonging to $C^k(\mathbb{R}^n)$.

Problem 2. This is the problem of existence of a linear extension operator $T: C^k(\mathbb{R}^n)|_F \longrightarrow C^k(\mathbb{R}^n)$.

We call these problems "Withney problems" because they have been formulated by H.Whitney in 1934. Whitney [8] has solved the both problems for the one-dimensional case (a generalization of that result can be found in [4]) but for the case n>1 the problems are still open. Let us note that the problems of the description of the trace space for the Zygmund class $\Lambda^k_\omega(\mathbb{R}^n)$ (which is close to the $C^k(\mathbb{R}^n)$ when $\omega(t)=t^k$) have been solved for n=1 in [3] ($\omega(t)=t^\alpha$), [2,6] (k=2) and [5] (and the extension operators constructed in these papers are linear). The case $\Lambda^2_\omega(\mathbb{R}^n)$ has been investigated in [7] and the corresponding linear extension operator has been constructed in [1].

---

*) the symbol ":=" means "equal by definition"

2. In this paper we give a solution of Problems 1 and 2 for the case k=1 and n arbitrary. This is the case of the trace space for the usual $\mathcal{C}^1(\mathbb{R}^n)$ space of differentiable functions.

Let us recall that $\mathcal{C}^1(\mathbb{R}^n)$ space consists of all bounded differentiable functions f with uniformly continuous gradient $\nabla f := \left(\frac{\partial f}{\partial x_1}, \dots, \frac{\partial f}{\partial x_n}\right)$. In other words, a function f belongs to $\mathcal{C}^1(\mathbb{R}^n)$ if and only if the ordinary modulus of continuity of gradient

$$\Omega(\nabla f;t) := \sup_{\|x-y\|\le t} \|\nabla f(x) - \nabla f(y)\|$$

converges to zero as $t \to 0$, where $\|\cdot\|$ denotes the usual Euclidean norm in $\mathbb{R}^n$. The norm on $\mathcal{C}^1(\mathbb{R}^n)$ is defined by

$$|f|_{\mathcal{C}^1(\mathbb{R}^n)} := \sup_{x\in\mathbb{R}^n} |f(x)| + \sup_{x\in\mathbb{R}^n} \|\nabla f(x)\|.$$

In order to formulate corresponding results we define a space which is close to $\mathcal{C}^1$, namely the space $\mathcal{C}^{1,\omega}(\mathbb{R}^n)$, where ω is a function of the modulus of continuity type. The space $\mathcal{C}^{1,\omega}(\mathbb{R}^n)$ is a subspace of $\mathcal{C}^1(\mathbb{R}^n)$ and consists of all smooth functions f with the modulus of continuity of gradient satisfying the following inequality

$$\Omega(\nabla f;t) \le \lambda\cdot\omega(t), \qquad t\ge 0.$$

We equipped the space $\mathcal{C}^{1,\omega}(\mathbb{R}^n)$ with the norm

$$|f|_{\mathcal{C}^{1,\omega}(\mathbb{R}^n)} := |f|_{\mathcal{C}^1(\mathbb{R}^n)} + \inf \lambda.$$

Let M be the set of functions ω of the modulus of continuity type; it is easy to show that $\mathcal{C}^1(\mathbb{R}^n) = \bigcup_{\omega\in M} \mathcal{C}^{1,\omega}(\mathbb{R}^n)$ and

$$|f|_{\mathcal{C}^1(\mathbb{R}^n)} \approx \inf_{\omega\in M} |f|_{\mathcal{C}^{1,\omega}(\mathbb{R}^n)}.$$

3. We formulate a hypothesis which we call "the hypothesis of finiteness".

Hypothesis. There is an integer N = N(n), depending only on n for which the following property holds: a function f belongs to $\mathcal{C}^1(\mathbb{R}^n)|_F$ if and only if for some $\omega\in M$ and every restriction $f|_{F'}$ to a subset $F'\subset F$, consisting of N(n) points, f can be extended to a function $f_{F'}$ lying in the unit ball of $\mathcal{C}^{1,\omega}(\mathbb{R}^n)$ with a fixed radius (that is $|f_{F'}|_{\mathcal{C}^{1,\omega}(\mathbb{R}^n)} \le \lambda$, where $\lambda = \lambda(f)$ depends only on f).

A similar hypothesis for the space $\mathcal{C}^{1,\omega}(\mathbb{R}^n)$ can be formulated. If the hypothesis holds, the problem of determination of the minimal constant N(n) arises.

For the formulation the hypothesis in terms of Banach spaces we introduce the following trace spaces. Let $\mathfrak{C}^1(F) := \mathfrak{C}^1(\mathbb{R}^n)|_F$ and $\mathfrak{C}^{1,\omega}(F) := \mathfrak{C}^{1,\omega}(\mathbb{R}^n)|_F$; we denote the usual trace norm in $\mathfrak{C}^1(F)$ by

$$|f|_{\mathfrak{C}^1(F)} := \inf \left\{ |\tilde{f}|_{\mathfrak{C}^1(\mathbb{R}^n)} : \tilde{f}|_F = f,\ \tilde{f} \in \mathfrak{C}^1(\mathbb{R}^n) \right\}.$$

The norm in $\mathfrak{C}^{1,\omega}(F)$ is defined in a similar manner. Next we introduce a new class of trace spaces closely connected with the hypothesis. This is the space $\mathfrak{C}^{1,\omega}(F;m)$, where m is an integer. It is defined by boundedness of the following norm

$$|f|_{\mathfrak{C}^{1,\omega}(F;m)} := \sup \left\{ |f|_{F'}|_{\mathfrak{C}^{1,\omega}(F')} : F' \subset F,\ \mathrm{card}\, F' \leq m \right\}.$$

In the case $\mathfrak{C}^1$ we define

$$|f|_{\mathfrak{C}^1(F;m)} := \inf_{\omega \in M} |f|_{\mathfrak{C}^{1,\omega}(F;m)}.$$

It is clear that

$$\mathfrak{C}^{1,\omega}(F) \subset \mathfrak{C}^{1,\omega}(F;m) \subset \mathfrak{C}^{1,\omega}(F;l), \qquad m>l$$

and for the space $\mathfrak{C}^1(F;m)$ the similar imbeddings take place as well. In virtue of these definitions the hypothesis can be reformulated in the following way

Hypothesis. $\mathfrak{C}^{1,\omega}(F) = \mathfrak{C}^{1,\omega}(F;N)$, $\mathfrak{C}^1(F) = \mathfrak{C}^1(F;N)$, $N=N(n)$.

Theorem 1. *The hypothesis holds true.*

We call this theorem "Theorem of finiteness".

Theorem 2. *The minimal constant N in the hypothesis is* $N(n) = 3 \cdot 2^{n-1}$.

4. Theorems 1 and 2 show that Problem 1 can be reduced to the case in which the subset F consists of N(n) points only. We believe it is a key moment in the solution of Problem 1.

So, for the complete solution of Problem 1 we must compute the trace norm $|f|_{\mathfrak{C}^1(F)}$, where card $F \leq N(n)$, only in terms of values of function f (and in terms of geometrical properties of F). We have the simple formula in the one-dimensional case and, naturally, it is equivalent to the Whitney criteria [4,8] (here N(1) = 3 and the formula contains only 3 values of f).

In order to demonstrate the corresponding result in the

two-dimensional case we consider the space Lip(F) which consists of all functions $f:F \to \mathbb{R}$ satisfying the following condition

$$|f|_{Lip(F)} := \sup_{x\in F} |f(x)| + \sup_{x,y\in F,\ x\neq y} \left|\frac{f(x)-f(y)}{\|x-y\|}\right|.$$

Next, for the set $X = (x_1,x_2,x_3) \subset \mathbb{R}^2$, $x_i \in \mathbb{R}^2$, we denote $\delta(X)$ as

$$\delta(X) := l(X)\cdot L(X)/\ \mathrm{mes}(\mathrm{conv}\ X)$$

where $l(X) := \min_{1\le i\neq j\le 3} \|x_i-x_j\|$, $L(X) := \max_{1\le i,j\le 3} \|x_i-x_j\|$. It is clear that in the case $\mathrm{mes}(\mathrm{conv}\ X) > 0$, $X \subset F$ there exists a unique polynomial $P_X(f)$ of degree ≤ 1, which interpolates f on X. Let the set $F \subset \mathbb{R}^2$ have affine dimension $\dim_{aff}F = 2$.

Theorem 3. *A function f belongs to $\mathfrak{C}^{1,\omega}(F)$ if and only if f belongs to $Lip(F)$ and there exists a constant $\lambda>0$ such that, for every three-points subsets $X,Y \subset F$,*

(1) $\|\nabla P_X(f)-\nabla P_Y(f)\| \le \lambda(\ \delta(X)\omega(L(X)) + \delta(Y)\omega(L(Y)) + \omega(diam(X\cup Y))\ )$.

The $\mathfrak{C}^{1,\omega}(F)$-norm of f is equivalent to $|f|_{Lip(F)} + \inf \lambda$.

Theorem 5. *For some compact subset $F\subset\mathbb{R}^2$ there does not exist a $f \in Lip(F)$ and there exists a function $\omega\in M$ such that inequality (1) holds with $\lambda=1$. The $\mathfrak{C}^1(F)$-norm of f is equivalent to norm $|f|_{Lip(F)}$.*

Let us note that the terms used in the case $n>1$ are very complicated and the problem of description of the trace space norm in simpler terms is still open.

5. **Problem 2.** This problem does not require new terms as it happens with Problem 1; the question is clear - "Yes" or "No"? It is important to note that Problem 2 in the multidimensional case essentially differs from the one-dimensional case. In the one-dimensional case closed sets have comparatively simple structure and the construction of linear extension operator can be received simultaneously with the characterization of the trace space $\mathfrak{C}^1(\mathbb{R}^1)|_F$. However, in the multidimensional case Problem 2 requires essentially new ideas. It is related with the specific nature of the multidimensional case, where the very existence of a linear extension operator is highly problematical.

As we have mentioned, H.Whitney [8] has shown that the answer to Problem 2 is "Yes" in the one-dimensional case. It was very unexpected that the answer to Problem 2 for the case $\mathfrak{C}^1(\mathbb{R}^n)$ with $n>1$

is "No"! Namely,

THEOREM 5. *For a some compact subset $F \subset \mathbb{R}^2$ does not exist a continuous linear extension operator $T: \mathcal{E}^1(\mathbb{R}^2)|_F \to \mathcal{E}^1(\mathbb{R}^2)$.*

Thus, an interesting problem arises from Theorem 5: describe those subsets F for which there exists corresponding linear extension operator.

However, the situation is essentially different in the case of $\mathcal{E}^{1,\omega}(\mathbb{R}^n)$, where ω is a given function and n is arbitrary. The answer in this case is "Yes"!

THEOREM 6. *Let F be an arbitrary closed subset of $\mathbb{R}^n$. Then there exists a continuous linear extension operator $T_\omega: \mathcal{E}^{1,\omega}(\mathbb{R}^n)|_F \to \mathcal{E}^{1,\omega}(\mathbb{R}^n)$ (depending on ω). The operator norm of T_ω is less than some constant $\gamma=\gamma(n)$ depending only on n.*

REFERENCES

[1] JU. A. BRUDNYI, P. A. SHVARTSMAN, *A linear extension operator for a space of smooth functions defined on a closed subset of $\mathbb{R}^n$*, Soviet Math. Dokl. 31 (1985), 48–51.

[2] V.K.DZYADYK, I. A. SHEVCHUK, *Extension of functions which are traces of functions on arbitrary subset of the line with given second modulus of continuity*, Izv. Akad. Nauk SSSR, Ser. Mat. 47 (1983), 248–267.

[3] A.JONSSON, *The trace of the Zygmund class $\Lambda^k(\mathbb{R})$ to closed sets and interpolating polinomials*, J. Approx. Theory 44 (1985), 1–13.

[4] J.MERRIEN, *Prolongateurs de fonctions differentiables d'une variable reelle*, J. Math. Pures Appl. 45 (1966), 291–309.

[5] I.A.SHEVCHUK, *On traces of functions of the class H_k^φ on the line*, Dokl. Akad. Nauk SSSR 273 (1983), 313–314.

[6] P.A.SHVARTSMAN, *On traces of functions of two variables satisfying the Zygmund conditions*, Studies in the Theory of Functions of Several Real Variables, Yaroslav. Gos. Univ., Yaroslavl 1982, 145–168.

[7] P.A.SHVARTSMAN, *Lipschitz section of multivalued mappings and the traces of functions of the Zygmund class to an arbitrary compact*, Dokl. Akad. Nauk SSSR 276 (1984), 3.

[8] H.WHITNEY, *Differentiable functions defined in closed sets*, Trans. Amer. Math. Soc. 36 (1934), 369–387.

ON THE ANALYTIC RADON-NIKODYM PROPERTY

A.V.Bukhvalov

Leningrad, USSR

Abstract. The operators with the analytic Radon-Nikodym property are introduced and investigated. The analytic Radon-Nikodym property is applied to the question when the interpolation spaces constructed by means of the two complex methods of Calderon coincide. Some open problems are formulated.

1. Introduction.

This paper is a continuation of the author's investigations of the analytic Radon-Nikodym property (ARN) of Banach spaces (see [8,11,12]). The necessary definitions connected with the (ARN)-property are given in this Introduction. A new class of operators with the analytic Radon-Nikodym property is introduced and investigated in Section 2. The third section is devoted to the question of coincidence of the interpolation spaces $[X_0,X_1]_\theta$ and $[X_0,X_1]^\theta$ constructed by means of two different complex methods of Calderon. Here the (ARN)-property is used. In Section 4 we work with the (ARN)-property in spaces of vector-valued functions.

Throughout the paper X,Y (may be with indexes) denote complex Banach spaces; X^* is the Banach conjugate of X. All operators are assumed to be linear and continuous.

In 1975 (cf. [8]) the author introduced the class of Banach spaces with the (ARN)-property. The initial definition was given in terms of the possibility to generalize some classical theorems on the existence of boundary values to the case of analytic functions with values in these spaces. Independently, this class was introduced under another name by A.A.Danilevich in his thesis (Moscow, 1976). To emphasize the connection of (ARN)-property with the ordinary

Radon-Nikodym property (RN) of Banach spaces we give a definition of the (ARN)-property in terms of vector measures. This definition was known to the author already at the end of the 70's (see the dissertation of the author for the degree of doctor of physical and mathematical sciences, Leningrad, 1984), but in explicit form it appears only in [19,32].

Let $\mathcal{B}$ be the σ-field of Borel subsets of $[0,2\pi]$. A vector measure $\vec{m}:\mathcal{B} \to X$ of bounded variation is called *analytic* if

$$\int_0^{2\pi} e^{-int} d\vec{m}(t) = 0 \qquad \text{for all integers} \quad n < 0.$$

A complex Banach space X possesses *the analytic Radon-Nikodym property* ($X \in (ARN)$) if for every analytic vector measure $\vec{m}:\mathcal{B} \to X$ of bounded variation there exists a (strongly) measurable function $\vec{f} \in L^1([0,2\pi],X)$, such that

$$\text{(1)} \qquad \vec{m}(A) = \int_A \vec{f}(t)dt, \qquad A \in \mathcal{B}.$$

This definition differs from the ordinary Radon-Nikodym property ($X \in (RN)$) in one point: in the definition of (RN)-property the representation (1) is true for an arbitrary measure $\vec{m}$ of bounded variation which is absolutely continuous with respect to the Lebesgue measure, while in the definition of the (ARN)-property the demand of absolute continuity is superfluous because of the vector-valued generalization of the F. and M.Riesz theorem [43]. If we assume this definition of the (ARN)-property, it is evident that $X \in (RN)$ implies $X \in (ARN)$. The interest of the class (ARN) is caused by the fact that the class (ARN) is essentially wider that the class (RN) (cf.[12]).

To formulate the result from [12], we have to mention some definitions from the theory of Banach lattices [44]. Throughout the paper we denote by E, W some complex Banach spaces, considered as complexifications of real Banach lattices. A Banach lattice E is said to be a *KB-space* provided

$$(0 \le e_n \uparrow, \ \sup \|e_n\| < \infty) \implies (\exists\ e \in E: \|e_n - e\| \to 0).$$

The main result of [12] is the following theorem.

Theorem A. *A Banach lattice E belongs to (ARN) iff E is a KB-space.*

From Theorem A we see that any L^1- space is an (ARN)-space since it is, evidently, a KB-space. On the other hand, it is well known that $L^1(0,1) \notin (RN)$. We conclude from these facts that the following Banach spaces do not have the (RN)-property but have the (ARN)-property:

$$L^1,\ (L^\infty)^*,\ C(K)^*,\ W_1^l.$$

The Sobolev space W_1^l (either l is an integer or not) has the (ARN)-property, since it can be isometrically embedded into a finite sum of L^1-spaces and the (ARN)-property is hereditary. The number of similar examples may be enlarged if we consider other metrics. For example, the spaces with the mixed norm $L^1(L^2)$ and $L^2(L^1)$ are of the same type.

The (ARN)-property has been extensively studied in the recent years in the various directions (vector-valued analytic martingales, geometry of Banach spaces, noncommutative Banach function spaces etc.; cf. [1,4,5,6,17,19,20,22,24,27-34,36,38-40]).

2. ARN-operators.

In analogy to the definition of operators with the Radon-Nikodym property (RN-operators) (cf. [41]), we define operators with the analytic Radon-Nikodym property (ARN-operators): An operator $U:X \to Y$ is called *ARN-operator* if it transforms each analytic measure $\vec{m}:\mathcal{B} \to X$ of bounded variation to a measure $U(\vec{m}):\mathcal{B} \to Y$, which has a representation with a density $\vec{g} \in L^1([0,2\pi],Y)$, i.e.

$$U(\vec{m})(A) = \int_A \vec{g}(t)dt, \qquad A\in\mathcal{B}. \tag{2}$$

Note that the measure $U(\vec{m})$ is automatically analytic and of bounded variation.

It is not hard to show that (ARN)-operators can be characterized in terms of the boundary values of the vector-valued analytic functions. Let

$$\mathbb{D} = \{z\in\mathbb{C}: |z| < 1\}.$$

By $H^p(\mathbb{D},X)$ $(1 \le p \le \infty)$ we denote the Hardy space of all analytic functions $\vec{F}: \to X$ such that

$$\|\vec{F}\|_p = \sup_{0<r<1} \Big(\frac{1}{2\pi} \int_0^{2\pi} \|\vec{F}(re^{i\theta})\|^p d\theta\Big)^{1/p} < \infty.$$

By $\tilde{H}^p(\mathbb{D},X)$ we denote the subspace of all functions $\vec{F} \in H^p(\mathbb{D},X)$ satisfying one of the following equivalent conditions:

(i) There exists $\vec{f} \in L^p([0,2\pi],X)$ such that $\vec{F}$ is the Poisson integral of $\vec{f}$, that is

$$\vec{F}(z) = \vec{F}(re^{i\theta}) = \frac{1}{2\pi} \int_0^{2\pi} P_r(\theta-\sigma)\vec{f}(\sigma)d\sigma; \tag{3}$$

(ii) The functions $\vec{F}_r(e^{i\theta}) := \vec{F}(re^{i\theta})$ are convergent in norm of the space X as $r \to 1$ for almost every (a.e.) θ.

(iii) (for $p \neq \infty$) The functions $\vec{F}_r$ are convergent in norm of the space $L^p([0,2\pi],X)$ as $r \to 1$.

If $\vec{f}$ is the limiting function from (ii) and (iii) respectively, then $\vec{f}$ satisfies (3).

This statement is proved in [8]. In [8,12] the equality

$$H^p(\mathbb{D},X) = \tilde{H}^p(\mathbb{D},X)$$

(independent of p, $1 \le p \le \infty$) is used as the definition of $X \in$ (ARN). This equivalence has an analogue for the operator case.

Proposition 1. *An operator $U:X \to Y$ is an (ARN)-operator iff*

$$U(H^p(\mathbb{D},X)) \subset \tilde{H}^p(\mathbb{D},X) \tag{4}$$

for all p, $1 \le p \le \infty$ (equivalently for one such p).

Proof. Let us translate (4) into the language of vector measures for p=1, obtaining additionally the independence of (4) of p. From [8] we know that the boundary values of $\vec{F} \in H^1(\mathbb{D},X)$ always exist in the following weak sense. Namely, there exists X^*-scalarly measurable function $\vec{f}:[0,2\pi] \to X^{**}$, such that

(i) $\vec{F}$ is the weak Poisson integral of $\vec{f}$;

(ii) The Fourier coefficients of $\vec{f}$ possesses the following property: $c_n(\vec{f}) \in X$, $n \ge 0$; $c_n(\vec{f}) = 0$, $n<0$;

(iii) the norm of $\vec{F}$ in $H^1(\mathbb{D},X)$ is calculated by the formula

$$\|\vec{F}\|_1 = \frac{1}{2\pi} \int_0^{2\pi} \|\vec{f}(\theta)\|_{X^{**}}\, d\theta.$$

Let us consider the vector measure

$$\vec{m}([a,b]) = \int_a^b \vec{f}(\theta)d\theta; \qquad [a,b] \subset [0,2\pi]. \tag{5}$$

A priori the values of $\vec{m}$ lie in X^{**} but we shall show that $\vec{m}([a,b]) \in X$. By the Hardy theorem on Fourier coefficients of H^1-functions one can easily see that the integrated Fourier series of $\vec{f}$ is weakly unconditionally convergent (to $\vec{m}([a,b])$) but it is not sufficient for the proof in that case of an arbitrary space X. To avoid this difficulty we recall some facts about the factorization of the real-valued analytic functions (see Proposition 2 in [12]). Since the function $z \to \langle \vec{F}(z),x^* \rangle$ $(x^* \in X^*,\ \|x^*\| \le 1)$ is an analytic scalar-valued function from H^1 with the boundary values $\langle \vec{f}(\theta),x^* \rangle$, we have

$$\frac{1}{2\pi} \int_0^{2\pi} \log \|\vec{f}(\theta)\|_{X^{**}}\, d\theta \ge \frac{1}{2\pi} \int_0^{2\pi} \log |\langle \vec{f}(\theta),x^* \rangle|\, d\theta > -\infty.$$

Then we can define the external function as usual by the following formula

$$F_e(z) = \exp\left[\frac{1}{2\pi}\int_0^{2\pi}\frac{e^{i\theta}+z}{e^{i\theta}-z}\log\|\vec{f}(\theta)\|_{X^{**}}\,d\theta\right].$$

So F_e is a scalar-valued analytic function such that $|F_e(e^{i\theta})| = \|\vec{f}(\theta)\|_{X^{**}}$ a.e. Applying Jensen inequality to the function $\langle\vec{F}(z),x^*\rangle$ we have

$$\log|\langle F(re^{i\theta}),x^*\rangle| \le \frac{1}{2\pi}\int_0^{2\pi}\log|\langle\vec{f}(\sigma),x^*\rangle|\cdot P_r(\theta-\sigma)\,d\sigma$$

$$\le \frac{1}{2\pi}\int_0^{2\pi}\log\|\vec{f}(\theta)\|_{X^{**}}\,P_r(\theta-\sigma)\,d\sigma = \log|F_e(re^{i\theta})|.$$

Hence, $\|\vec{F}(z)\|_X \le |F_e(z)|$ $(z\in\mathbb{D})$ and we can define the internal function

$$\vec{F}_i(z) = \frac{\vec{F}(z)}{F_e(z)}, \qquad z\in\mathbb{D}.$$

It is clear that $\vec{F}_i \in H^\infty(\mathbb{D},X)$. So we have the factorization

$$\vec{F}(z) = F_e(z)\vec{F}_i(z),$$

where $F_e \in H^1(\mathbb{D})$, $\vec{F}_i \in H^\infty(\mathbb{D},X)$. It is evident that $\vec{F} \in H^p(\mathbb{D},X)$ iff $F_e\in H^p$, and $\vec{F} \in \tilde{H}^p(\mathbb{D},X)$ iff $F_e\in H^p$ and $\vec{F}_i \in \tilde{H}^\infty(\mathbb{D},X)$. These facts give the independence of (4) of p.

Let us return to the proof that $\vec{m}([a,b]) \in X$. Let B be the Blaschke product for F_e and represent

$$\vec{F}(z) = F_1(z)\vec{F}_2(z),$$

where

$$F_1(z) = B(z)\left(\frac{F_e(z)}{B(z)}\right)^{1/2},$$

$$\vec{F}_2(z) = \left(\frac{F_e(z)}{B(z)}\right)^{1/2}\vec{F}_i(z).$$

Then $F_1 \in H^2(\mathbb{D})$, $\vec{F}_2 \in H^2(\mathbb{D},X)$. Consider the Fourier series

$$\vec{F}_2(e^{i\theta}) = \sum_{n=0}^{\infty}\vec{c}_n e^{i\theta},$$

where $\vec{c}_n \in X$. Let us check that

$$\vec{m}([a,b]) = \sum_{n=0}^{\infty}\vec{c}_n\int_a^b F_1(e^{i\theta})e^{in\theta}\,d\theta$$

in the sense of the convergence in norm in X. For $x^*\in X^*$, $\|x^*\| \le 1$, we have

$$\left|\left\langle\sum_{n=N}^{M}\vec{c}_n\int_a^b F_1(e^{i\theta})e^{in\theta}\,d\theta,\ x^*\right\rangle\right| \le$$

$$\leq \left(\sum_{n=N}^{M} |\langle \vec{c}_n, x^* \rangle|^2 \right)^{1/2} \left(\sum_{n=N}^{M} \left| \int_a^b F_1(e^{i\theta}) e^{in\theta}\, d\theta \right|^2 \right)^{1/2}$$

$$\leq \left(\frac{1}{2\pi} \int_0^{2\pi} |\langle \vec{F}_2(e^{i\theta}),\ x^* \rangle|^2\, d\theta \right)^{1/2} \left(\sum_{n=N}^{M} |a_n|^2 \right)^{1/2}$$

$$\leq \|\vec{F}_2\|_2 \left(\sum_{n=N}^{M} |a_n|^2 \right)^{1/2},$$

where $\{a_n\}$ is the sequence of the Fourier coefficients of the function $F_1(e^{i\theta})\chi_{(a,b)}(\theta)$ from $L^2([0,2\pi])$. Since x^* is arbitrary we obtain the norm estimate and we have proved the convergence of the series for $\vec{m}$ in X. Thus, $\vec{m}$ is an analytic X-valued measure of bounded variation. Conversely, if we take such a measure $\vec{m}$ and construct the function $\vec{f}:[0,2\pi] \to X^{**}$ satisfying (5), then $\vec{f}$ has automatically all the other properties mentioned above. Thus, we identify the space $H^1(\mathbb{D},X)$ with the space of analytic X-valued measures of finite variation by means of formula (5).

The analogue of formula (5) allows to identify the space $\tilde{H}^1(\mathbb{D},Y)$ with the space of Y-valued measures represented by means of Bochner integral.

Now the statement of Proposition 1 is obvious.

Proposition 2. *If $U:X \to Y$ is an ARN-operator then U does not fix a copy of c_0 (i.e. there exists no subspace X_0 of X, isomorphic to c_0, such that $U|_{X_0}$ is an isomorphism).*

Proof. It is an immediate corollary of the well-known fact that c_0 fails to have (ARN)-property.

As it was proved in [12], the converse of Proposition 2 is false in the case of general Banach spaces even for the identity operator.

To formulate a hypothesis to generalize Theorem A to the case of the (ARN)-operators on Banach lattices we recall some facts from [26] about the operators on Banach lattices which does not fix a copy of c_0. Let E be a Banach lattice and Y be a Banach space; $U:E \to Y$ be a linear operator. The following result is proved in [26].

Theorem B. 1. *The following assertions are equivalent:*

(i) U does not fix a copy of c_0;

(ii) $0 \leq e_n \uparrow$ in E and $\sup \|e_n\| < \infty$ implies $\{Ue_n\}$ is norm convergent in Y.

2. *If U does not fix a copy of c_0 then there exist a Banach lattice W with order continuous norm, a lattice homomorphism $R:E \to W$*

and an operator $Q:W \to Y$ with $U = QR$ such that R does not fix a copy of c_o.

We see that the identity operator id:E $\to$ E does fix a copy of c_o iff E is a KB-space. So it is a natural class of operators to formulate a generalization of Theorem A. From Proposition 2 we deduce that any ARN-operator does not fix a copy of c_o. Unfortunately, we have not succeeded to prove the complete converse to this statement.

Concerning this problem we may add the following remarks. First of all it is a natural to consider the problem of factorizing ARN-operators from a Banach lattice to a Banach space trough a lattice with the (ARN)-property, that is a KB-space. It is evident from Theorem A that every operator, which factors trough a KB-space is an ARN-operator. From the results of [26] and Proposition 2 we easily see that the converse is true in the following cases: the norm of the Banach lattice is order continuous, or the operator is a dual one. However, in general, the converse is not true. There exist ARN-operators on Banach lattices which do not factorize through a KB-space. Indeed, in [25] N.Ghoussoub and W.B.Johnson constructed a Banach lattice E and a lattice homomorphism U from E onto c_o which satisfy:

(i) U does not fix an isomorphic copy of c_o;

(ii) U is a RN-operator and, consequently, an ARN-operator;

(iii) U maps weak Cauchy sequences into norm convergent sequences;

(iv) If U is written as a product of two operators, then one of them fixes a copy of c_o.

From (iv) we see that U cannot be factorized through a KB-space (and even through a Banach space with the (ARN)-property), but U is an ARN-operator. This shows that the answer in the case of operators cannot be deduced from Theorem A.

I think that the following result holds true.

CONJECTURE. *Let E be a Banach lattice and let Y be a Banach space. An operator $U:E \to Y$ is an ARN-operator if and only if U does not fix a copy of c_o.*

For to prove the Conjecture it is sufficient to obtain some new form of weighted estimate for the maximal radial function of H^1-function (cf. [21]) depending on a parameter. We need not theorem about some condition of Muckenhoupt type but some kind of existence theorem for an arbitrary weight (cf. [15,23,42]). Maybe the separation theorem "without compactness" from [13] will be useful here. Of course, it is just Theorem B makes the use of the maximal radial function possible.

Remark. We can introduce the class of the HRN-operators, connected with boundary values of harmonic vector-valued functions, by means of the following analogue of formula (4):

$$U(h^p(\mathbb{D},X)) \subset \tilde{h}^p(\mathbb{D},Y), \qquad 1<p\le\infty,$$

either for all such p or for one (the spaces of harmonic functions h and $\tilde{h}$ are defined in full analogy with H and $\tilde{H}$). One can prove in the same way as in [12] that U is a HRN-operator iff U is a RN-operator.

3. The analytic Radon-Nikodym property and interpolation.

Let (X_o,X_1) be an interpolation pair of Banach spaces. The two well-known complex interpolation methods of Calderon lead to the spaces $[X_o,X_1]_\theta$ and $[X_o,X_1]^\theta$, $0<\theta<1$ (see [14,3]). It is proved in [2] that the natural embedding

$$(6) \qquad [X_o,X_1]_\theta \subset [X_o,X_1]^\theta$$

is an isometry. It is easy to find examples ([14]) when the inclusion (6) is proper. A.Calderon [14] proved that if X_o is reflexive then

$$(7) \qquad [X_o,X_1]_\theta = [X_o,X_1]^\theta$$

J.Peetre [37] and V.Vodopianov (Thesis, Kazan. Univ., 1980) independently proved that (7) is true if $X_o \in (RN)$. The equality (7) is useful, for example, in the consideration of interpolation of analytic families of linear operators [16]. In this connection it is interesting to find new sufficient conditions for the validity of (7).

To formulate the main result we have to recall some definitions and introduce a new class of Banach spaces. Let (Ω,Λ,μ) be a σ-finite measure space; let $S = S(\Omega,\Lambda,\mu)$ be the space of all (complex) measurable functions. A Banach ideal space on (Ω,Λ,μ) is a linear subset E of S which is a Banach space satisfying

$$(|e_1|\le|e_2|,\ e_1\in S,\ e_2\in E) \implies (e_1\in E,\ \|e_1\|\le\|e_2\|).$$

With E' we denote the Banach ideal space dual to E, which can be identified with the space of all functionals possessing an integral representation (these are exactly the order continuous functionals on E), that is

$$E' = \left\{e'\in S:\ \|e'\|_{E'} = \sup\left\{\int |ee'|d\mu:\ \|e\|\le 1\right\} < \infty\right\}.$$

If E is a KB-space then $E^* = E'$.

We say that a Banach space X, which is a linear subset of $S(\Omega,\Lambda,\mu)$, is well embedded into $S(\Omega,\Lambda,\mu)$ ($X\in(WE)$) provided there exists a Banach ideal space W on (Ω,Λ,μ) such that the identity embedding of X into W is continuous. Note that the Sobolev spaces, the Besov spaces and BMO are continuously embedded into some L^p and so they are in the class (WE) as the most part of the other spaces considered in analysis. Note that we may assume without loss of generality that W is a weighted space L^1. Indeed, it is sufficient to take an arbitrary $\wp\in W'$, $\wp>0$ a.e. (such $\wp$ exists because of the σ-finiteness of μ) and take $W_1 = L^1(\wp d\mu)$. Clearly $W \subset W_1$. Note also that not every Banach space, which is continuously embedded into S, is well embedded.

Now we can formulate the main result.

Theorem 3. *Let $X_0 = E$ be a KB-space on (Ω,Λ,μ) and let X_1 be a Banach space which is well embedded into $S(\Omega,\Lambda,\mu)$. Then the equality (7) holds.*

Proof. Let us remind the definition of the space $[X_0,X_1]^\theta$ [14,3]. We denote by $\mathfrak{G} = \mathfrak{G}(X_0,X_1)$ the space of all (X_0+X_1)-valued functions ψ which are continuous in the open strip $\Pi = \{z\in\mathbb{C}:\ 0\le \mathrm{Re}\le 1\}$ and analytic in the open strip $\Pi_0 = \{z\in\mathbb{C}:\ 0<\mathrm{Re}<1\}$ such that

(i) $\|\psi(z)\|_{X_0+X_1} \le c(1+|z|)$, $z\in\Pi$;

(ii) For j=0,1 we have $\psi(j+it_1) - \psi(j+it_2) \in X_j$, $\forall t_1,t_2\in\mathbb{R}$ and

$$\|\psi\|_{\mathfrak{G}} = \max_{j=0,1}\ \sup_{t_1\ne t_2} \left\| \frac{\psi(j+it_2) - \psi(j+it_1)}{t_2 - t_1}\right\|_{X_j} < \infty.$$

The second complex interpolation method assigns to the pair (X_0,X_1) the interpolation space $[X_0,X_1]^\theta$, $0<\theta<1$, consisting of all $x \in X_0+X_1$ such that $x = \psi'(\theta)$ for some $\psi \in \mathfrak{G}(X_0,X_1)$, with the norm

$$\|x\| = \inf\{\|\psi\|_{\mathfrak{G}}:\ x=\psi'(\theta),\ \psi\in\mathfrak{G}\}.$$

Let us turn to the proof of (7). Obviously we have to show the converse inclusion to (6). Take $x \in [X_0,X_1]^\theta$ and $\psi \in \mathfrak{G}(X_0,X_1)$ such that $x=\psi'(\theta)$. Set

$$\vec{m}([a,b]) := \psi(ib) - \psi(ia), \tag{8}$$

where $[a,b] \subset [0,1]$. Then $\vec{m}$ extends to a X_0-valued Borel measure of bounded variation which is absolutely continuous with respect to the Lebesgue measure (it follows from the condition (ii) on ψ). According to the statement 9.5 in [14] (see also Lemma 4.3.2 [3]) it is sufficient to prove that the measure (8) is representable as a Bochner integral (1) with respect to a X_o-valued measurable function. In order to obtain (1) we need new ideas in comparison with [37,45], since in the general case $X_o \notin (RN)$.

The natural candidate to play the role of $\vec{f}$ in (1) would be the function $\psi'(it)$ if it exists as measurable X_o-valued function. However, the existence of the strong derivative for an arbitrary X_o-valued Lipschitz function is equivalent to $X_o \in (RN)$. Nevertheless, just the investigation of the boundary behaviour of $\psi'(z)$ is the key.

First of all, note that $X_o+X_1 \in (WE)$. We embed X_o+X_1 in a weighted L^1-space W according to the remark preceding Theorem 3. From Theorem A we know that $W \in (ARN)$, which is essential for what follows. To use this fact, we consider a conform mapping $\alpha:\mathbb{D} \to \Pi_o$ which is defined also everywhere on the circle $\{z: |z|=1\}$ except two fixed points corresponding to the 'upper and lower infinity' of the strip Π. Since the function $\psi':\Pi_o \to X_o+X_1$ is analytic and bounded, $\psi':\Pi_o \to W$ and $\psi'\circ\alpha:\mathbb{D} \to W$ have the same properties; hence $\psi'\circ\alpha \in H^{\infty}(\mathbb{D},W)$. Since $W \in (ARN)$ we conclude that the non-tangential limits of $\psi'\circ\alpha$ exists a.e. on the circle (in the norm of W). Then

$$(9) \qquad \psi'(it+\tfrac{1}{n}) \longrightarrow \varphi(it), \qquad n \to \infty$$

in the norm of W for a.e. $t \in (-\infty,\infty)$. We will prove that

$$(10) \qquad \vec{m}([a,b]) = i\int_a^b \varphi(it)dt.$$

From $E = X_o \subset X_o + X_1 \subset W$ we obtain $E' = X_o' \supset W'$ and W' is a total set of linear functionals on $E = X_o$. If $y^* \in W'$, then from (9) we have

$$\langle\psi'(it+\tfrac{1}{n}),y^*\rangle \longrightarrow \langle\varphi(it),y^*\rangle \qquad \text{a.e. } t;\ n \to \infty.$$

Besides it,

$$|\langle\psi'(it+\tfrac{1}{n}),y^*\rangle| \le \|\psi'(it+\tfrac{1}{n})\|_W\cdot\|y^*\|_{W'}$$
$$\le c\|\psi'(it+\tfrac{1}{n})\|_{X_o+X_1}\cdot\|y^*\|_{W'} \le c_1\|y^*\|_{W'}.$$

Consequently, by Lebesgue's theorem we have

$$\int_a^b\langle\psi'(it+\tfrac{1}{n}),y^*\rangle dt \longrightarrow \int_a^b\langle\varphi(it),y^*\rangle dt, \qquad n \to \infty.$$

On the other hand, since $(X_o+X_1)^* \supset W'$, we have

$$i\cdot\int_a^b\langle\psi'(it+\tfrac{1}{n}),y^*\rangle dt = \int_a^b\frac{d}{d\tau}\ \langle\psi(i\tau+\tfrac{1}{n}),y^*\rangle|_{\tau=t}dt$$

$$= \langle\psi(ib+\tfrac{1}{n}) - \psi(ia+\tfrac{1}{n}),\ y^*\rangle \longrightarrow \langle\psi(ib) - \psi(ia),\ y^*\rangle = \langle\vec{m}([a,b]),y^*\rangle,$$

as $n \to \infty$. Comparing these two limits we obtain (10).

Since $\varphi:[0,1] \to W$ is measurable the integral (10) is a.e. differentiable in norm of W. In particular, for a.e. t

$$e_n(t) = \frac{\psi(it+\frac{i}{n}) - \psi(it)}{(\frac{i}{n})} \longrightarrow \varphi(it), \qquad n \to \infty$$

in measure μ. Since the sequence $\langle e_n(t)\rangle$ is norm bounded for any t in the KB-space $E = X_o$ (this follows from the definition of $\psi \in \mathfrak{G}$), we have $\varphi(it) \in X_o = E$ for a.e. t. As the function $\varphi:[0,1] \to W$ is measurable we may assume that the function $\varphi(it)(\omega)$ is measurable in the two variables (t,ω) [7]. We proved that the values of φ lie in the KB-space X_o. These two facts together give that φ is a measurable X_o-valued function [7]. This concludes the proof of Theorem 3.

For general Banach pairs (X_o,X_1) the proof of Theorem 3 gives the following result.

Theorem 4. *Let (X_o,X_1) be a Banach pair such that X_o+X_1 is continuously embedded into some Banach space $Y\in(ARN)$. Suppose that*
(i) $x_n \to x$ $(x_n\in X_o,\ x\in Y)$ in the norm of Y and $\sup\|x_n\|_{X_o} < \infty$ implies $x\in X_o$;
(ii) any function $\varphi:[0,1] \to X_o$, which is measurable as an Y-valued function is scalarly equivalent with respect to the restrictions of the functionals from Y^* to some X_o-valued measurable function.
Then (7) holds true.

Remarks. (i) There is some result on the validity of the equality (7) connected with (ARN)-property in [31] but it deals with the simple situation $X_1 \subset X_o \in$ (ARN). Evidently, it is a particular case of Theorem 4 with $X_o=Y$.

(ii) The assumption (ii) of Theorem 4 holds true if X_o is weakly compactly generated. It is an easy consequence of some well known properties of such spaces (see the references in [10]).

(iii) It is proved in [17] (the result belongs to G.Pisier) that even in the case when both spaces X_o, X_1 have the (ARN)-property then it may occur that $[X_o,X_1]_\theta = [X_o,X_1]^\theta$ fails to have this property.

(iv) It is unlikely to obtain necessary and sufficient conditions for the validity of equality (7) in a comprehensible form. For example, $[X,X]_\theta = [X,X]^\theta$ for every Banach space X. Nevertheless, we can attempt to find some properties of Banach pairs (X_o,X_1) for which (7) holds. For example, it is likely that if (7) holds true for a Banach pair (X_o,X_1) for some $\theta_o \in (0,1)$, then (7) holds true for all

$\theta \in (0,1)$.

Now we consider the case when (E_0,E_1) is a pair of Banach spaces on (Ω,Λ,μ). In this situation the complex method's interpolation spaces are closely connected with some special Banach ideal space $E(\theta) = E_0^{1-\theta}E_1^{\theta}$ which was introduced by A.Calderon [14]. The properties of this space were investigated in details by G.Ya.Lozanowskii (see [35], for example). The following Proposition 5 is an immediate corollary of the results from [14,35].

Proposition 5. *Let E_0, E_1 be a Banach ideal spaces on (Ω,Λ,μ).*

(i) If $E(\theta)$ is a KB-space then

(11) $$[E_0,E_1]_\theta = [E_0,E_1]^\theta = E(\theta).$$

(ii) If E_0 is a KB-space then $E(\theta)$ is a KB-space too. Consequently, the equality (11) holds and, moreover, all the spaces in (11) have the (ARN)-property.

(iii) If E_0 is arbitrary and $E_1=E_0'$ then $E(\theta)$ is reflexive. Consequently, the equality (11) holds.

Remark. From the assertion (ii) we see that the main interest of the Theorem 3 lies in the fact that X_1 may not be a Banach ideal space. This case is not covered by the technique from [14,35].

It is proved in [45] that RN-operators have an improvement property, i.e. if an operator $U:(X_0,X_1) \to (Y_0,Y_1)$ is such that $U:X_0 \to Y_0$ is an RN-operator, then U acts from $[X_0,X_1]^\theta \longrightarrow [Y_0,Y_1]_\theta$. It is interesting to consider an open question whether it is possible to generalize Theorem 3 to ARN-operators in a similar manner.

4. The (ARN)-property in spaces of vector-valued functions.

In [32,20] it was established that for $1 \le p < \infty$ we have the following equivalence:

(12) $$L^p(X) \in (ARN) \iff X \in (ARN).$$

In [11] independently a more general result (see Theorem 6 bellow) was announced.

Let E be a Banach ideal space on (Ω,Λ,μ), X be a Banach space. By $E(X)$ we denote the Banach space of all (strongly) measurable functions $\vec{f}:\Omega \to X$ such that $\|\vec{f}(\cdot)\|_X \in E$, provided with the norm

$\|\vec{f}\| = \| \ \|\vec{f}(\cdot)\|_X\|_E.$

Theorem 6. *Let E be a Banach ideal space on (Ω,Λ,μ) and X be a Banach space. Then*

(12) $$E(X) \in (ARN) \iff (E \in (ARN) \ \& \ X \in (ARN)).$$

Remark 1. For the (RN)-property the analogous result was established in [10].

Remark 2. It was convenient in [20] to check the existence of the boundary values in $H^p(L^p(X))$ by using Fubini's theorem. In the case when $E \neq L^p$ this approach is not suitable because of the generalized Kolmogorov-Nagumo theorem [9].

Proof. Since E and X can be embedded into E(X) isomorphically as subspaces and the (ARN)-property is hereditary, the implication '$\Longrightarrow$' in (13) is trivial.

Let us prove the implication '$\Longleftarrow$' in (13) Note that E is a KB-space by Theorem A. By means of the approach from the beginning of the proof of Theorem 1 in [10] we may assume that E and X are separable. This is convenient concerning questions of measurability of functions.

Take $\vec{F} \in H^{\infty}(\mathbb{D},E(X))$. Just as in [8] we obtain that the function

$$\vec{f}(\omega,z) := \vec{F}(z)(\omega)$$

is an X-valued analytic function for a.e. $\omega\in\Omega$, and that

(14) $$e_o(\omega) = \sup_{0<r<1} \int_0^{2\pi} \|\vec{f}(\omega,re^{i\theta})\|_X d\theta$$

belongs to E. Hence, for a.e. ω the function $\vec{f}(\omega,z)$ belongs to $H^1(\mathbb{D},X)$. Since $X\in(ARN)$ the boundary values $\vec{g}(\omega,e^{i\theta})$ of the function $\vec{f}(\omega,z)$ exist for a.e. ω and

(15) $$e_o(\omega) = \int_0^{2\pi} \|\vec{g}(\omega, \ e^{i\theta})\|_X d\theta.$$

The end of the proof is based on the same ideas as the proofs of Theorem A in [12] and Theorem 3 of the present paper. However, some delicate questions concerning measurability occur in this case.

We obtain that for a.e. ω

(16) $$[0,2\pi] \ni \theta \longrightarrow \vec{g}(\omega,e^{i\theta}) \in X$$

is a measurable X-valued function;

(17) $$\|\vec{f}(\omega,re^{i\theta}) - \vec{g}(\omega,e^{i\theta})\|_X \longrightarrow 0, \quad r \to 1$$

for a.e. θ (the excluded set of θ's depend on ω);

(18) $$\int_0^{2\pi} \|\vec{f}(\omega, re^{i\theta}) - \vec{g}(\omega, e^{i\theta})\|_X d\theta \longrightarrow 0, \quad r \to 1;$$

(19) $$\vec{f}(\omega, re^{i\theta}) = \frac{1}{2\pi}\int_0^{2\pi} P_r(\theta-\sigma)\cdot \vec{g}(\omega, e^{i\sigma})d\sigma$$

for a.e. θ (the integral on the right has to be understood as X-valued Bochner integral for fixed ω).

We have to construct a measurable function $\vec{\varphi}:[0,2\pi] \to E(X)$, such that

(20) $$\vec{F}(re^{i\theta}) = \frac{1}{2\pi}\int_0^{2\pi} P_r(\theta-\sigma)\cdot \vec{\varphi}(e^{i\sigma})d\sigma.$$

Taking into account the similarity of (19) and (20), we see that the problem lies in the correction of the function $\vec{g}$ in a suitable way.

For a.e. ω the function $\vec{f}(\omega,z)$ is analytic. Hence, for a.e. ω the function $\vec{f}(\omega, re^{i\theta})$ is continuous in the norm of X in the two variables (r,θ). Then for any r and a.e. ω the function

$$\theta \longrightarrow \vec{f}(\omega, re^{i\theta}) \in X$$

is measurable. Passing to a weighted L^1-space we may assume that $E \subset L^1$. From (14), (17) and the Lebesgue dominated convergence theorem we have

$$\int_\Omega\int_0^{2\pi} \|\vec{f}(\omega, r_n e^{i\theta}) - \vec{f}(\omega, r_m e^{i\theta})\|_X \, d\theta \, d\mu(\omega) \longrightarrow 0, \quad n,m \to \infty.$$

Since the space $L^1([0,2\pi]\times\Omega, X)$ is complete, there exists a function $\vec{\varphi} \in L^1([0,2\pi]\times\Omega, X)$ such that

(21) $$\int_\Omega\int_0^{2\pi} \|\vec{f}(\omega, r_n e^{i\theta}) - \vec{\varphi}(\omega, e^{i\theta})\|_X \, d\theta \, d\mu(\omega) \longrightarrow 0, \quad n \to \infty.$$

We show that $\vec{\varphi}$ is the desired function. For this purpose we have to prove two things, i.e. $\vec{\varphi}:[0,2\pi] \to E(X)$ is measurable and equality (20) holds. First of all we establish the correspondence between $\vec{\varphi}$ and $\vec{g}$. From (21) we see (passing to a subsequence if it is necessary) that for a.e. ω

(22) $$\|\vec{f}(\omega, r_n e^{i\theta}) - \vec{\varphi}(\omega, e^{i\theta})\|_X \longrightarrow 0, \quad n \to \infty$$

for a.e. θ. From (17) and (22) we obtain that for a.e. ω

(23) $$\vec{g}(\omega, e^{i\theta}) = \vec{\varphi}(\omega, e^{i\theta})$$

for a.e. θ. On the other hand, from (21) we can conclude that for a.e. θ

(24) $$\|\vec{f}(\omega, r_n e^{i\theta}) - \vec{\varphi}(\omega, e^{i\theta})\|_X \longrightarrow 0, \quad n \to \infty$$

for a.e. ω. Since $\vec{F} \in H^\infty(\mathbb{D}, E(X))$ the sequence of the functions

$\omega \to \|\vec{f}(\omega, r_n e^{i\theta})\|_X$ is bounded in E-norm for any θ. Since E is a KB-space, this fact together with (24) implies that the function $\omega \to \vec{\varphi}(\omega, e^{i\theta})$ belongs to E(X) for a.e. θ. We have assumed that E and X are separable, so E(X) is separable too. In order to prove the measurability of $\vec{\varphi}:[0,2\pi] \to E(X)$ it is sufficient to check the scalar measurability of $\vec{\varphi}$ on a total set of functionals on E(X) (see, for example Lemma 1 in [10]).

Take $e^*\in E'$, $x^*\in X^*$ and consider the following expression

$$\langle \vec{\varphi}(\cdot, e^{i\theta}), e^*\otimes x^*\rangle = \int \langle \vec{\varphi}(\omega, e^{i\theta}), x^*\rangle \, e^*(\omega)\, d\mu(\omega). \tag{25}$$

From (21) we see that the function $(\omega,\theta) \to \vec{\varphi}(\omega, e^{i\theta}) \in X$ is measurable which proves the measurability in θ of the integral (25).

It remains to prove (20). It is sufficient to check this equality on the functionals $e^*\otimes x^*$. However this can be easily seen using (19) and (23).

For a Banach lattice E and a Banach space X the tensor product $E\hat{\otimes}X$ (in V.Levin's sense) is defined (see [44]), which is a variant of the construction of the space E(X).

Remark. Let E, W be Banach lattices and X, Y Banach spaces. Let $U:E \to W$, $V:X \to Y$ be ARN-operators. It is likely that $U\otimes V: E\hat{\otimes}X \to W\hat{\otimes}Y$ is an ARN-operator.

Theorem 6 permits to extend the proof of Theorem 3 to spaces of vector-valued functions. The method of the proof of Theorem 3 gives the following result.

Proposition 7. *Let E_o be a Banach ideal space on (Ω,Λ,μ) which is a KB-space; let $X\in(ARN)$ be a Banach space; let Y be a Banach space, continuously embedded into W(X), where W is a Banach ideal space on (Ω,Λ,μ). Then*

$$[E_o(X),Y]_\theta = [E_o(X),Y]^\theta .$$

Acknowledgement: The part of this paper has been written while the author was at the University of Tübingen on basis of the scientific exchange program with Deutscher Akademischer Austauschdienst. The author wants to thank Prof. R.Nagel for his hospitality and Dr. F.Räbiger for interesting discussions.

REFERENCES

[1] A.B.Aleksandrov, *Esseys on non locally convex Hardy classes*, Lect. Notes Math. 864 (1981), 1-89.

[2] J.Bergh, *On the relation between two complex methods of interpolation*, Indiana J. Math. 28 (1979), 775-778.

[3] J.Bergh, J.Löfström, *Interpolation Spaces. An Introduction*. Springer Verlag, Berlin-Heidelberg-New York, 1976.

[4] O.Blasco, *Boundary values of functions in vector-valued Hardy spaces and geometry on Banach spaces*, J. Funct. Anal. 78 (1988), 346-364.

[5] S.Bu, *Quelques remarques sur la propriété de Radon-Nikodym analytique*, Compt. Rend. Acad. Sci. Paris 306 (1988), 757-760.

[6] S.Bu, B.Khoulani, *Une caracterisation de la propriété de Radon-Nikodym analytique pour les espaces de Banach isomorphes a leur carre*, to appear.

[7] A.V.Bukhvalov, *On spaces with mixed norm*, Vestnik Leningrad. Univ. (1973), no. 19, 5-12. English translation: Vestnik Leningrad. Univ. Math. 6 (1979), 303-311.

[8] A.V.Bukhvalov, *Hardy spaces of vector-valued functions*, Zap. Nauchn. Sem. LOMI 65 (1976), 5-16. English translation: J. Soviet Math. 16 (1981), 1051-1059.

[9] A.V.Bukhvalov, *A generalized Kolmogorov-Nagumo theorem on tensor products*, Kachestv. i Pribl. Metody Issled. Operat. Uravn., Jaroslavl' 4 (1979), 48-65 (in Russian).

[10] A.V.Bukhvalov, *Radon-Nikodym property in Banach spaces of measurable vector-valued functions*, Mat. Zametki 26 (1979), 875-884. English translation: Math. Notes 26 (1979), 939-944 (1980).

[11] A.V.Bukhvalov, *Some properties of the interpolation methods*, Proc. XII-th School on the Theory of Operators in Function Spaces, Part I, Tambov, 1987, p.30 (in Russian).

[12] A.V.Bukhvalov, A.A.Danilevich, *Boundary properties of analytic and harmonic functions with values in Banach space*, Math. Zametki 31 (1982), 203-214. English translation: Math. Notes 31 (1982), 104-110.

[13] A.V.Bukhvalov, G.Ya.Lozanovskii, *On sets closed in mesure in spaces of measurable functions*, Trudy Maskov. Mat. Ob-va 34, 129-150. English translation: Trans. Moscow Math. Soc. (1978), issue 2, 127-148.

[14] A.P.Calderon, *Intermediate spaces and interpolation, the complex method*, Studia Math. 24 (1964), 113-190.

[15] A.Cordoba C.Fefferman, *A weighted norm inequality for singular integrals*, Studia Math. 57 (1976), 97–101.

[16] M.Cwikel, S.Janson, *Interpolation of analytic families of operators*, Studia Math. 79 (1984), 61–71.

[17] S.J.Dilworth, *Complex convexity and the geometry of Banach spaces*, Math. Proc. Cambr. Phil. Soc. 99 (1986), 495–506.

[18] P.G.Dodds, *o-weakly compact mappings of Riesz spaces*, Trans. Amer. Math. Soc. 214 (1975), 389–402.

[19] P.N.Dowling, *Representable operators and the analytic Radon-Nikodym property in Banach spaces*, Proc. Royal Irish Acad. 85A (1985), 143–150.

[20] P.N.Dowling, *The analytic Radon-Nikodym property in Lebesque-Bochner function spaces*, Proc. Amer. Math. Soc. 99 (1987), 119–122.

[21] P.L.Duren, *Teory of H^p Spaces*, Academic Press, New York, 1970.

[22] G.A.Edgar, *Analytic martingale convergence*, J. Funct. Anal. 69 (1986), 268–280.

[23] J.Garcia-Cuerva, J.L.Rubio de Francia, *Weighted Norm Inequalities and Related Topics*, North Holland, Amsterdam, 1985.

[24] D.J.H.Garling, *On martingales with values in a complex Banach space*, Math. Proc. Cambr. Phil. Soc., to appear.

[25] N.Ghoussoub, W. B. Johnson, *Counterexamples to several problems on the factorization of bounded linear operators*, Proc. Amer. Math. Soc. 92 (1984), 233–238.

[26] N.Ghoussoub, W. B. Johnson, *Factoring operators through Banach lattices not containing C(0,1)*, Math. Z. 194 (1987), 153–171.

[27] N.Ghoussoub, J.Lindenstrauss, B. Maurey, *Analytic martingales and plurisubharmonic barries in complex Banach spaces*, Contemporary Math., to appear.

[28] N.Ghoussoub, J. Maurey, *Plurisubharmonic martingales and barries in complex quasi-Banach spaces*, to appear.

[29] N.Ghoussoub, J. Maurey, *An non-linear method for constructing certain basic sequences in Banach spaces*, to appear.

[30] N.Ghoussoub, J. Maurey, W.Schachermayer, *Pluriharmonically dentable complex Banach spaces*, to appear.

[31] U.Haagerup, G.Pisier, *Factorization of analytic functions with values in non-commutative L_1-spaces*, to appear.

[32] W.Hensgen, *Hardy-Räume vektorwertiger Funktionen*, Dissertation, Univ. München, 1986.

[33] W.Hensgen, *Operatoren $H^1 \to X$*, Manuscr. Math. 59 (1987), 399–422.

[34] N.J.Kalton, *Differentiability properties of vector valued functions*, Lect. Notes in Math. 1221 (1986), 141–181.

[35] G.Ya.Lozanovskii, *On some Banach lattices*, Sibirsk. Mat. Zh. 10 (1969), 584–599. English translation: Siberian Math. J. 10 (1969), 419–431.

[36] M.Nawrocki, *The Fréchet envelopes of vector-valued Smirnov classes*, Studia Math. 94 (1989), 61–75.

[37] J.Peetre, *H^{∞} and complex interpolation*, Lund, 1981.

[38] G.Pisier, *Factorization of operator valued analytic functions*, to appear.

[39] Quanhua Xu, *Inégalités pour les martingales de Hardy et renormage des espaces quasinormés*, Compt. Rend. Acad Sci. Paris 306 (1988), 601–604.

[40] Quanhua Xu, *Applications du theoreme de factorisation pour des fonctions a valeurs operateurs*, to appear.

[41] O.I.Reynov, *Operators of type RN in Banach spaces*, Sibirsk. Mat. Zh. 19 (1978), 857–865. English translation: Siberian Math. J. 19 (1978), 606–612.

[42] J.Rubio de Francia, *Weighted norm inequalities and vector valued inequalities*, Lect. Notes Math. 908 (1982), 86–101.

[43] R.Ryan, *The F. and M.Riesz theorem for vector measures*, Proc. Acad. Sci. Amsterdam A66 (1963), 408–412.

[44] H.H.Schaefer, *Banach Lattices and Positive Operators*, Springer Verlag, Berlin-Heidelberg-New York, 1974.

[45] V.V.Vodopianov, *Radon-Nikodym property in Calderon's interpolation constructions*, Izv. VUZ Mat. (1987), no. 11, 8–10. English translation: Soviet Math. (Iz. VUZ) 31 (1987), no. 11.

ON THE LIONS-SCHECHTER COMPLEX INTERPOLATION METHOD

M. J. Carro, Joan Cerdà

Barcelona, Spain

Abstract. We describe the basic properties of the complex interpolation method associated to a distribution of finite support.

Some examples related to the domain of positive operators and L^p spaces are given.

Introduction.

Lions [8] and Schechter [10] study a variant of the complex interpolation method for compatible pairs of Banach spaces associated to certain functionals T that in the case $T = \delta(\theta)$ is the first interpolation method of Calderón. This method can be applied also to complex interpolation families of Banach spaces of [5].

The functional T acts on the space $\mathcal{F}$ of vector-valued analytic functions introduced by Calderón. Thus, it is convenient to consider continuous linear functionals on the space $\mathcal{H}$ of analytic functions on the strip in the case of interpolation pairs, and on the disk in the case of complex interpolation families. In Section 1 we include some results about these functionals.

We describe some of the basic properties of this complex interpolation method when T is of finite type (see (2)): interpolation of operators, duality and connection with the real method. Examples about L^p spaces and domains of fractional powers of positive operators are included.

Supported in part by DGICYT Grant PS87-0027

1. Analytic functionals.

Suppose that Ω is a domain of the complex plane $\mathbb{C}$, like the strip $S = \{z: 0 < \Re z < 1\}$ and the disc $D = \{z: |z| < 1\}$.

The space $\mathcal{H}(\Omega)$ of all holomorphic functions defined on Ω is a locally convex space with the topology of uniform convergence on compact subsets of Ω.

If $\mathbb{C}$ is the complex sphere, $\mathcal{H}(\mathbb{C}\setminus\Omega)$ will be the vector space of all holomorphic functions defined on open neighborhoods of $\bar{\mathbb{C}}\setminus\Omega$ and which vanish at ∞. Two such functions are identified if they agree on a neighborhood of $\bar{\mathbb{C}}\setminus\Omega$.

An analytic functional on Ω is a continuous linear functional $T \in \mathcal{H}'(\Omega)$ on $\mathcal{H}(\Omega)$. By defining

$$h_T(\zeta) = \langle T, \frac{1}{\zeta - z} \rangle$$

(here $\frac{1}{\zeta-z}$ is a function of z for every $\zeta \in \bar{\mathbb{C}}$, equal to 0 if $\zeta=\infty$) we obtain a function $h_T \in \mathcal{H}(\bar{\mathbb{C}}\setminus\Omega)$ such that

$$(1) \qquad \langle T,f\rangle = \frac{1}{2\pi i}\int_\gamma h_T(z)f(z)dz$$

and the value of (1) is independent of the "Cauchy curve" γ in $\Omega\cap G$, where G is the domain of h_T.

Fantapiè's representation (1) defines an isomorphism between $\mathcal{H}'(\Omega)$ and $\mathcal{H}(\bar{\mathbb{C}}\setminus\Omega)$. See reference [7] for the details.

An analytic functional T on Ω is said to be of *finite type* if it is the restriction to $\mathcal{H}(\Omega)$ of a distribution of finite support on Ω. Thus, in this case,

$$(2) \qquad T = \sum_{j=1}^{n}\sum_{k=0}^{m(j)} a_{j,k}\delta^{(k)}(z_j)$$

where $a_{j,k} \in \mathbb{C}$, $a_{j,m(j)}\neq 0$, $\langle\delta^{(k)},f\rangle = f^{(k)}(z)$ and $z_1,\dots z_n$ are n different points of Ω. From the Cauchy integral formula it follows that for the corresponding Fantapiè's representation we have

$$h_T = \frac{P_T}{Q_T},$$

where P_T and Q_T are two polynomials such that Q_T has zeroes z_j with multiplicity m(j), and P_T has no zeroes in $\{z_1,\dots,z_n\}$.

If E is a complex Banach space and $\mathcal{H}(\Omega;E)$ is the corresponding space of vector-valued holomorphic functions on Ω, (1) allows to extend any $T \in \mathcal{H}'(\Omega)$ to a continuous linear operator $T:\mathcal{H}(\Omega;E) \to E$. If T is of finite type we can use (2) for this extension.

Finally observe that, if $G \in \mathcal{H}(\Omega)$ and $T \in \mathcal{H}'(\Omega)$, we can define GT in the sense of the distributions

$$\langle GT,f\rangle = \langle T,Gf\rangle, \qquad f \in \mathcal{H}(\Omega;E)$$

and, if $u \in E'$,

$$\langle uT,f\rangle = \langle T,\langle u,f\rangle\rangle, \qquad f \in \mathcal{H}(\Omega;E).$$

2. The interpolated spaces

Let (A_o,A_1) be an interpolation pair of complex Banach spaces, and $\Sigma = A_o+A_1$ the sum space. We suppose that $\Delta = A_o \cap A_1$ is dense in A_j $(j = 0,1)$.

Like in [8] and [10], for a fixed analytic functional T on S we can define the interpolated space

$$[A_o,A_1]_T = \{x=T(f): f\in\mathcal{F}\},$$

a Banach space with $\|x\|_T = \inf\{\|f\|_{\mathcal{F}}: T(f)=x, f\in\mathcal{F}\}$. Here $\mathcal{F} = \mathcal{F}(A_o,A_1) \subset \mathcal{H}(S;\Sigma)$ is the class of vector-valued analytic functions of the first complex interpolation method of Calderón (see [1]). In the case $T = \delta(\theta)$, with $0<\theta<1$, we have $[A_o,A_1]_T = [A_o,A_1]_\theta$.

In the same way we can consider the case of a *complex interpolation family* (c.i.f.) $A = \{A_\gamma: \gamma\in\Gamma\}$ $(\Gamma = \partial D)$ in the sense of [5], with the containing space $\mathcal{U}$ and the log-intersection space $\mathcal{A}$. We suppose that $\mathcal{A}$ is dense in every A_γ.

If $T \in \mathcal{H}'(D)$, we can define the interpolated space

$$A[T] = \{x=T(f): f\in\mathcal{F}\}$$

with $\|x\|_T = \inf\{\|f\|_{\mathcal{F}}: T(f)=x, f\in\mathcal{F}\}$ as before, but now $\mathcal{F} = \mathcal{F}(A) \subset \mathcal{H}(D;\mathcal{U})$ is the family of analytic functions introduced in [5].

We observe that Δ ($\mathcal{A}$ in the case of a c.i.f.) is dense in the interpolated space because

$$\mathcal{G} = \{g\in\mathcal{F}: \dim[\mathrm{Range}(g)] < \infty \text{ and } \mathrm{Range}(g) \subset \Delta \text{ or } \mathcal{A}\}$$

is a dense subspace of $\mathcal{F}$ and T is a continuous function from $\mathcal{F}$ onto $[A_o,A_1]_T$ ($A[T]$ for a c.i.f.).

It is convenient also to note that the finite type case (2) can often be reduced to the case $T = \delta^{(m)}(z)$ because

$$[A_o,A_1]_T = \sum_{j=1}^{n}[A_o,A_1]_{\delta^{(m(j))}(z_j)},$$

and the same is true for a c.i.f.

There is no problem to prove the interpolation theorem:

THEOREM 1. *If $(A_0,A_1),(B_0,B_1)$ are two interpolation pairs and $L:A_0\cap A_1 \to B_0\cap B_1$ is linear with*

$$\|L(a)\|_j \le M_j\|a\|_j \qquad (j=0,1),$$

then for any $T \in \mathcal{H}'(S)$ we have $\|L(a)\|_T \le M\|a\|_T$ with $M = \max(M_0,M_1)$.

In the same way, if

$$H_z[g] = \frac{1}{2\pi}\int_0^{2\pi} g(\gamma)\,\frac{e^{i\gamma}+z}{e^{i\gamma}-z}\,d\gamma$$

is the Herglotz transform, we have [2]

THEOREM 1 BIS. *Let $A = \{A_\gamma: \gamma\in\Gamma\}$ and $B = \{B_\gamma: \gamma\in\Gamma\}$ be two c.i.f., $T \in \mathcal{H}'(D)$ and $L:\mathcal{A} \to \bigcap_{\gamma\in\Gamma} B_\gamma$ a linear operator such that*

$$\|L(a)\|_\gamma \le M(\gamma)\|a\|_\gamma \qquad a.e.\ \gamma\in\Gamma.$$

Then

$$\|L(a)\|_T \le \|a\|_{GT}$$

for $G(z) = \exp(-H_z[\log M])$.

3. Duality.

Calderón introduced his second interpolation method to study duality. This is based on its second space of vector-valued analytic functions

$$\bar{\mathcal{F}} = \bar{\mathcal{F}}(A_0',A_1') \subset \mathcal{H}(S;A_0'+A_1').$$

Then

$$(3) \qquad [A_0,A_1]_\theta' = \{u\in\Delta^*: u = h'(\theta),\ h \in \bar{\mathcal{F}}\}.$$

If $T = \delta(\theta)$, $u \in A_0'+A_1' = (A_0\cap A_1)'$ and $g\in\mathcal{G}$, it follows from the definitions that $\langle uT,g\rangle = \langle h'(\theta),g(\theta)\rangle$ and condition (3) is equivalent to

$$uT = h'T, \qquad h \in \bar{\mathcal{F}}.$$

But for $T = \delta^{(m)}(\theta)$ we need a more restricted class:

$$\bar{\mathcal{F}}(T) = \{h\in\bar{\mathcal{F}}: \langle T,\langle h',f\rangle\rangle = 0 \text{ if } T(f)=0,\ f\in\mathcal{F}\},$$

where $\langle h',f\rangle(z) := \lim_n \langle h'(z),g_n(z)\rangle$ for any sequence $g_n \in \mathcal{G}$ such that $\lim g_n = f$ in $\mathcal{F}$.

In the case of a c.i.f. $A = \{A_\gamma: \gamma\in\Gamma\}$, to the space $\mathcal{W} \subset \mathcal{H}(D;\mathcal{U})$ of [5] and $T \in \mathcal{H}'(D)$ of finite type we associate the new class

$$\mathcal{W}(T) = \{H\in\mathcal{W}\colon \langle R,\langle f,H\rangle\rangle = 0 \text{ if } T(f)=0,\ f\in\mathcal{F}\},$$

where $R \in \mathcal{X}'(S)$ is such that $h_R = \frac{1}{Q_T}$ if $h_T = \frac{P_T}{Q_T}$.

The corresponding duality theorem reads as follows:

Theorem 2. *If T is of finite type,*

$$A[T]' = \{u\in\mathcal{A}^*\colon uT = HT \text{ for some } H \in \mathcal{W}(T)\}$$

and

$$\|u\| \cong \inf\{\|H\|_{\mathcal{W}}\colon uT = HT,\ H \in \mathcal{W}(T)\}.$$

Remark. If T is $\delta(z)$ we have $\mathcal{W}(T) = \mathcal{W}$ and this theorem coincides with the duality theorem of [5]. But this equality is not true in general.

4. The connection with the real method.

Recall that a Banach space E is of *Fourier type p*, with $1 \le p \le 2$, if the Fourier transform $\mathcal{F}$ maps $L^p_E(\mathbb{R})$ continuously in $L^q_E(\mathbb{R})$ for $\frac{1}{p} + \frac{1}{q} = 1$.

Let us consider an interpolation pair (A_o,A_1). The relationship with the real method (see [1])

$$(A_o,A_1)_{\theta,1} \subset [A_o,A_1]_\theta \subset (A_o,A_1)_{\theta,\infty},$$

is sharpened if A_j is of type p_j $(j=0,1)$. In this case we have

$$(A_o,A_1)_{\theta,p} \subset [A_o,A_1]_\theta \subset (A_o,A_1)_{\theta,q},$$

with $1/p + 1/q = 1$ [9], where $1/p = (1-\theta)/p_o + \theta/p_1$.

If we consider the new complex interpolation method associated to $T = \delta^{(n)}(\theta)$, new real methods are required.

Denote

$$f(t) = t^\theta(n + |\log t|)^n \qquad (n\in\mathbb{N}).$$

These functions are in the class B_K of [6] and for such a function, Kalugina and Gustavsson consider the two parameters interpolation method

$$(A_o,A_1)_{f,p} = \{x\in\Sigma\colon \frac{K(\cdot,x)}{f} \in L^p_*\},$$

Where $L^p_* = L^p(dt/t)$ and $K(t,x)$ is the usual K-functional, with the norm

$$\|x\| = \left\| \frac{K(\cdot,x)}{f} \right\|_{L^p_*}.$$

Theorem 3. *For every $0 < \theta < 1$, and $n \in \mathbb{N}$,*

$$(A_0,A_1)_{f,1} \subset [A_0,A_1]_{\delta_\theta^{(n)}} \subset (A_0,A_1)_{f,\infty}.$$

The next result improves the previous one when the spaces are of Fourier type greater than 1.

Define $(A_0,A_1)_{f,p_0,p_1}$ to be the set of all those $a \in A_0 + A_1$ such that there exists a measurable function $u:\mathbb{R}^+ \to A_0 \cap A_1$ with

$$a = \int_0^\infty u(t)\, dt/t, \qquad \|u(t)\|_{A_0}/f(t) \in L_*^{p_0} \quad \text{and} \quad \|tu(t)\|_{A_1}/f(t) \in L_*^{p_1}$$

Theorem 4. *If A_0 is of type p_0 and A_1 of type p_1, then*

$$(A_0,A_1)_{f,p_0,p_1} \subset [A_0,A_1]_{\delta_\theta^{(n)}} \subset (A_0,A_1)_{f,q_0,q_1}.$$

For more details see [4].

Remark. If $n=0$, $f(t) = t^\theta$ and we have

$$(A_0,A_1)_{f,p_0,p_1} = (A_0,A_1)_{\theta,p_0,p_1} = (A_0,A_1)_{f,p}.$$

As far as we know, the equality $(A_0,A_1)_{f,p_0,p_1} = (A_0,A_1)_{f,p}$ is an open problem when $n>0$.

5. Examples.

(I) Interpolation of L^p spaces.

Interpolation results of L^p spaces like

(a) $L^p(X;[A_0,A_1]_\theta) = [L^p(X;A_0),L^p(X;A_1)]_\theta$.

(b) $[L^p(\omega_0),L^p(\omega_1)]_\theta = L^p(\omega)$ with $\omega = \omega_0^{1-\theta}\omega_1^\theta$ (interpolation with change of measures).

(c) $[L^{p_0},L^{p_1}]_\theta = L^p$ (Riesz–Thorin theorem)

have an extension for the Lions–Schechter interpolation method:

(a') $L^p(X;[A_0,A_1]_{\delta^{(n)}(\theta)}) = [L^p(X;A_0),L^p(X;A_1)]_{\delta^{(n)}(\theta)}$. The proof is quite similar to the one of (a).

(b') $[L^p(\omega_0),L^p(\omega_1)]_{\delta'(\theta)} = L^p(\omega)$ with

$$\omega = \omega_0^{1-\theta}\omega_1^\theta \left(1 + |\log \frac{\omega_0}{\omega_1}|\right)^{-p}$$

as stated in [8].

(c') $[L^{p_0},L^{p_1}]_{\delta'(\theta)} = \{f_0+f_1\log|f_1|: f_j \in L^p\}$ with the "natural" topology.

Both results (b') and (c') are particular cases of interpolation of c.i.f.

$$L^p(\omega) = \{L^{p(\gamma)}(\omega_\gamma): \gamma \in \Gamma\}$$

of L^p spaces with change of measures contained in [2]. For n=1

$$L^p(\omega)[\delta'(z_o)] = \{f_o+f_1(K_{p,\omega}+\log|f_1|): f_j \in L^{p(z_o)}(\omega_{z_o})\}.$$

Here $1/p(z) = P_z[1/p]$, $\omega_z = \exp\{p(z)H_z[(\log\omega)/p]\}$ and

$$K_{p,w} = \left(\omega_{z_o}^{H_z[1/p]}\omega_z^{P_z[1/p]}\right)'_{z=z_o}$$

The corresponding result for $L^p(\omega)[\delta^{(n)}(z_o)]$ in the case of an interpolation pair without change of measure is

$$[L^{p_o},L^{p_1}]_{\delta^{(n)}(\theta)} = \left\{\sum_{j=o}^{n} f_j(\log|f_j|)^j: f_j \in L^p\right\} = L_\Phi,$$

where Φ is the Orlicz function $\Phi = (\phi^{-1})^{1/p}$ such that $\phi(t) = t(n+|\log t|)^n$.

(II) Domains of fractional powers of a positive operator.

Let L be a linear closed operator with dense domain in E. L is said to be positive if $(-\infty,0] \subset \mathrm{Res}(L)$ (resolvent set) and there exists $C \geq 0$ such that

$$\|(L+t)^{-1}\| \leq C(1+t)^{-1} \qquad \forall\ t \in [0,\infty).$$

Any self-adjoint positive-definite operator on a Hilbert space is a positive operator, and so is $-L$ if L is the infinitesimal generator of a strongly continuous semigroup $\{G(t): t \geq 0\}$ such that $\beta<0$ (where $\|G(t)\| \leq Me^{\beta t}$).

Functions of L like L^α or logL are easily defined with the use of the spectral decomposition of L in the Hilbert space case, and the study of fractional powers has been extended to positive operators on a Banach space E.

In this case, for any positive integer m, L^{-m} is a bounded operator (we have $-m \in \mathrm{Res}(L)$) and L^m is easily defined by induction on

$$D(L^m) = \{x\in D(L^{m-1}): L^{m-1}(x) \in D(L)\}.$$

It is a one-to-one closed operator whose inverse is L^{-m}, and it follows that $D(L^m)$ is dense in A.

A general formula for L^α on a dense subspace F of E can be found in [11]. When $\Re\alpha \in [0,1]$ we can take $F = D(L^3)$ and

$$L^\alpha a = C(\alpha) \int_0^\infty t^\alpha L^2 (L+t)^{-3} a \, dt$$

with $C(\alpha) = 2/\Gamma(1+\alpha)\Gamma(2-\alpha)$, and L^α is the clausure of $a \longmapsto L^\alpha a$.

It is known that $[A,D(L)]_\theta = D(L^\theta)$ if the positive operator L satisfies the following condition (see [11], 1.15.3):

L^{it} is uniformly bounded for $t\in\mathbb{R}$.

In fact it is enough to have $\|L^{it}\|_{B(A)} \leq C$ for $-\varepsilon < t < \varepsilon$. In the Hilbert space case L has this property, but this is not always true in the general case. We shall impose this extra condition to L.

Let $a \in D(L^3)$. The function $F(\xi) = L^\xi(a)$ is analytic in S and

$$F'(0) = 2 \int_0^\infty \left[\log t + \frac{C'(0)}{2}\right] L^2(L+t)^{-3} a \, dt.$$

In analogy with what it happens in the Hilbert space case, we define

$$\log La = F'(0),$$

we have a clausurable and densely defined operator and, in this way, log L is defined as its closure. One can prove that

$$(\log L)^n a = F^{(n)}(0) \qquad \forall\ n\in\mathbb{N}.$$

If $\mathcal{R}_j = R[(\log L)^j_{|D(L^\theta)}]$ and $\|x\|_{\mathcal{R}_j} = \inf \{\|x_1\|_{D(L^\theta)}: (\log L)^j x_1 = x\}$, in [3] it is proved that the spaces $[A,D(L)]_{\delta_\theta^{(n)}}$ and $\sum_{j=0}^n \mathcal{R}_j$ with the sum norm are equivalent.

In the Hilbert space case, if L is definite-positive and $1 \notin \sigma(L)$, $[A,D(L)]_{\delta_\theta^{(n)}} = \mathcal{R}_n$.

If L is the operator $Lf = \mathcal{F}^{-1}(1+\|x\|^2)^{1/2}\mathcal{F}(f)$, $\mathcal{F}$ the Fourier transform, these results are applied in [3] to the interpolation of Sobolev spaces. If $LF = \omega f$, L^p interpolation with change of measures is obtained.

REFERENCES

[1] J. Bergh, J. Löfström, *Interpolation Spaces, An Introduction*, Springer Verlag, Berlin-Heidelberg-New York, 1976.

[2] M.J.Carro, *Interpolación Compleja de Operadores Lineales*, Ph. D. Thesis, University of Barcelona (1988).

[3] M.J.Carro, J.Cerdá, *Fractional powers of linear operators and complex interpolation*, Math. Nachr. (to appear).

[4] M.J.Carro, J.Cerdá, J.Sueiro, *On Fourier type and uniform convexity of interpolation spaces*, Bolletino U.M.I. (to appear).

[5] R.Coifman, M. Cwikel, R. Rochberg, S. Sagher, G. Weiss, *A theory of complex interpolation for families of Banach spaces*, Advances in Math. 43 (1982), 203–229.

[6] J.Gustavsson, *A function parameter in connection with interpolation of Banach spaces*, Math. Scand. 42 (1978), 289–305.

[7] G.Köthe, *Toplogical Vector Spaces I*, Springer-Verlag, Berlin-Heidelberg-New York, 1969.

[8] J.L.Lions, *Quelques procédés d'interpolation d'operateurs linéaires et quelques applications*, Séminaire Schwartz II (1960/61), 2–3.

[9] J.Peetre, *Sur la transformation de Fourier des fonctions à valeurs vectorielles*, Rend. Sem. Univ. Padova 42 (1969), 15–26.

[10] M. Schechter, *Complex interpolation*, Comp. Math. 18 (1967), 117–147.

[11] H.Triebel, *Interpolation Theory, Function Spaces, Differential Operators*, North-Holland, Amsterdam-New York-Oxford, 1978.

REITERATION FOR AND EXACT RELATIONS BETWEEN SOME REAL INTERPOLATION SPACES

CARLOS E. FINOL, LECH MALIGRANDA, LARS ERIC PERSSON

Caracas, Venezuela; Luleå, Sweden

ABSTRACT. For the real interpolation spaces $(A_o,A_1)_{f,q}$ with a parameter function f we prove a general reiteration result, where we need not assume some separation condition between the corresponding parameter functions. As one application we obtain a sharp embedding result between the spaces $(A_o,A_1)_{(\theta,b),q}$ obtained by using the function parameter $f(t) = t^{\theta}(1+|\log\ t|)^{b}$. This result may be regarded as a generalization of some well-known embeddings between Lorentz-Zygmund spaces.

INTRODUCTION. Let $\theta,\theta_1,\theta_o,q,q_o,q_1$ and q_θ denote real numbers satisfying $0 < \theta,\theta_o,\theta_1 < 1$, $0 < q,q_o,q_1 \leq \infty$ and $1/q_\theta = (1-\theta)/q_o+\theta/q_1$, where $1/\infty = 0$. Let (A_o,A_1) be a compatible pair of quasi-Banach spaces. The Lions-Peetre real interpolation spaces $(A_o,A_1)_{\theta,q}$ are well understood and widely used in different kinds of applications, see e.g. the books [2],[3],[5],[7],[17] (and compare with our Section 1). In particular, we note that *the reiteration formula*

$$((A_o,A_1)_{\theta_o,q_o},(A_o,A_1)_{\theta_1,q_1})_{\theta,q} = (A_o,A_1)_{\eta,q},$$

where $\eta = (1-\theta)\theta_o+\theta\theta_1$ holds if $\theta_o \neq \theta_1$ (see e.g. [3,p.50]) or if $q = q_\theta$ (see e.g. [5,p.186] or our Theorem 1.1). Unfortunately, if none of these conditions is satisfied, then the situation is much more complicated. In this paper we present exact descriptions of the more general (parameter function) spaces

$$((A_o,A_1)_{f_o,q_o},(A_o,A_1)_{f_1,q_1})_{f,q} \tag{0.1}$$

in cases which do not include the complicated cases described above. (The real interpolation spaces $(A_o,A_1)_{f,q}$ with a parameter function f are described in Section 1). Moreover we introduce the spaces $(A_o,A_1)_{(\theta,b),q}$ which generalizes the usual Lorentz-Zygmund spaces

$L^{p,q}(\log L)^b$ (see [1]) in a natural way. We prove a sharp embedding theorem between these spaces which in particular generalizes the well-known embedding between the $L^{p,q}(\log L)^b$-spaces (see [1,p.31]).

The paper is organized in the following way: In Section 1 we give some preliminaries including some necessary theory about real interpolation. In Section 2 we present the announced descriptions of the spaces (0.1). We remark that the proofs are easy consequences of a fundamental estimate of Brudnyi-Krugljak (see (1.1)) and some fairly new descriptions of real interpolation spaces between weighted L^p and Lorentz-spaces (see [9] and [15]). In Section 3 we introduce the spaces $(A_o,A_1)_{(\theta,b),q}$ and prove the announced embedding theorem between these spaces. Moreover we prove that this embedding is in a sense the sharpest possible.

Conventions: C denotes any positive constant (not the same in different appearances). The equivalence symbol $f(t) \approx g(t)$ means that $af(t) \le g(t) \le bf(t)$ for some constants $a,b > 0$ and all $t>0$. Two quasi-normed spaces A and B are considered as equal, and we write A = B, whenever their quasi-norms are equivalent.

1. Preliminaries.

We consider the Lebesgue space $L^q_* = L^q(0,\infty,dt/t)$. The class Q_ε, $\varepsilon>0$, consists of the functions $\psi:(0,\infty) \to (0,\infty)$ satisfying

$$\|\psi\|_{L^1_*} = 1, \quad t^{-\varepsilon}\psi(t) \text{ is decreasing and } t^{\varepsilon}\psi(t) \text{ is increasing.}$$

Let (Ω,Σ,μ) be a σ-finite measure space and let $\omega = \omega(x)$ be a weight function on Ω, i.e. let ω be a measurable and positive function on Ω. Let $E = E(\mu)$ be an ideal quasi-normed subspace of the space $S(\mu)$ of all μ-measurable functions, which are finite almost everywhere. The weighted ideal quasi-normed space $E(\omega) = E(\omega,\mu)$ consists of all $a \in S(\mu)$ satisfying

$$\|a\|_{E(\omega)} = \|a\omega\|_E < \infty.$$

Let a^* denote the nonincreasing rearrangement of a. The weighted Lorentz space $L^{\varphi q}(\omega,\mu)$ consists of all $a \in S(\mu)$ for which $(a\omega)^*$ belongs to the space $L^q_*(\varphi)$ endowed with quasi-norm

$$\|a\|_{L^{\varphi q}(\omega,\mu)} = \|(a\omega)^*\|_{L^q_*(\varphi)},$$

see e.g. [9]. (The function $a\omega$ is rearranged with respect to the measure μ.) If $\omega=1$, then $L^{\varphi q}(\omega,\mu) = L^{\varphi q}(\mu) = L^{\varphi q}$. In particular, if $\varphi(t) = t^{1/p}(1+|\log t|)^b$, then $L^{\varphi q}$ coincides with the usual

Lorentz-Zygmund spaces $L^{p,q}(\log L)^b$ investigated by Bennett and Rudnick in [1].

Let $\bar{A} = (A_o, A_1)$ denote a quasi-Banach pair. The K-functional $K(t,a)$ is defined for every $a \in A_o+A_1$ and $t>0$ as

$$K(t,a) = K(t,a,\bar{A}) = \inf_{a=a_o+a_1} (\|a_o\|_{A_o} + t\|a_1\|_{A_1}).$$

Let E be an ideal quasi-normed subspace of $S(\mu)$, where $\mu = dt/t$ and $\Omega = (0,\infty)$. The real interpolation space $\bar{A}_E$ is defined as the set of all $a \in A_o+A_1$ satisfying

$$\|a\|_{\bar{A}_E} = \|K(t,a,\bar{A})\|_E < \infty.$$

In particular if $E = L^q_*(\frac{1}{f})$, where f is a parameter function, we obtain the real interpolation spaces $\bar{A}_{f,q}$ with a parameter function f. The necessary hypothesis on the parameter function f can be given in several essentially equivalent ways. In this paper we use the Matuszewska-Orlicz indices α_f and β_f defined for every $f \in B$, where B denotes the class of all continuous functions $f:(0,\infty) \longrightarrow (0,\infty)$ such that, for every $t>0$,

$$\bar{f}(t) = \sup_{s>0} (f(st)/f(s)) < \infty.$$

The definition of α_f and β_f are as follows

$$\alpha_f = \sup_{0<t<1} \frac{\log \bar{f}(t)}{\log t} = \lim_{t \to 0+} \frac{\log \bar{f}(t)}{\log t},$$

$$\beta_f = \inf_{t>1} \frac{\log \bar{f}(t)}{\log t} = \lim_{t \to \infty} \frac{\log \bar{f}(t)}{\log t}.$$

It is well-known that $-\infty < \alpha_f \le \beta_f < \infty$. See [8] for more information about these and other indices. In this paper we assume that the parameter functions f belong to the class $B_o = \{f \in B: 0 < \alpha_f \le \beta_f < 1\}$.

Example 1.1. Let $f(t) = t^\theta(1+|\log t|)^b$, $0 < \theta < 1$, $b \in \mathbb{R}$. Then $\bar{f}(t) = t^\theta(1+|\log t|)^{|b|}$ and $\alpha_f = \beta_f = \theta$ and, thus, $f \in B_o$. In this case we denote the spaces $A_{f,q}$ by $A_{(\theta,b),q}$. For the case b=0 we have the usual parameter spaces $A_{\theta,q}$.

More information concerning real interpolation with a parameter function between quasi-Banach pairs can be found in [14] and the references given there. For the Banach case these spaces are in fact special cases of the spaces $(A_o,A_1)_\Phi{}^K$ already studied by Peetre in [11].

Later on we need the following fundamental estimate by Brudnyi-Krugljak (see [4],[5] and also Nilsson [10]):

Let (A_0,A_1) *be a quasi-Banach pair and* (E_0,E_1) *any pair of interpolation spaces between* ℓ^∞ *and* $\ell^\infty((2^{-i})_i)$. *Then, for any* $a \in A_0+A_1$ *and* $t>0$,

(1.1) $$K(t,a,\bar{A}_{E_0},\bar{A}_{E_1}) \approx K(t,(K(2^i,a,\bar{A}))_i,E_0,E_1).$$

2. Reiteration.

First of all we note that using the crucial estimate (1.1) with $E_j = \ell^{q_j}(\omega_j)$, where $\omega_j = \{1/f_j(2^i)\}_i$, $i \in Z$, $j=0,1$, we obtain that

(2.1) $$(\bar{A}_{f_0,q_0},\bar{A}_{f_1,q_1}) = \bar{A}_E, \text{ where } E = \left[L_*^{q_0}(\tfrac{1}{f_0}),L_*^{q_1}(\tfrac{1}{f_1})\right]_{f,q}$$

In the general case it is difficult to give simple descriptions of the spaces E (compare e.g. with our Lemma 2.2) but in some special cases such simple descriptions are known. For example, in **the diagonal cases** $q_0 = q_1 = q$, $0 < q \le \infty$ and $0 < q_0,q_1 < \infty$, $q = q_\theta$, $f(t) = t^\theta$ we have

(2.2) $$E = L_*^q(1/f_2) \quad \text{with} \quad f_2 = f_0 f(f_1/f_0).$$

See [9,Theorem 2] or [14,Lemma 3.1] and [3,p.115]. Moreover, if the functions f_1 and f_2 are **separated** from each other by some suitable index condition (corresponding to the condition $\theta_0 \neq \theta_1$ in the parameter case) we can use the following lemma:

Lemma 2.1. *Let* ω_0 *and* ω_1 *be positive weight functions on* Ω, $\omega_{01} = \omega_0/\omega_1$ *and* $\omega = \omega_0/f(\omega_{01})$. *Let* $0 < p_0,p_1,q \le \infty$, $f \in B_0$, $\omega_0,\omega_1,\omega_{01} \in B$ *with* $\alpha_{\omega_i}>0$, $i=0,1$ *and* $\alpha_{\omega_{01}}>0$ *or* $\beta_{\omega_{01}}<0$. *If* $b = b(t)$ *is a positive and continuous function on* $(0,\infty)$ *such that* $b(t)\,t^c$ *is increasing or decreasing for some constant* c, *then*

$$\|b\|_{(L_*^{p_0}(\omega_0),L_*^{p_1}(\omega_1))_{f,q}} \approx \|b\|_{L_*^q(\omega)}.$$

Lemma 2.1 is proved in [9,p.771]. See also [15]. The following theorem generalizes the usual parameter versions of the reiteration theorem to the function parameter case.

Theorem 2.1. *Let* $0 < q,q_0,q_1 \le \infty$, $f,f_0,f_1 \in B_0$, $f_{10} = f_1/f_0$ *and* $f_2 = f_0 f(f_1/f_0)$. *Then*

(2.3) $$(\bar{A}_{f_0,q_0},\bar{A}_{f_1,q_1})_{f,q} = \bar{A}_{f_2,q}$$

if one of the following conditions holds:

(a) $q_0 = q_1 = q$.

(b) $f(t) = t^{\theta}$, $q = q_{\theta}$, $0 < q_o, q_1 < \infty$.

(c) $\alpha_{f_{10}} > 0$ or $\beta_{f_{10}} < 0$.

Proof. Let (a) or (b) be satisfied. Then (2.2) holds and, thus, according to (2.1), (2.3) is satisfied. Moreover, using Lemma 2.1 with $\omega_i = 1/f_i$, i=1,2 and $b(t) = K(t,a,A_o,A_1)$ we find that also the assumption (c) together with (2.1) imply (2.3) and the proof is complete.

Remark. Concerning Theorem 2.1 (a) and (b) see also [5,p.186] and [14,Examples 4.1–4.2] or [9,Example 1]. Other proofs of Theorem 2.1 (c) can be found in [14,p.211] and (at least for the case $q < \infty$) in [6,Theorem 2.1].

In order to be able to treat the general off-diagonal case $q_o = q_1 \neq q$ we need the following lemma:

Lemma 2.2. *Let $f \in B_o$, $0 < p,q \leq \infty$, $\gamma = 1/p-1/q$ and $\varepsilon_o = \min(\alpha_f, 1-\beta_f)/|\gamma|$, $\gamma \neq 0$. If ω_o and ω_1 are weight functions on Ω, $\omega_{01} = \omega_o/\omega_1$ and $\omega = \omega_o/f(\omega_{01})$, then, for any $\varepsilon \in (0,\varepsilon_o)$,*

$$(L^p(\omega_o), L^p(\omega_1))_{f,q} = \begin{cases} \bigcap_{\psi \in Q_\varepsilon} L^p(\omega(\psi \circ \omega_{01})^{\gamma}) & \text{if } q > p, \\ \bigcup_{\psi \in Q_\varepsilon} L^p(\omega(\psi \circ \omega_{01})^{\gamma}) & \text{if } q < p. \end{cases}$$

A proof of Lemma 2.2 can be found in [9,p.769]. See also [15].

Theorem 2.2. *Let $f, f_o, f_1 \in B_o$, $f_{10} = f_1/f_o$, $f_2 = f_1 f(f_{10})$, $0 < p,q \leq \infty$, $\gamma = 1/q-1/p$, $\gamma \neq 0$ and $\varepsilon_o = \min(\alpha_f, 1-\beta_f)/|\gamma|$. Then, for any ε, $0 < \varepsilon < \varepsilon_o$,*

$$(\overline{A}_{f_o,p}, \overline{A}_{f_1,p})_{f,q} = \begin{cases} \bigcap_{\psi \in Q_\varepsilon} \overline{A}_{f_{2,\psi},p} & \text{if } q > p, \\ \bigcup_{\psi \in Q_\varepsilon} \overline{A}_{f_{2,\psi},p} & \text{if } q < p. \end{cases}$$

where $f_{2,\psi} = f_2(\psi \circ f_{10})^{\gamma}$.

Proof. We use (2.1) together with Lemma 2.2 where $\Omega = (0,\infty)$, $d\mu = dt/t$, $\omega_i = 1/f_i$, $i = 0,1$, and the proof follows.

Remark. If $A_o = L(\Omega)$ and $A_1 = L^{\infty}(\Omega)$, then

$$K(t,a,A_o,A_1) = \int_c^t a^*(u)du, \quad 0 < t < \infty, \tag{2.4}$$

see [11] or [3,p.109]. We conclude that in this case the spaces $(A_o,A_1)_{f,q}$ can be identified with generalized Lorentz spaces of the type $L^{\varphi q}$. Therefore, by combining Theorem 2.1(c) with Theorem 2.2 we obtain a new proof of Theorem 3.1 in [13].

We close this section by stating a description of the spaces (0.1) also for the most complicated off-diagonal case $q_o \neq q_1$, $q_\theta \neq q$.

Theorem 2.3. *Let* $f, f_o, f_1 \in B_o$, $0 < q, q_o, q_1 \leq \infty$, $q_\theta \neq q$, $q_o \neq q_1$,

$$f_2 = f_o^{1/(q_1/q_o-1)} f_1^{1/(q_o/q_1-1)}, \quad f_3 = (f_1/f_o)^{1/(1/q_o-1/q_1)}$$

and $\varphi(t) = t^{1/q_o}/f(t^{1/q_o-1/q_1})$. *Then*

$$(\bar{A}_{f_o,q_o}, \bar{A}_{f_1,q_1})_{f,q} = \bar{A}_E,$$

where $E = L^{\varphi q}(f_2, f_3 dt/t)$.

Proof. The proof is similar to those of Theorems 2.1 and 2.2. In this case we use Theorem 4 in [9] (with $\omega_i = 1/f_i$, $i = 0,1$, and $d\mu = dt/t$) instead of (2.1)- Lemma 2.1 and Lemma 2.2, respectively.

3. A sharp embedding between the spaces $\bar{A}_{(\theta,b),q}$.

First of all we point out the following obvious embeddings:

$$\bar{A}_{(\theta,b),q_o} \subset \bar{A}_{(\theta,b),q_1} \quad \text{if } q_o \leq q_1 \tag{3.1}$$

and

$$\bar{A}_{(\theta,b_o),q} \subset \bar{A}_{(\theta,b_1),q}, \quad \text{if } b_o \leq b_1.$$

The next theorem may be regarded as a complement of (3.1)-(3.2) which gives us precise information about the importance of the quantity 1/q-b in the definition of the spaces $A_{(\theta,b),q}$.

Theorem 3.1. *Let* $0 < q_1 < q_o \leq \infty$ *and* $-\infty < b_o, b_1 < \infty$. *If* $1/q_o - b_o > 1/q_1 - b_1$, *then*

$$\bar{A}_{(\theta,b_o),q_o} \subset \bar{A}_{(\theta,b_1),q_1}. \tag{3.3}$$

Proof. We choose a and θ_o such that $1-\theta < a < 1$ and $\theta-1+a < \theta_o a < \min(\theta,a)$. Let $f(t) = t^{\theta_o}$, $f_o(t) = t^{\theta-\theta_o a}(1+|\log t|)^{b_o}$ and $f_1(t) = t^{\theta+(1-\theta_o a)}(1+|\log t|)^{b_o}$ Then $f_{1o}(t) = f_1(t)/f_o(t) = t^a$ and $f_2(t) = f_o(t)f(f_{1o}(t)) = t^\theta(1+|\log t|)^{b_o}$. Therefore, according to Theorem 2.1(c) and Theorem 2.2, we find that the assumption $q_o > q_1$ implies that, for any fixed ε, $0 < \varepsilon < \varepsilon_o$,

(3.4) $$\overline{A}_{(\theta,b_o),q_o} = \bigcap_{\psi\in Q_\varepsilon} \overline{A}_{f_{2,\psi}q_1},$$

where $f_{2,\psi} = \psi^\gamma(t^a)t^\theta(1+|\log t|)^{b_o}$; $\gamma = 1/q_o-1/q_1$. We assume that $a \in \overline{A}_{(\theta,b_o),q_o}$ and choose ε such that $0 < \varepsilon < \varepsilon_o$ and $\psi \in Q_\varepsilon$ such that $\psi(t) \approx (1+|\log t|)^{-1-\varepsilon}$. Then, by (3.4), we find that $a \in A_{(\theta,b),q_1}$ for

$$b = b_o-(1+\varepsilon)(1/q_o-1/q_1) > b_o+1/q_1-1/q_o.$$

Thus (3.3) holds for every b_1 satisfying

$$b_o+1/q_1-1/q_o < b_1 \le b_o+(1+\varepsilon_o)(1/q_1-1/q_o).$$

Hence, in view of (3.2) we conclude that (3.3) holds for every $b_1 > b_o+1/q_1-1/q_o$ and the proof is complete.

Next we prove a statement showing that Theorem 3.1 is a way best possible.

Proposition 3.2. *Let $0 < q_1 < q_o \le \infty$ and $-\infty < b_o,b_1 < \infty$. If $1/q_o-b_o = 1/q_1-b_1$, then (3.3) does not hold in general.*

Proof. Let $A_o = L^r(\Omega)$, $0 < r < \infty$, and $A_1 = L^\infty(\Omega)$. Then

$$K(t,a,A_o,A_1) \approx \left(\int_0^{t^r} (a^*(u))^r du\right)^{1/r},$$

see [3,p.109]. Therefore, according to a suitable variant of Hardy's inequality (see e.g. Lemma 3.2 in [14] applied with $\psi(t) = t^{1/r}$ and $f(t) = 1/t^\theta(1+|\log t|)^b$, we obtain that

$$\|a\|_{\overline{A}_{(\theta,b),q}} \le C \left(\int_0^\infty (t^{-\theta}(1+|\log t|)^{-b})^q \left(\int_0^{t^r} (a^*(u))^r du\right)^{q/r} \frac{dt}{t}\right)^{1/q}$$

$$\le C \left(\int_0^\infty (t^{-\theta/r}(1+|\log t|)^{-b})^q \left(\int_0^{t} (a^*(u))^r du\right)^{q/r} \frac{dt}{t}\right)^{1/q}$$

$$\le C \left(\int_0^\infty (t^{(1-\theta)/r}(1+|\log t|)^{-b})^q (a^*(t))^q \frac{dt}{t}\right)^{1/q}$$

Since this inequality trivially holds in the opposite direction (with another constant) we have that

(3.5) $$\overline{A}_{(\theta,b),q} = L^{p,q}(\log L)^{-b}, \quad p = r/(1-\theta).$$

Let $q_o > 0$ and consider the function

$$f(t) = \begin{cases} t^{-1/p}\left(1+\log \frac{1}{t}\right)^{b_1-\frac{1}{q_1}}\left(1+\log\log \frac{e}{t}\right)^{-a}, & 0 < t \le 1, \\ 0, & t > 1, \end{cases}$$

where a is a real number to be chosen later on. According to (3.5) and the assumption $b_1 = b_o+1/q_1-1/q_o$ we find that

(3.6)
$$\|a\|_{\bar{A}_{(\theta,b_1),q_1}} = \|a\|_{L^{p,q_1}(\log L)^{-b_1}}$$
$$= \int_o^1 \frac{1}{t(1+\log 1/t)(1+\log\log 1/t)^{aq_1}}\,dt$$

and

(3.7)
$$\|a\|_{\bar{A}_{(\theta,b_o),q_o}} = \|a\|_{L^{p,q_o}(\log L)^{-b_o}}$$
$$= \int_o^1 \frac{1}{t(1+\log 1/t)(1+\log\log 1/t)^{aq_o}}\,dt.$$

We choose a satisfying $1/q_o < a \le 1/q_1$ and use (3.6)-(3.7) to obtain that

(3.8)
$$a \in \bar{A}_{(\theta,b_o),q_o} \quad \text{but} \quad a \notin \bar{A}_{(\theta,b_1),q_1}.$$

We only need to modify the arguments used above to see that (3.8) holds for the case $q_o = \infty$ too. The proof is complete.

Corollary 3.3. *Let $0 < p < \infty$, $0 < q_1 < q_o \le \infty$ and $-\infty < b_o, b_1 < \infty$.*

a) If $1/q_o+b_o > 1/q_1+b_1$, then

(3.9)
$$L^{p,q_1}(\log L)^{b_1} \supset L^{p,q_o}(\log L)^{b_o}$$

b) If $1/q_o+b_o = 1/q_1+b_1$, then the inclusion (3.9) does not hold in general.

Proof. We choose r, $0 < r < \infty$, and θ, $0 < \theta < 1$, such that $p = r/(1-\theta)$. Let $A_o = L^r(\Omega)$, $0 < r < \infty$, and $A_1 = L^\infty(\Omega)$. Then, in view of (3.5), we find that (3.3) implies (3.9). The statement in b) was proved in Proposition 3.2.

Remark. Another proof of (3.9) can be found in [1,p.31]. See also [13].

Remark. As seen in this paper it is difficult to describe the spaces

(3.10)
$$(\bar{A}_{\theta_o,q_o}, \bar{A}_{\theta_o,q_1})_{\theta,q}, \quad q_o \ne q_1,$$

in off-diagonal cases $q \ne q_\theta$. These difficulties appear usually as well in special cases e.g. when we are concerned with scales of Besov, Lorentz, weighted L^p or operator ideal spaces, etc. (see e.g. [9],[13],[15],[16] and the references given in these papers). One possibility to avoid this type of troubles can be to replace the scale (3.10) by the scale

(3.11) $$(\bar{A}_{\theta_0,q_0},\bar{A}_{\theta_0,q_1})_{(\theta,b),q_\theta}, \quad b = \frac{1}{q} - \frac{1}{q_\theta}.$$

On the other hand, according to Theorem 3.1 (and Proposition 3.2), we see that the scales in (3.10) and (3.11) are very closed related and on the other hand we note that our Theorem 2.1(b) gives a fairly uncomplicated description of the reiteration spaces (3.11).

REFERENCES

[1] C.Bennett, K. Rudnik, *On Lorentz-Zygmund spaces*, Dissertationes Math. 175 (1980), 1-67.

[2] C.Bennett, R.Sharpley, *Interpolation of Operators*, Academic Press, 1988.

[3] J.Bergh, J. Löfström, *Interpolation Spaces. An Introduction*, Grundlehren der Matematischen Wissenschaften 223, Springer Verlag, Berlin-Heidelberg-New York, 1976.

[4] Ju. A. Brudnyi, N. Ja. Krugljak, *Real interpolation functors*, Dokl. Akad. Nauk. SSSR, 256 (1981), 14-17 (Russian); Soviet Math. Dokl., 23, 1981, 5-8.

[5] Ju. A. Brudnyi, N. Ja. Krugljak, *Real interpolation functors*, Book manuscript, Jaroslavl, 1981, 212 pp (Russian).

[6] H.P.Heinig, *Interpolation of quasi-normed spaces involving weights*, Can. Math. Conf. Proc. 1 (1981), 245-267.

[7] S.G.Krein, Yu. I. Petunin, E. M. Semenov, *Interpolation of linear operators*, Nauka Moscow, 1978 (Russian); English translation A.M.S., Providence, 1982.

[8] L.Maligranda, *Indices and interpelation*, Dissertationes Math. 234 (1985), 1-54.

[9] L.Maligranda, L-E.Persson, *Real interpolation between weighted L^p and Lorentz spaces*, Bull. Acad. Polon. Math. 35 (1987), 685-832.

[10] P.Nilsson, *Reiteration theorems for real interpolation and approximation spaces*, Ann. Mat. Pura Appl. 32 (1982), 291-330.

[11] J.Peetre, *A theory of interpolation of normed spaces*, Notas de Matematica 39 (1968), 1-86.

[12] L-E.Persson, *An exact description of Lorentz spaces*, Acta Sci. Math. (Szeged) 46 (1983), 177-195.

[13] L-E.Persson, *Exact relations between some scales of spaces and interpolation*, Taubner-Texte zur Mathematik 103 (1988), 112-122.

[14] L-E.PERSSON, *Interpolation with a parameter function*, Math. Scand. 59 (1986), 199–22.

[15] L-E.PERSSON, *Real interpolation between cross-sectional L^p-spaces in quasi-Banach bundles*, Research report 1, Dept. of Math., Luleå University, 1986, 1–17.

[16] L-E.PERSSON, *Real interpolation between some operator ideals*, Lecture Notes in Math. 1302 (1986), 347–362.

[17] H.TRIEBEL, *Interpolation Theory. Function Spaces. Differential Operators*, North-Holland, 1978.

A SET OF K-FUNCTIONALS FOR A FIXED BANACH COUPLE

N.Ja.Krugljak

Jaroslavl, USSR

Let $\vec{X} = (X_o, X_1)$ be a Banach couple and

$$K(t,x;\vec{X}) = \inf_{x=x_o+x_1} \{\|x_o\|_{X_o} + t\|x_1\|_{X_1}\} \qquad (t>0)$$

be a K-functional of element $x \in \Sigma(\vec{X}) = X_o + X_1$. Since $K(\cdot,x;\vec{X})$ is a nonnegative concave function on $\mathbb{R}_+ = (0,+\infty)$, so the set of K-functionals of all elements $x \in \Sigma(\vec{X})$ ($\vec{X}$ – is fixed) forms a subset $K_{\vec{X}}$ of cone $\mathcal{C}onv$ – nonnegative concave functions on $\mathbb{R}_+$.

The aim of this report is to show that it is important to study the properties of the set

$$K_{\vec{X}} = \{f \in \mathcal{C}onv: f = K(\cdot,x;\vec{X}),\ x \in \Sigma(\vec{X})\}.$$

Problem 1. Characterize the set $K_{\vec{X}}$ (exactly or within to equivalence).

In particular, the next theorem shows that not every subset of $\mathcal{C}onv$ may coincide with $K_{\vec{X}}$ for some $\vec{X}$.

Theorem 1. *(Brudnyi, Krugljak) If $f \in K_{\vec{X}}$ and*

$$f = \sum_{i=1}^{\infty} \varphi_i \qquad (\varphi_i \in \mathcal{C}onv,\ i=1,2,\dots).$$

Then for every $\varepsilon>0$ there exist such functions $f_i \in K_{\vec{X}}$ that

$$f \le \sum_{i=1}^{\infty} f_i \qquad f_i \le (8+\varepsilon)\varphi_i.$$

Remark. The constant $8+\varepsilon$ in Theorem 1 was obtained by M.Cwikel.

Problem 2. For which Banach couples, under the conditions of Theorem 1, we can find such elements $f_i \in K_{\vec{X}}$, that $f \le \sum_{i=1}^{\infty} f_i$ and

$c^{-1}\varphi_i \le f_i \le c_i\varphi_i$ $(i = 1,2...)$, where the constant $c>0$ is not depended on f and φ_i $(i = 1,2,...)$.

Below we shall denote by $\mathfrak{Conv}_0$ the subcone of the cone $\mathfrak{Conv}$ which consists of functions $f \in \mathfrak{Conv}$ for which

$$\lim_{t\to 0} f(t) = \lim_{t\to\infty} \frac{f(t)}{t} = 0.$$

The function $f \in \mathfrak{Conv}_0$ we shall call *quasipower*, if for some $\gamma>0$ we have $Sf \le \gamma f$ where

$$(Sf)(t) = \int_0^\infty \min(1,t/s)f(s)\,\frac{ds}{s} \tag{1}$$

1. The converse problem of theory of approximation.

Let X be a Banach space, A_n $(n = 0,1,2,...)$ monotone (i.e., $A_n \subset A_{n+1}$) family of subsets of X for which $A_0 = \{0\}$ and $\overline{\bigcup_n A_n} = X$.

Let us define the best approximation of $x \in X$ by the formula

$$E_n(x) = \inf_{y\in A_n} \|x-y\|_X.$$

Problem 3. Characterize the set E of all sequences of best approximations

$$E = \{\{E_n(x)\}_{n=0}^\infty : x \in X\}.$$

The classical result is

Theorem 2. *(S.N.Bernstein) If A_n is a linear subspace of X $(n = 1,2,...)$ and $\dim A_n = n$, then the sequence $\{\alpha_n\}_{n=0}^\infty \subset E$, if and only if*

$$\alpha_n \ge \alpha_{n+1} \quad (n = 0,1,2,...) \quad \text{and} \quad \lim \alpha_n = 0. \tag{2}$$

We must note that the condition "A_n is linear subspace of X" is not fulfilled in some important cases. For example, this condition is not fulfilled if A_n consists of 2π-periodic functions, which have non zero Fourier coefficients not more n; or A_n consists of rational functions of degree $\le n$.

So, it arises the problem of generalization of the theorem of Bernstein. This problem was analyzed by Brudnyi [1] under additional restrictions:

1) $A_n+A_m \subset A_{n+m}$ $(n,m = 0,1,2,...)$,

ii) $\lambda A_n \subset A_n$ $(n = 0,1,\ldots,\ \lambda \in \mathbb{R})$,

iii) the sets A_n is not merge, i.e., there exists the constant $\gamma>0$ such that for every n we have

$$\sup_{x\in A_{n+1}\cap B_X} E_n(x) \geq \gamma \qquad (B_X = \{x\in X:\ \|x\|_X \leq 1\}).$$

Theorem 3. *(Ju.A.Brudnyi) If the sequence $(\alpha_n)_{n=0}^{\infty}$ is convex and satisfies (2), then we can find such element $x \in X$ that*

$$E_n(x) \geq \alpha_n; \qquad \varliminf_{n\to\infty} \frac{E_n(x)}{\alpha_n(x)} \leq \gamma_1 < +\infty,$$

where γ_1 is dependend only on γ.

The proof of Theorem 3 is based on the connection of sequence $\{E_n(x)\}_{n=0}^{\infty}$ with K-functional of x in some specially constructed pair of Abelian groups. Then Brudnyi used to such pair the analogy of

Theorem 4. *(N.Ja.Krugljak) If $K_{\vec{X}}$ contains at least one quasipower function, then for every function $g \in \mathcal{C}onv_0$ we can find such function $f_g \in K_{\vec{X}}$ that*

$$c^{-1} f_g \leq g \leq c f_g,$$

where $c>0$ is not depended on g.

Theorem 4 can also be used in the problem of describing the set of the modulus of continuity. In solving this problem it is important that for many domains $\Omega \subset \mathbb{R}^n$ we have:

$$\omega_k(t^{1/k},f)_p \approx K(t,f;L_p(\Omega),W_p^k(\Omega)),$$

where $\omega_k(t,f)_p$ is the k-th modulus of continuity of f in L_p.

2. Geometry of Banach spaces.

The classical result of Kadec-Pełczyński says that if $2 < p < \infty$ and a closed subspace U of the space $L_p(0,1)$ is not Hilbert, then U contains the space V which is complemented in L_p and isomorphic to ℓ_p.

This theorem can be easy obtained from the remarkable theorem of M.Levy.

Theorem 5. *(M.Levy)*

i) Let X be a subspace of the space $\vec{X}_{\theta p}$ $(\theta \in (0,1),\ p \in (1,\infty))$

and X is not closed in $\Sigma(\vec{X})$. Then for every $\varepsilon>0$ there exists a subspace Y_ε of the space X (Y_ε is not closed in $\Sigma(\vec{X})$), which is $(1+\varepsilon)$ isomorphic to ℓ_p and $(1+\varepsilon)$ complemented in $\vec{X}_{\theta p}$.

ii) The same is true for $X = \vec{X}_{\theta p}$ if $\Delta(\vec{X}) = X_0 \cap X_1$ is not closed in $\Sigma(\vec{X})$.

In the above theorem, $\vec{X}_{\theta p}$ denotes the linear subspace of $\Sigma(\vec{X})$ which consists of such elements $x \in \Sigma(\vec{X})$ for which the following norm is finite:

$$\|x\|_{\theta p} = \left[\int_0^\infty (t^{-\theta}K(t,x;\vec{X}))^p \frac{dt}{t}\right]^{1/p} < +\infty.$$

The question, which we shall discuss below is: will the Levy theorem remain true if we replace the space $\vec{X}_{\theta p}$ by the space $\vec{X}_{\omega p}$, where

$$\|x\|_{\omega p} = \left[\int_0^\infty (\omega^{-1}(t)K(t,x;\vec{X}))^p \frac{dt}{t}\right]^{1/p} < +\infty.$$

Theorem 6. *(N.Ja.Krugljak) If $\omega \in \mathcal{C}onv_0$, then the Levy theorem remains true if we replace $\vec{X}_{\theta p}$ by $\vec{X}_{\omega p}$.*

The proof of this theorem uses the Levy technic and the next property of $K_{\vec{X}}$.

Theorem 7. *(N.Ja.Krugljak) If $\Delta(\vec{X})$ is not closed in $\Sigma(\vec{X})$, then the function*

$$F(t) = \sup \{f(t): f = K(\cdot,x,\vec{X}),\ \|x\|_{\Sigma(\vec{X})} \le 1\}$$

has at least one of the limits $\lim\limits_{t\to 0} F(t)$, $\lim\limits_{t\to\infty} \frac{F(t)}{t}$ greater than zero.

From Theorem 7 it follows that if $\Delta(\vec{X})$ is not closed in $\Sigma(\vec{X})$ then

$$\Sigma^\circ(\vec{X}) \neq K_{L_\infty^\omega}(\vec{X})$$

for any $\omega \in \mathcal{C}onv_0$.

The proofs of cited results can be find in [1].

REFERENCES

[1] Ju. A. Brudnyi, N. Ja. Krugljak, *Interpolation spaces and interpolation functors.1*, to appear in North Holland Press.

ANISOTROPIC SOBOLEV SPACES ON RIEMANNIAN SYMMETRIC MANIFOLDS

LESZEK SKRZYPCZAK

Poznań, Poland

In this paper we define and study anisotropic function spaces of Sobolev type on Riemannian symmetric manifolds. The spaces are defined in terms of infinitesimal translations, i.e., infinitesimal generators of one-parameter groups of translations of a symmetric manifold. The paper is organized as follows. In Section 1 we recall some basic facts about Riemannian symmetric manifolds and prove a theorem concerning zero points of infinitesimal translation which will be needed later on. The definition of the spaces and their basic properties are given in Section 2. In Section 3 we discuss the local smoothness of the functions belonging to the spaces being under consideration.

1. Infinitesimal translations of Riemannian symmetric manifolds.

Let M be a connected m-dimensional Riemannian symmetric manifold, and let S_x denote a symmetry with respect to a point $x \in M$. For further details concerning this definition and other facts mentioned below we refer to [5] and [7]. The symmetry S_x, $x \in M$, is an isometry of the Riemannian manifold M. We fix a point $o \in M$. It will be called a basic point of M. In this section we are especially interested in a group G(M) of translations of the manifold M. The group G(M) is a subgroup of the group I(M) of all isometries of M generated by isometries of the form S_xS_y, $x,y \in M$. It is the smallest group of isometries of M acting transitively on M and invariant with respect to the automorphism $\delta: g \longrightarrow S_o g S_o$.

O.Loos extended the notation of the symmetry of points of M to the differential operators defined on M (cf. [7]). Let D and D' be differential operators on M. Let D⊗D' denote the differential operator on M×M composed of the action of D on the first variable and the action of D' on the second one. Then we define

$$[(D\cdot D')f](x) := [(D\otimes D')(f\circ\mu)](x,x), \qquad f \in C^{\infty}(M),$$

where

$$\mu: M\times M \ni (x,y) \longrightarrow S_x y \in M.$$

The operator $D\cdot D'$ is a differential operator on M. If D or D' is an identity operator (e.g. D = id) then we will write $id\cdot D' = x\cdot D'$. In this notation $S_x y = x\cdot y$.

A mapping $F:N \to M$ of two symmetric manifolds N and M is called *a homomorphism of symmetric manifolds* if $F(x\cdot y) = F(x)\cdot F(y)$. If M is connected then every isometry $F:M \to M$ is an automorphism of the symmetric manifold M.

For every vector $v \in T_oM$, a vector field

$$\tilde{v}(x) = 1/2\ v\cdot(o\cdot x), \qquad x \in M,$$

is an infinitesimal translation of M, i.e., an infinitesimal generator of a one-parameter group of translations. The group generated by the field $\tilde{v}$ can be described in the following way

$$\varphi(t) = S_{\exp_o(tv/2)}S_o, \qquad t \in \mathbb{R},$$

(exp stands for an exponential mapping of the Riemannian manifold M). Let sM denote the vector space spanned by the vector fields $\tilde{v}$, $v \in T_oM$. The spaces T_oM and aM are lineary isomorphic and the isomorphism is given by $v \to \tilde{v}$. The Lie algebra lG of the group G(M) is isomorphic to the direct sum of the space sM and the Lie algebra lH of the subgroup H of isotropy of the point o

$$lG = sM \oplus lH,$$

moreover,

$$[lH,sM] = sM, \qquad [sM,sM] = lH.$$

It can be proved that the manifold M is diffeomorphic to the homogeneous space G(M)/H.

A submanifold N of M is called *a symmetric subspace of M* if $N\cdot N \subset N$. The symmetric manifold M is called *abelian* if the group G(M) is abelian. A closed connected abelian symmetric subspace of a compact symmetric manifold is isomorphic as a symmetric manifold to the standard torus T^k, so it will be called *a torus of a symmetric manifold*.

Let vol denote the Riemannian measure of the manifold M. The main purpose of this paragraph is to prove the following theorem.

Theorem 1.1. *Let M be a connected m-dimensional Riemannian symmetric manifold with a basic point o. Let*

$$Zero_o M = \{x \in M: \text{ there exists } v \in T_o M,\ v \neq 0 \text{ such that } \tilde{v}(x) = 0\}.$$

Then

$$vol(Zero_o M) = 0.$$

Proof. The proof is long, so we sketch it only, leaving the details to the reader. According to the famous Cartan theorem every connected, simply connected Riemannian symmetric manifold M is isomorphic as a symmetric and Riemannian manifold to the product $M_1 \times M_2 \times M_3$ of symmetric manifolds of Euclidean, compact and non-compact types respectively (cf. [5],[7]). Using this theorem we can divide the proof into several steps. It is clear that if M is of the Euclidean type then the set $Zero_o M$ is empty. We prove that $Zero_o M = \emptyset$ also for non-compact type.

Step 1. If M is a manifold of the non-compact type then a symmetry with respect to any point $x \in M$ is a geodesic symmetry and for $x,y \in M$ there is exactly one maximal geodesic γ joining these points. The set $Zero(\tilde{v}) = \{x \in M: \tilde{v}(x) = 0\}$, $v \in T_o M$, is empty or it is a connected totally geodesic submanifold of M (cf. [6]).

Let $\tilde{v}(x) = 0$, $x \in M$. It is not hard to prove that

$$\widetilde{(-v)}(S_o x)f = \tilde{v}(x)(f \circ S_o), \qquad f \in C^{\infty}(M).$$

Since $d_o S_o(v) = -v$ and $\widetilde{(-v)} = -\tilde{v}$, we get $\tilde{v}(S_o x) = 0$. The geodesic γ joining the points x and $S_o x$ must be contained in $Zero(\tilde{v})$. But it leads to a contradiction since $o \in \gamma$ and $\tilde{v}(o) = v \neq 0$.

Step 2. Let M be a manifold of the compact type. If M is abelian then $Zero_o M = \emptyset$. Let M be a non-abelian and A be a maximal torus of M. Let

$$V_A = \{x \in A: \text{ there exists } \tilde{v} \in \mathfrak{s}M \text{ such that } \tilde{v}(x) \in 0\}.$$

There exists a finite set P of linear forms λ, $\lambda \neq 0$, defined on $T_o A$, and for every $\lambda \in P$ there are subspaces $\mathfrak{s}M_\lambda$ of $\mathfrak{s}M$ and $\mathfrak{l}H_\lambda$ of $\mathfrak{l}H$ such that (cf. [7,Ch.6]):

– $\mathfrak{s}M = \mathfrak{s}A \oplus \sum_{\lambda \in P} \mathfrak{s}M_\lambda$, $\qquad \mathfrak{s}A = \{\tilde{v} \in \mathfrak{s}M: \tilde{v}(o) \in T_o A\}$,

$\mathfrak{l}H = \mathfrak{l}H^{\mathfrak{s}A} \oplus \sum_{\lambda \in P} \mathfrak{l}H_\lambda$, $\qquad \mathfrak{l}H^{\mathfrak{s}A} = \{x \in \mathfrak{l}H: [X, \mathfrak{s}A] = 0\}$,

and $\dim \mathfrak{s}M_\lambda = \dim \mathfrak{l}H_\lambda$;

– $\mathfrak{l}H^{x} = \mathfrak{l}H^{\mathfrak{s}A} \oplus \sum_{x \in U_\lambda} \mathfrak{l}H_\lambda$, $\qquad \mathfrak{l}H^{x} = \{X \in \mathfrak{l}H: X(x) = 0\}$

$U_\lambda = \{x \in A: x = \exp_o v,\ e^{2\pi\sqrt{-1}\lambda(v)} = 1\}$, and

$$sM_x = \sum_{\lambda\in P_x} sM_\lambda, \qquad P_x = \{\lambda\in P\colon x\cdot o \in U_\lambda \text{ and } x \notin U_\lambda\}.$$

Thus

$$V_A = \bigcup_{\lambda\in P_x} V_\lambda \qquad V_\lambda = \{x\in A\colon x\cdot o \in U_\lambda \text{ and } x \in U_\lambda\}.$$

The set $W_\lambda = \{v\in T_oA\colon \lambda(v) \in Z\}$ is a set of measure zero since $\lambda \neq 0$. Consequently, the set U_λ is also of measure zero since $U_\lambda = \exp_o W_\lambda$. We may identify A with T^k, $k = \dim A$, in the way that the point o is identified with the unit of T^k and $x\cdot o = x^2$. Let

$$F_o = \{x\in T^k\colon x = (\cos t_1 + \sqrt{-1}\sin t_1, \dots, \cos t_k + \sqrt{-1}\sin t_k),$$
$$t_j=0 \text{ for at least one } j,\ 1 \le j \le k\}.$$

The set $F_o \cap V_\lambda$ is of measure zero. Since the square root is a smooth function on $T^k\setminus F_o$, the set $V_\lambda \cap (T^k\setminus F_o)$ is also of measure 0. Thus V_A is a set of measure 0.

It is well known that $M = \bigcup_{h\in H} h(A)$. We will show that

$$\mathrm{Zero}_o M = \bigcup_{h\in H} h(V_A).$$

For any maximal torus A' of M there is $h \in H$ such that $A' = h(A)$. If $v \in T_oM$ and $v' = (d_oh)v$ then

$$\tilde{v}'(h(x))(f) = \tilde{v}(x)(f\circ h) \quad \text{for any } f \in C^\infty(M)$$

(cf. 2.(1)). Thus $h(V_A) \subset V_{A'}$. On the other hand $A = h^{-1}(A')$, so $h^{-1}(V_{A'}) \subset V_A$.

Let $H^A = \{h\in H\colon h(x)=x \text{ for every } x\in A\}$. The group H^A is a compact Lie subgroup of the group H. It is not difficult to prove that $H/_{H^A}$ is a homogeneous space of dimension $n = \dim M - \dim A$. Let

$$G\colon H/_{H^A}\times A \ni (h,a) \longrightarrow h_1^{-1}(a) \in M,$$

where h_1 is an element of the equivalence class h. The mapping G is well defined. Moreover, it is a smooth surjection and

$$G(H/_{H^A}\times V_A) = \mathrm{Zero}_o M.$$

Since $H/_{H^A}\times V_A$ is a set of measure zero, the set $\mathrm{Zero}_o M$ is also of measure zero.

Step 3. Let M be simple connected. Then according to the Cartan theorem, M is isomorphic to $M_1\times M_2\times M_3$, where M_1, M_2, M_3 are symmetric manifolds of Euclidean, non-compact and compact type respectively. Let $v \in T_oM$. Then there are vectors $v_i \in T_{o_i}M$, $o = (o_1,o_2,o_3)$, such that $v = \sum_{i=1}^3 v_i$ and $\tilde{v} = \sum_{i=1}^3 \tilde{v}_i$. Thus $\tilde{v}(x) = 0$ if and only if $\tilde{v}_i(x) = 0$ for every $i = 1,2,3$. Since $\mathrm{Zero}_{o_1}M_1 = \emptyset$ and $\mathrm{Zero}_{o_2}M_2 = \emptyset$, the vector fields $\tilde{v}_i \equiv 0$ for $i = 1,2$. Consequently,

$$\mathrm{Zero}_o M = M_1 \times M_2 \times \mathrm{Zero}_{o_3}M_3.$$

Thus $\mathrm{Zero}_o M$ is a set of measure zero.

Step 4. Let M be any connected Riemannian symmetric manifold. Then there is a connected, simple connected Riemannian symmetric manifold N which is a universal covering manifold of M. Moreover, the covering mapping $\Pi:N \to M$ is an smooth homomorphism of symmetric manifolds. Let us choose a point $\bar{o} \in N$ such that $\Pi(\bar{o}) = o$. Since Π is regular at the point $\bar{o}$, the tangential map $d_{\bar{o}}\Pi$ defines an isomorphism of the spaces sN and sM by:

$$sN \ni \tilde{\bar{v}} \to \tilde{\bar{v}}(\bar{o}) = \bar{v} \xrightarrow{d_{\bar{o}}-\Pi} v \longrightarrow \tilde{v} \in sM.$$

Moreover, it is not difficult to prove that

$$\tilde{v}(x)f = \tilde{\bar{v}}(\bar{x})(f\circ\Pi), \qquad f \in C^{\infty}(M),\ x = \Pi(\bar{x}).$$

Thus $\mathrm{Zero}_o(M) \subset \Pi(\mathrm{Zero}_{\bar{o}}(N))$.

2. Anisotropic Sobolev spaces on symmetric manifolds.

Let M be a m-dimensional connected Riemannian symmetric manifold with a basic point o. Let $\bar{v} = (v_1,\dots,v_m)$, $v_i \in T_oM$, be a system of vectors orthonormal in the sense of Riemannian metric tensor g. Each of the vectors v_i appoints an infinitesimal translation

$$\tilde{v}_i(x) = 1/2\ v_i\cdot(o\cdot x), \qquad x \in M,$$

such that $\tilde{v}_i(o) = v_i$. Moreover, let $\varphi_i(t)$ be a one-parameter group of translation corresponding to $\tilde{v}_i$.

THEOREM 2.1. *Let M be a connected m-dimensional Riemannian symmetric manifold with a basic point o. Then, for every system $(v_1,\dots,v_m)$ of linearly independent vectors $v_i \in T_oM$, the set consisted of all points $x \in M$ such that the vectors $\tilde{v}_i(x)$, $i = 1,\dots,m$, are linearly dependent, is a closed nowhere-dense set of Riemannian measure zero.*

Proof. Let

$$F = \{x \in M\text{: the vectors } \tilde{v}_i(x),\ i=1,\dots,m \text{ are linearly dependent}\}.$$

Since the vector spaces T_oM and sM are isomorphic and the isomorphism is given by $v \to \tilde{v}$, it should be clear that $F = \mathrm{Zero}_o M$. Thus, by Theorem 1.1 the Riemannian measure of the set F equals zero. To finish the proof, it is sufficient to show that the set F is closed. A mapping

$$\Phi:\dot{\mathbb{R}}^m\times M \ni ((a_1,\dots,a_m),x) \longrightarrow \sum_{i=1}^{m} a_i\tilde{v}_i(x) \in TM, \quad \dot{\mathbb{R}}^m = \mathbb{R}^m\setminus\{(0,\dots0)\},$$

is smooth since the vector fields $\tilde{v}_i$ are smooth. So the set $G = \Phi^{-1}(0)$ is closed. Moreover,

$$G = \bigcup_{x \in F} G_x, \qquad G_x = \{(a_1,\dots,a_m) \neq 0: \sum_{i=1}^{m} a_i \tilde{v}_i(x) = 0\},$$

and the set F is equal to the projection of G onto the second component. Let $x_n \to y$ in M, $(x_n) \subset F$. For each x_n there exists $a(x_n) \in G_{x_n}$ with the Euclidean norm one. Let $a(x_{n_k})$ be a convergent subsequence of $(a(x_n))$. Then the sequence $(a(x_{n_k}), x_{n_k})$ is convergent in $\mathbb{R}^m \times M$ and its limit belongs to G. Thus $y \in F$.

To simplify our future notation, we introduce the following symbols:

a) for a multi-index $\bar{n} = (n_1,\dots,n_k)$ the symbol $\delta(\bar{n})$ stands for the set of all sequences of the length $\bar{n}$ consisted of n_1 numbers 1, n_2 numbers 2,..., n_k numbers k;

b) if $0 < n_i \le m$ then $\tilde{v}_{\bar{n}}$ stands for the operator

$$(\tilde{v}_{n_k}(\tilde{v}_{n_{k-1}}(\dots(\tilde{v}_{n_1}())\dots),$$

and $\tilde{v}_i^k$ stands for the operator $\tilde{v}_{\bar{n}}$ with $n_1=n_2=\dots=n_k=i$.

Analogously as in the Euclidean case, we will regard two norms:

$$\|f\,|W_p^{\bar{n},\bar{v}}\| = \|f\|_p + \sum_{i=1}^{m} \sum_{k=1}^{n_i} \|\tilde{v}_i^k f\|_p,$$

and

$$\|f\,|V_p^{\bar{n},\bar{v}}\| = \|f\|_p + \sum_{\bar{k}:\left(\frac{k_1}{n_1}+\dots+\frac{k_m}{n_m}\right)\le 1} \ \sum_{i \in \delta(\bar{k})} \|\tilde{v}_{\bar{i}} f\|_p,$$

where $\|\cdot\|_p$, $1 \le p < \infty$, is the norm of the space $L_p(M,\mathrm{vol})$ and $\bar{n} = (n_1,\dots n_m)$, $n_i = 0,1,2,\dots$.

Definition 2.2. Let $\bar{v} = (v_1,\dots v_m)$, $v_i \in T_oM$, $g(v_i,v_j) = \delta_{i,j}$. Moreover, let $\bar{n} = (n_1,\dots n_m)$, $n_i \in \mathbb{N}$, $1 \le p < \infty$. Then

$$W_p^{\bar{n},\bar{v}}(M) = \{f \in L_p(M,\mathrm{vol}): \|f\,|W_p^{\bar{n},\bar{v}}\| < \infty\},$$

$$V_p^{\bar{n},\bar{v}}(M) = \{f \in L_p(M,\mathrm{vol}): \|f\,|V_p^{\bar{n},\bar{v}}\| < \infty\}.$$

Remarks: a) If $M = \mathbb{R}^m$ and $\bar{v} = \left(\frac{\partial}{\partial x_1},\dots,\frac{\partial}{\partial x_m}\right)$ then the spaces $W_p^{\bar{n},\bar{v}}(\mathbb{R}^m)$ and $V_p^{\bar{n},\bar{v}}(\mathbb{R}^m)$ coincide with the classical anisotropic Sobolev spaces on $\mathbb{R}^m$. In that case $V_p^{\bar{n},\bar{v}}(\mathbb{R}^m) = W_p^{\bar{n},\bar{v}}(\mathbb{R}^m)$ if $1 < p < \infty$ (cf. [12]).

b) As a direct consequence of Definition 2.2 we have:

$$V_p^{\bar{n},\bar{v}}(M) \hookrightarrow W_p^{\bar{n},\bar{v}}(M),$$

$$V_p^{\bar{n},\bar{v}}(M) \hookrightarrow V_p^{\bar{k},\bar{v}}, \qquad W_p^{\bar{n},\bar{v}}(M) \hookrightarrow W_p^{\bar{k},\bar{v}}(M)$$

provided that $\bar{k} \le \bar{n}$, i.e., $k_i \le n_i \quad i = 1,\dots,m$.

We will write $V_p^{n,\bar{v}}(M)$, $W_p^{n,\bar{v}}(M)$ if $\bar{n} = (n,..,n)$.

Proposition 2.3. *Let $\bar{\upsilon} = (\upsilon_1,\dots,\upsilon_m)$ and $\bar{\omega} = (\omega_1,\dots,\omega_m)$ be orthonormal basis of the space T_oM. Then, for every $n \in \mathbb{N}$ and $1 \le p < \infty$*

$$V_p^{n,\bar{v}}(M) = V_p^{n,\bar{w}}(M)$$

and the corresponding norms are equivalent.

Proposition 2.3 is a simple consequence of the isomorphism of the spaces sM and T_oM and, therefore, the proof is omitted.

Let ψ be an element of the isotropy group $I_o(M)$ of the point o, i.e., the group of those isometries of M which do not move the point o. We will investigate how the operator

$$T_\psi : f \to f \circ \psi^{-1}$$

acts on the spaces $W_p^{\bar{n},\bar{v}}(M)$ and $V_p^{\bar{n},\bar{v}}(M)$.

Theorem 2.4. *If $\psi \in I_o(M)$ then the operator T_ψ maps the space $W_p^{\bar{n},\bar{v}}(M)$ isometrically onto $W_p^{\bar{n},\bar{w}}(M)$ and the space $V_p^{\bar{n},\bar{v}}(M)$ isometrically onto $V_p^{\bar{n},\bar{w}}(M)$, where $\bar{\omega} = ((d_o\psi)\upsilon_i)_{i=1}^m$.*

Proof. The system $\bar{w}$ is orthonormal since ψ is an isometry of M. Moreover, ψ being an isometry is an automorphism of symmetric manifold M, i.e., $\psi(S_x y) = S_{\psi(x)}\psi(y)$.

For every $f \in C^\infty(M)$ and every i, $1 \le i \le m$, we have:

$$\tilde{w}_i(\psi(x))(f) = \frac{1}{2}[w_i \cdot (o \cdot \psi(x))](f) = \frac{1}{2}[w_i \cdot \psi(o \cdot x)](f)$$

$$= \frac{1}{2}[(d_o\psi)v_i \cdot \psi(o \cdot x)](f) = \frac{1}{2}[(d_o\psi)v_i \otimes \psi(o \cdot x)](f \circ \mu)$$

$$= \frac{1}{2}[v_i \otimes (o \cdot x)](f \circ \mu \circ (\psi \times \psi)) = \frac{1}{2}[v_i \otimes (o \cdot x)](f \circ \psi \circ \mu)$$

$$= \frac{1}{2}[v_i \cdot (o \cdot x)](f \circ \psi) = \tilde{v}_i(x)(f \circ \psi),$$

where $(\psi \times \psi)(x,y) = (\psi(x),\psi(y))$. Thus

$$(1) \qquad \tilde{w}_i(x)(f \circ \psi^{-1}) = \tilde{v}_i(\psi^{-1}(x))(f), \qquad f \in C^\infty(M).$$

Now, we can prove by induction that

$$(2) \qquad \tilde{w}_I(x)(f \circ \psi^{-1}) = \tilde{v}_{i_k}(\psi^{-1}(x))(\tilde{v}_{i_{k-1}} \dots \tilde{v}_{i_1}(\cdot)(f))$$

for every multi-index $\bar{\iota} = (i_1,\dots,i_k)$. Using standard arguments we can extend this identity to the functions belonging to the space $L_p(M,\mathrm{vol})$.

The Riemannian measure vol is invariant with respect to isometries of M, so it follows from (2) that

$$\|\tilde{w}_{\bar{\iota}}(f\circ\psi^{-1})\|_p = \|\tilde{v}_{\bar{\iota}}(f)\|_p$$

for every multi-index $\bar{\iota}$. Thus, the operator T_ψ maps $W_p^{\bar{n},\bar{v}}(M)$ $(V_p^{\bar{n},\bar{v}}(M))$ isometrically into $W_p^{\bar{n},\bar{w}}(M)$ $(V_p^{\bar{n},\bar{w}}(M))$. Surjectivity of the operator follows from the fact that

$$T_\psi \circ T_{\psi^{-1}} = \mathrm{id} \qquad \text{and} \qquad (d_o\psi^{-1})w_i = v_i.$$

We recall that a Riemannian manifold M is called *isotropic* if for every $x \in M$ the group $I_x(M)$ acts transitively on spheres $S(x,r) = \{y\in M\colon \rho(x,y) = r\}$.

Corollary 2.5. *Let M be an isotropic connected Riemannian symmetric manifold. Let $\bar{v} = (v_1,\dots,v_m)$ and $\bar{w} = (w_1,\dots,w_m)$ be an orthonormal basis in T_oM. Then there is a permutation $\bar{w}'$ of $\bar{w}$ such that, for every multi-index $\bar{n}$, the space $W_p^{\bar{n},\bar{v}}(M)$ $(V_p^{\bar{n},\bar{v}}(M))$ is isometrically isomorphic to the space $W_p^{\bar{n},\bar{w}'}(M)$ $(V_p^{\bar{n},\bar{w}'}(M))$.*

Proof. Let $IL(o) = \{d_o\psi\colon \psi \in I_o(M)\}$. The group $IL(o)$ acts transitively on the sphere $S(0,\varepsilon) \subset T_oM \approx \mathbb{R}^m$, $\varepsilon < i(M)$, since M is an isotropic manifold and $d_o\psi = \exp_o^{-1}\circ\psi\circ\exp_o$ (here $i(M)$ denotes an injectivity radius of M). Thus the rotation group $O(\mathbb{R}^m)$ is contained in $IL(o)$ and, in consequence, there exist $\psi \in I_o(M)$ and a permutation $\bar{w}'$ of $\bar{w}$ such that $\bar{w}' = ((d_o\psi)v_i)_{i=1}^m$. Now the corollary follows from the previous theorem.

Corollary 2.6. *If M is an isotropic manifold then, for arbitrary orthonormal systems $\bar{v}$ and $\bar{w}$, the spaces $W_p^{n,\bar{v}}(M)$ and $W_p^{n,\bar{w}}(M)$ are isometric, $1 \le p < \infty$, $n \in \mathbb{N}$.*

Proposition 2.7. *If an isometry $\psi \in I(M)$ commutes with the groups $\{\varphi_i(t)\}$, $i = 1,\dots,m$, then the operator T_ψ is an isometry of $W_p^{\bar{n},\bar{v}}(M)$ $(V_p^{\bar{n},\bar{v}}(M))$, $1 \le p < \infty$, $\bar{n} \in \mathbb{N}^m$.*

The proof of Proposition 2.7 is similar to the proof of Theorem 2.2 and it is omitted here.

Theorem 2.8. *The spaces* $W_p^{\bar{n},\bar{\nu}}(M)$ *and* $V_p^{\bar{n},\bar{\nu}}(M)$ *are Banach spaces for every possible* $\bar{n}$, $\bar{\nu}$ *and* p.

Proof. In the first place we prove the completeness of the space $W_p^{\bar{n},\bar{\nu}}(M)$. Since

$$W_p^{\bar{n},\bar{\nu}}(M) = \bigcap_{i=1}^{m} W_p^{n_i,\nu_i}(M),$$

where $W_p^{n_i,\nu_i}(M) = W_p^{(0,\dots,0,n_i,0,\dots,0),\bar{\nu}}(M)$, and the norm $\|\cdot|W_p^{\bar{n},\bar{\nu}}\|$ is equivalent to $\sup\{\|\cdot|W_p^{n_i,\nu_i}(M)\|: i = 1,\dots,m\}$, it is sufficient to prove the completeness of $W_p^{n_i,\nu_i}(M)$ (cf. [3,Lemma 2.3.1]).

Let $(f_n)_{n=1}^{\infty}$ be a Cauchy sequence in $W_p^{n_i,\nu_i}(M)$. Then there are functions $h,h^{(1)},\dots,h^{(n_i)} \in L_p(M,vol)$ such that $f_n \to h$ and $\tilde{\nu}_i^j f_n \to h^{(j)}$ in $L_p(M,vol)$ if $n \to \infty$.

Let $\{(U_k,\Psi_k)\}_k$ be a covering of the manifold $M\setminus Zero(\tilde{\nu}_i)$ by straightening charts of the vector field $\tilde{\nu}_i$. Let $f_{n,k} = f_n\circ\Psi_k^{-1}$, $h_k = h\circ\Psi_k^{-1}$, $h_k^{(j)} = h^{(j)}\circ\Psi_k^{-1}$. Then

$$\int_{\Psi_k(U_k)} (h_k - f_{n,k}) \frac{\partial^l}{\partial x_1^l} \varphi \, dx = (-1)^l \int_{\Psi_k(U_k)} \left(\frac{\partial^l}{\partial x_1^l} h_k - \frac{\partial^l}{\partial x_1^l} f_{n,k}\right) \varphi \, dx,$$

for $\varphi \in C_0^{\infty}(\Psi_k(U_k))$, $l \in \mathbb{N}$, $0 < l \le n_i$. But

$$\int_{\Psi_k(U_k)} \left|\frac{\partial^l}{\partial x_1^l} f_{n,k} - h_k^{(l)}\right| dx \longrightarrow 0$$

because

$$\int_{U_k} |\tilde{\nu}_i^l f_n - h_k^{(l)}|^p \, dx \longrightarrow 0,$$

and the set $\Psi_k(U_k)$ is a set of finite measure. It follows from the inequality

$$\left|\int_{\Psi_k(U_k)} \left(\frac{\partial^l}{\partial x_1^l} h_k - \frac{\partial^l}{\partial x_1^l} f_{n,k}\right) \varphi \, dx - \int_{\Psi_k(U_k)} \left(\frac{\partial^l}{\partial x_1^l} h_k - h_k^{(l)}\right) \varphi \, dx\right|$$

$$\le \int_{\Psi_k(U_k)} \left|h_k^{(l)} - \frac{\partial^l}{\partial x_1^l} f_{n,k}\right| |\varphi| \, dx \longrightarrow 0$$

that

$$(3) \quad \int_{\Psi_k(U_k)} \left(\frac{\partial^l}{\partial x_1^l} h_k - \frac{\partial^l}{\partial x_1^l} f_{n,k}\right) \varphi \, dx \longrightarrow \int_{\Psi_k(U_k)} \left(\frac{\partial^l}{\partial x_1^l} h_k - h_k^{(l)}\right) \cdot \varphi \, dx.$$

On the other hand,

$$\int_{\Psi_k(U_k)} |h_k - f_{n,k}| \, dx \longrightarrow 0$$

since

$$\int_{U_k} |f_n-h|^p \, d\,\mathrm{vol} \longrightarrow 0.$$

Consequently,

$$(4) \qquad \int_{\Psi_k(U_k)} (h_k - f_{n,k}) \frac{\partial^l}{\partial x_1^l} \varphi \, dx \longrightarrow 0.$$

It follows from (3) and (4) that

$$\int_{\Psi_k(U_k)} \left(\frac{\partial^l}{\partial x_1^l} h_k - h_k^{(l)}\right) \varphi \, dx = 0.$$

Therefore

$$\vartheta_i^l h = h^{(l)} \quad \text{on } U_k.$$

The set Zero(ϑ_i) is of Riemannian measure zero, so $\vartheta_i^l f_n \to v_i^l h$ in $L_p(M,\mathrm{vol})$ and the completeness of $W_p^{\bar{n},\bar{v}}(M)$ is proved.

Now, let (f_n) be a Cauchy sequence in $V_p^{\bar{n},\bar{v}}(M)$. The operators $\vartheta_i : f \to \vartheta_i f$, $D(\vartheta_i) = \{f \in L_p(M,\mathrm{vol}) : \vartheta_i f \in L_p(M,\mathrm{vol})\}$, $1 \le i \le m$, are closed in $L_p(M,\mathrm{vol})$ since the space $W_p^{1,v_i}(M)$ is complete. Therefore, it follows from the definition of $V_p^{\bar{n},\bar{v}}(M)$ that $\vartheta_i f_n \to \vartheta_i f$ in $L_p(M,\mathrm{vol})$ provided $f_n \to f$. Similarly, it follows from $\vartheta_i f_n \to \vartheta_i f$ that $\vartheta_j \vartheta_i f_n \to \vartheta_j \vartheta_i f$, etc. Thus

$$\|f_n - f \,|\, V_p^{\bar{n},\bar{v}}\| \to 0.$$

Theorem 2.9. *The expression*

$$\|f\,|\,W_p^{\bar{n},\bar{v}}(M)\|^{(1)} = \|f\|_p + \sum_{i=1}^{m} \|v_i^{n_i} f\|_p$$

is a norm in $W_p^{\bar{n},\bar{v}}(M)$ *equivalent to the norm* $\|\cdot\,|\,W_p^{\bar{n},\bar{v}}\|$.

Proof. It is sufficient to prove that

$$\|f\|_p + \|\vartheta_i^{n_i} f\|_p$$

is an equivalent norm in $W_p^{n_i,v_i}(M)$, $1 \le i \le m$. Since

$$\|f\|_p + \|\vartheta_i^{n_i} f\|_p \le \|f\,|\,W_p^{n_i,v_i}\|$$

and $W_p^{n_i,v_i}(M)$ is a Banach space, the theorem follows from the open mapping theorem provided the space $(W_p^{n_i,v_i}(M), \|\cdot\|^{(1)})$ is complete. But in the proof of the previous theorem we have showed that if $f_n \to f$ and $\vartheta_i^l f_n \to h$ in $L_p(M,\mathrm{vol})$ then $h = \vartheta_i^l f$.

3. Local smoothness.

We are going to discuss the local smoothness of the function belonging to $W_p^{n,\bar{v}}(M)$. Let $\psi \in C_o^\infty(M)$ and let M_ψ denotes the following operator

$$M_\psi : f \longrightarrow \psi f.$$

Theorem 3.1. *Let $\psi \in C_o^\infty(M)$ and $\operatorname{supp}\psi \cap \mathrm{Zero}_o M = \emptyset$. Then the operator M_ψ is a continuous linear operator mapping the space $W_p^{n,\bar{v}}(M)$ into the Sobolev–Aubin space $W_p^n(M)$, $1 < p < \infty$, $n \in \mathbb{N}$.*

Remark. The Sobolev spaces $W_p^n(M)$ on Riemannian manifolds were defined by T.Aubin in terms of covariant derivatives (cf. [1],[2]). Later H.Triebel defined the scale $F_{p,q}^s(M)$ of function spaces on Riemannian manifolds with bounded geometry, $-\infty < s < \infty$, $0 < p < \infty$, $0 < q < \infty$ or $p = q = \infty$. He proved also that $W_p^n(M) = F_{p,2}^n(M)$ if $1 < p < \infty$ (cf. [3],[11]). Since Riemannian symmetric manifolds are manifolds with bounded geometry, we can use this identification in our proof.

Proof of Theorem 3.1. Since the set $\operatorname{supp}\psi$ is compact and the set $\mathrm{Zero}_o M$ is closed, there is a finite covering of the set $\operatorname{supp}\psi$ by a family of geodesic balls $\{B(y_i,\delta_i)\}_{i=1}^k$, $\max\{\delta_i : i=1,\dots,k\} < i(M)/8$, such that $B(y_i,\delta_i) \cap \mathrm{Zero}_o M = \emptyset$ for every i. Let $\{\chi_i\}_{i=1}^k$ be a family of functions such that

$$\chi_i \in C_o^\infty(M), \qquad \operatorname{supp}\chi_i \subset B(y_i,\delta_i), \qquad 0 \le \chi_i \le 1,$$

$$\sum_{i=1}^k \chi_i(x) = 1 \quad \text{if} \quad x \in \operatorname{supp}\psi.$$

Let us fix i, $1 \le i \le k$. We show that $(\chi_i\psi f)\circ\exp_{y_i} \in W_p^n(\mathbb{R}^m)$.

Let $\sum_{i=1}^m v_{j,i}(x)\dfrac{\partial}{\partial x_i}$ be a representation of the vector field $\bar{v}_j$, $1 \le j \le m$, in the chart $(B(y_i,\delta_i),\exp_{y_i}^{-1})$. Let $P_{j,n}(x,D)$ be a differential operator with the characteristic polynomial

$$P_{j,n}(x,z) = \left[\sum_{i=1}^m v_{j,i}(x)\, z_i\right]^n, \qquad j = 1,\dots,m.$$

Because the vector fields $\bar{v}_i$ are linearly independent on $B(y_i,\delta_i)$, the polynomials $P_{j,n}(x,z)$, $j = 1,\dots,m$, do not have a common zero except the origin provided $x \in B(0,\delta_i) \subset \mathbb{R}^m$. It follows from the theorem about coercivity (cf. [4]) that

$$\|h \,|\, \mathring{W}_p^n(B(0,\delta_i))\| \sim \|h\|_p + \sum_{j=1}^m \|P_{j,n}(x,D)h\|_p,$$

where $\mathring{W}^n_p(B(0,\delta_i))$ is a completion of the space $C^\infty_o(B(0,\delta_i))$ in the norm $\|h\|_p + \sum_{|\alpha|\le n} \left\| \frac{\partial^\alpha}{\partial x^\alpha} h \right\|_p$, and $\|\cdot\|_p$ denotes the L_p-norm with respect to the Lebesgue measure. Since supp $\chi_i\psi f \circ \exp_{y_i} \subset B(0,\delta_i)$, we have

$$\chi_i\psi f \circ \exp_{y_i} \in \mathring{W}^n_p(B(0,\delta_i)) \quad \text{iff} \quad \chi_i\psi f \circ \exp_{y_i} \in W^n_p(B(0,\delta_i)).$$

It should be obvious that if n = 1 then

$$\|\chi_i\psi f \circ W^{1,\bar{\nabla}}_p\| \sim \|\chi_i\psi f \circ \exp_{y_i} | W^1_p(B(0,\delta_i))\|.$$

Let us assume that this equivalence is also true for n = k. Then

$$\|\chi_i\psi f \circ \exp_{y_i} | W^{k+1}_p(B(0,\delta_i))\| \le \|\chi_i\psi f \circ \exp_{y_i}\|_p$$

$$+ \sum_{j=1}^{m} \|(P_{j,k+1} - \bar{\nabla}^{k+1}_j)(\chi_i\psi f \circ \exp_{y_i})\|_p + \|\bar{\nabla}^{k+1}_j(\chi_i\psi f \circ \exp_{y_i})\|_p.$$

But the operator $P_{j,k+1} - \nabla^{k+1}$ is a differential operator of smooth bounded coefficients in $B(0,\delta_i)$, therefore

$$\|(P_{j,k+1} - \bar{\nabla}^{k+1})(\chi_i\psi f \circ \exp_{y_i})\|_p \le C \cdot \|\chi_i\psi f \circ \exp_{y_i} | W^k_p(B(0,\delta_i))\|$$

$$\le C \cdot \|\chi_i\psi f | W^{k,\bar{\nabla}}_p(M)\|.$$

In consequence,

$$\|\chi_i\psi f | W^{n,\bar{\nabla}}_p(M)\| \sim \|\chi_i\psi f \circ \exp_{y_i} | W^n_p(B(0,\delta_i))\|. \tag{5}$$

As we mention above, the space $W^n_p(M)$ coincides with $F^n_{p,2}(M)$, so it is sufficient to prove that $\varphi_k\psi f \circ \exp_{x_i} \in F^n_{p,2}(\mathbb{R}^m)$, where $\{B(x_i,\delta)\}$ is a suitable covering and $\{\varphi_i\}$ suitable resolution of unity used in the definition of the scale $F^s_{p,q}(M)$ (cf. [11]). But this fact follows from (5) and the pointwise multipliers and diffeomorphic properties of the scale $F^s_{p,q}(\mathbb{R}^m)$ (cf. [10],[11]).

Corollary 3.2.. *If $1 < p,q < \infty$ and $n_1 - \frac{m}{p} = \frac{m}{q}$ then M_ψ is a continuous linear operator mapping the spaces $W^{n,\bar{\nabla}}_p(M)$ into $L_q(M)$.*

The proof of the above corollary is an immediate consequence of Theorem 3.1 and Theorem 2.10 in [2]. The next corollary follows from Theorems 1.1, 3.1 and properties of $W^n_p(M)$.

Corollary 3.3. *For almost every $x \in M$ there is an open neighborhood U_x of the point x such that if $f \in W^{n,\bar{\nabla}}_p(M)$ then $f|_{U_x} \in C^l(U_x)$, where*

$$l = \begin{cases} [n - m/p] & \text{if } m/p \notin \mathbb{N}, \\ n - m/p - 1 & \text{if } m/p \in \mathbb{N}. \end{cases}$$

Remark. As far as the spaces $V_p^{n,\bar{\nu}}(M)$ are regarded, we can also say something about smoothness of functions at points belonging to $\mathrm{Zero}_o M$. Namely, if $f \in V_p^{n,\bar{\nu}}(M)$, $p > 1$, and $\psi \in C_o^{\infty}(M)$ then $\psi f \in F_{p,2}^{n/2}(M)$ (cf. [8]).

REFERENCES

[1] T. Aubin, *Espaces de Sobolev sur les variétés Riemanniennes*, Bull. Sci. Math. 100 (1976), 149–173.

[2] T. Aubin, *Nonlinear Analysis on Manifolds. Monge – Ampere Equations.*, Springer Verlag, New York–Heidelberg–Berlin, 1982.

[3] J.Bergh, J. Löfström, *Interpolation Spaces. An Introduction.*, Springer Verlag, Berlin–Heidelberg–New York, 1976.

[4] O. V. Besov, *About coercivity in anisotropic Sobolev spaces*, Mat. Sb. 73 (1967), 585–600 (in Russian).

[5] S.Helgason, *Differential Geometry and Symmetric Spaces*, Academic Press, New York–London, 1962.

[6] S.Kobayashi, *Transformation Groups in Differential Geometry*, Springer Verlag, Berlin–Heidelberg–New York, 1972.

[7] O.Loos, *Symmetric Spaces, vol I,II.*, W.A.Benjamen INC., New York–Amsterdam, 1969.

[8] L.Skrzypczak, *Function spaces of Sobolev type on Riemannian symmetric manifolds*, Forum Math., to appear.

[9] H.Triebel, *Characterization of function spaces on a complete Riemannian manifold with bounded geometry*, Math. Nachr. 130 (1987), 321–346.

[10] H.Triebel, *Diffeomorphism properties and pointwise multiplier for function spaces*, Proceedings of the conference "Function Spaces" Poznań 1986, Teubner-Verlag, Leipzig 1988, 75–84.

[11] H.Triebel, *Spaces of Besov–Hardy–Sobolev type on complete Riemannian manifolds*, Arkiv der Mat. 24 (1986), 299–337.

[12] H.Triebel, *Theory of Function Spaces*, Birkhauser, Basel–Boston–Stuttgart, 1983.

LIST OF PARTICIPANTS

Abbreviations:

IM UAM = Institute of Mathematics, A.Mickiewicz University, Matejki 48/49, 60-769 Poznań, Poland

IM PP = Institute of Mathematics, Technical University, Piotrowo 3a, 61-138 Poznań, Poland

1.	Alghossein Ahmad,	Inst. of Math., University of Techrin, Latakie, Syria
2.	Alherk Ghassan,	University of Aleppo, Aleppo, Syria
3.	Banaś Józef,	Dept. of Math. Technical University of Rzeszów, W.Pola 2, 35-959 Rzeszów, Poland
4.	Basile Achille	Dipartimento di Matematica e Applicazioni, Universita degli Studi di Napoli, via Mezzocannone 8, 80-134 Napoli, Italy
5.	Beckenstein Edward,	St.John's University, 300 Howard Ave, Staten Island, New York 10301, USA
6.	Bereznoi Evgenij,	Jaroslavlskij Gos. Univ.,Jaroslavl, USSR
7.	Bernues Julio,	University of Zaragoza, Spain
8.	Blank Mariusz,	Inst. of Math., Teachers University, ul. Kaliskiego 7, 85-791 Bydgoszcz, Poland
9.	Bobrowski Dobiesław,	IM PP
10.	Bombal Fernando,	Departamento de Análysis Mat., Universidad Complutense, 28040 Madrid, Spain
11.	Borucka-Cieślewicz Anna,	Inst. of Math., Teachers University, Zielona Góra, Poland
12.	Brudnyi Ju.A.,	Jaroslavlskij Gos. Univ.,Jaroslavl, USSR
13.	Bukhvalov A.V.,	Leningrad Inst.of Finances and Economy, Giboedova Canal 30/32, USSR
14.	Cerdà Joan,	Dept. de Matematica Aplicada i Análisi, Universitat de Barcelona, Spain
15.	Cottin Claudia,	FB Mathematik, Universität Duisburg, Postfach 101503, D-4100 Duisburg 1, FRG
16.	Curberra Guillermo,	DPTO Analisis Matematico, Universidad de Sevilla, Apart. 1160, Sevilla, Spain
17.	Cwikel Michael,	Dept. of Math., Technion I.I.I. Haifa, Israel
18.	Dobrakov Ivan,	Matematický Ústav SAV, Obráncov mieru 49, 81473 Bratislava, Czechoslovakia
19.	Domański Paweł,	IM UAM
20.	Drewnowski Lech,	IM UAM
21.	Duran Antonio J.,	DPTO Analisis Matematico, Universidad de Sevilla, Sevilla, Spain

22. Finol Carlos, Departamento de Matemáticas, Facultad de Ciencias, Universidad Central de Venezuela, Apartado 40645, Caracas 1040-A, Venezuela.

23. Gniłka Stanisław, IM UAM

24. Gonzales Manuel, Departamento de Matematicas, Facultad de Ciencias, 39005 Santander, Spain

25. Greim Peter, Math. Dept. the Citadel, Charleston, SC 29409, USA

26. Grzybowski Jerzy IM UAM

27. Gurka Petr, Mathematical Inst. Chechoslovak Academy of Sciences, Žitná 25, 115 67 Praha 1, Czechoslovakia

28. Hanin L.G., 194021 Leningrad, Institutckij pr.29, apt.45, USSR

29. Harmand P., Universität Oldenburg, Fachbereich Mathematik, Ammerlander Heerstr. 114-118, Postfach 25 03, 2900 Oldenburg, FRG

30. Heberlein Krzysztof, Inst. of Math., Technical University Szczecin, Al. Piastów 48/49, 70-310 Szczecin, Poland

31. Heinig Hans Paul, Dept of Math. & Statistics, McMaster University Hamilton Ontario L8S - 4KI Canada

32. Hernández F.L., Universidad Complutense, Facultad de Matematicas, Deptart. Analisis Matematico, 28040 Madrid, Spain

33. Hudzik Henryk, IM UAM

34. Hüysmans C.B., Dept. of Math., University of Leiden, P.O.Box 95/2, 2300-RA- Leiden, The Netherlands

35. Jaramillo Jesus, Universidad Complutense, Facultad de Matematicas, Deptart. Analisis Matematico, 28040 Madrid, Spain

36. Jaroszewska M., IM UAM

37. Jędryka T.M. Teachers University Bydgoszcz, Żeglarska 67, 85-519 Bydgoszcz, Poland

38. Kasperski Andrzej, Silesian Technical University, Pstrowskiego 7, 44-101 Gliwice, Poland.

39. Kolomy Josef, Mathematical Inst. of Charles University, Sokolovská 83, 18600 Praha-Karlin, Czechoslovakia

40. Kozłowska Grażyna, Silesian Technical University, Pstrowskiego 7, 44-101 Gliwice, Poland.

41. Kremp Stefan Michael, Universität des Saarlandes, FB9 Mathematik, 6600 Saarbrücken, FRG

42. Krugljak N.Ya., Jaroslavl, Kalinina str.31, f.152, USSR

43. Kubiak Tomasz, IM UAM

44. Kufner Alois, Mathematical Institute, Czechoslovak Academy of Sciences, Žitná 25, Praha 1 - Nove Město, Czechoslovakia

45. Kun Soo Chang, Dept. of Math., Yonsei University, Seoul 120-149, Korea

46. Kurc Wiesław, IM UAM

47. Kutzarova Denka, Inst. of Math., Bulgarian Academy of Sciences, 1090 Sofia, P.O.Box 373, Bulgaria

48. Landes Thomas, Dept. of Economics, University-GHS Paderborn, Wartburgerstr. 100, D-4790 Paderborn, FRG

49. Laskowska Anna, Silesian Technical University, Pstrowskiego 7, 44-101 Gliwice, Poland.

50. Lee Peng Yee, Dept of Math., National University of Singapore, Kent Ridge, Rep. of Singapore 0511

51. Lewicki Grzegorz, Inst. of Math., Jagiellonian University, Reymonta 4, 30-059 Kraków, Poland

52. Liskowski Marian, IM PP

53. Llavona Jose G., Deptartamento de Analisis Matematico, Fac. Matematicas, Universitad Complutense, 28040 Madrid, Spain

54. Lusky Wolfgang, University of Paderborn, Fachbereich 17, Warburgerstr. 100, D-4790 Paderborn, FRG

55. Luxemburg Wilhelmus A.J., California Institute of Technology, Dept. of Math. 253-37 Pasadena, CA 91125, USA

56. Łenski Włodzimierz, IM UAM

57. Maligranda Lech, Departamento de Matemáticas, Instituto Venezolano de Investigationes Cientificas, Apartado 21827, Caracas 1020-A, Venezuela

58. Mastyło Mieczysław, IM UAM

59. Matkowski Janusz, Technical University Łódź, Bielsko Biała Branch, Findera 32, 43-300 Bielsko Biała, Poland

60. Mazur Tomasz, Technical University Radom, Malczewskiego 29, 26-600 Radom, Poland

61. Mikosz Mirosława, IM UAM

62. Musielak Anna, IM PP

63. Musielak Helena, IM UAM

64. Musielak Julian, IM UAM

65. Nawrocki Marek, IM UAM

66. Nekvinda Ales, Fac. Civil Engineering, Czech.Technical University, Thákorova 7, 166 29 Praha 6, Chechoslovakia

67. Nguyễn Hồng Thái, Belorussian State University, Minsk, USSR

68. Nikolova Ljudmila, Dept. of Math., Sofia University, Bul. A.Ivanov 5, Sofia 1229,

69. Novikov I.Ya., Voronezh State University, Voronezh, USSR

70. Nowak Marian, IM UAM

71. Operestejan B.A., Jaroslavl, pr.Oktjabrja 28-a, f.10, USSR

72. Pasternak-Winiarski Z. Inst. of Math., Technical University of Warsaw, Pl. Jedności Robotniczej 1, 00-661 Warsaw, Poland.

73.	Paúl Pedro J.,	E.S.Ingenieros Industriales, Av.Reina Mercedes s/n, 41012 Sevilla, Spain
74.	Persson Lars Erik,	Dept. of Math., Luleå University, 95187 Luleå, Sweden
75.	Pick Luboš	Mathematical Inst. Chechoslovak Academy of Sciences, Žitná 25, 115 67 Praha 1, Czechoslovakia
76.	Płuciennik Ryszard,	IM PP
77.	Pych-Taberska P.,	IM UAM
78.	Rodin V.A.,	Leninskij pr. 129, f.29, Voronezh, USSR
79.	Ruiz Cezar Bermejo,	Universidad Complutense de Madrid, Facultad de Matematicas, Departamento de Analisis Matematico, 28040 Madrid, Spain
80.	Schempp Walter,	Lehrstuhl für Mathematik I, University of Siegen, Siegen, FRG
81.	Shvartzman P.M.,	Jaroslavlskij Gos. Univ.,Jaroslavl, USSR
82.	Sickel Winfried,	Sektion Mathematik der FSU Jena UHH, 17.OG, Jena 6900, GDR
83.	Skrzypczak Leszek,	IM UAM
84.	Stachowiak-Gniłka D.,	IM UAM
85.	Staenlund Michael,	Dept. of Math., Luleå University, 95187 Luleå, Sweden
86.	Stoiński Stanisław,	IM UAM
87.	Suárez Antonio G.,	Universidad Complutense de Madrid, Facultad de Matematicas, Departamento de Analisis Matematico, 28040 Madrid, Spain
88.	Szelmeczka J.	Catholic University of Lublin, Lublin, Poland
89.	Taberski Roman,	IM UAM
90.	Urbański Ryszard,	IM UAM
91.	Véntura Echandia Liendo,	Departamento de Matemáticas, Instituto Venezolano de Investigationes Cientificas, Caracas, Venezuela
92.	Wang Tingfu,	Dept. of Basic Sci., Harbin University Sci.Tech., Harbin, P.R.China
93.	Weber Hans,	Instituto di Matematica, Facolta Scienze MM.FF.NN. Università della Basilicata, 85100 Potenza, Italy
94.	Werner Dirk,	I.Math.Inst., FU Berlin, Arnimallee 3, D-1000 Berlin 33, FRG
95.	Wisła Marek,	IM UAM
96.	Wnuk Witold,	Mathematical Institute, Polish Academy of Sciences, Branch Poznań, Mielżyńskiego 27/29, 61-725 Poznań, Poland
97.	Wójtowicz Marek,	Inst. of Math., Technical University of Zielona Góra, Podgórna 50, 65-246 Zielona Góra, Poland
98.	Zabrejko P.P.,	Minsk, Kuzniechnaja str. 3, f.80, USSR